PETROLOGY FOR STUDENTS

PETROLOGY

FOR STUDENTS

S. R. NOCKOLDS

Fellow of Trinity College, Cambridge

R. W. O'B. KNOX

Petrographer at the Institute of Geological Sciences, Leeds

G. A. CHINNER

Lecturer in Petrology, Cambridge University

CAMBRIDGE UNIVERSITY PRESS

CAMBRIDGE

LONDON · NEW YORK · MELBOURNE

Published by the Syndics of the Cambridge University Press
The Pitt Building, Trumpington Street, Cambridge CB2 1RP
Bentley House, 200 Euston Road, London NW1 2DB
32 East 57th Street, New York, NY 10022, USA
296 Beaconsfield Parade, Middle Park, Melbourne 3206, Australia

First published 1978
Reprinted 1979

Printed in the United States of America
First printed by
Cox & Wyman Ltd, London, Fakenham and Reading
Reprinted by
Hamilton Printing Co., Rensselaer, New York

Library of Congress cataloguing in publication data

Nockolds, Stephen Robert
Petrology for students

Includes index
1. Petrology I. Knox, Robert William O'Brien, 1942– joint author.
II. Chinner, G. A., 1932– joint author. III. Title
QE431.2.N6 552 77-3635
ISBN 0 521 21443 6 hard covers
ISBN 0 521 29184 4 paperback

CONTENTS

PREFACE

When Cambridge University Press decided not to reprint Dr A. Harker's well-known *Petrology for Students* it was suggested that a replacement be provided on the same lines.

The present book, in which the section on igneous rocks has been written by Nockolds, that on sedimentary rocks by Knox and that on metamorphic rocks by Chinner, is the result. Dr Harker's book, with its emphasis on carefully drafted drawings of thin sections, has for many years provided an essentially petrographic text for elementary classes in petrology. We have attempted to continue in this rôle, sufficient reference being made to the relevant scientific literature for the student to relate petrography to petrogeny.

Through the courtesy of CUP we have been able to use any of Harker's thin-section drawings, and in the igneous section the majority of the drawings are his. Knox has drawn all the figures (except fig. 24.1) for the sedimentary section, and Chinner those (except fig. 36.5) for the metamorphic section. We are grateful to Mr G. M. Part for drawing figs. 3.1, 3.2. and 3.6, and to Dr Germaine Joplin for drawing figs. 9.1, 16.1 and 16.2. Figs. 12.1, 12.3 and 20.1 are reproduced from Dr Joplin's *A Petrography of Australian Igneous Rocks* by kind permission of the author and the publishers, Angus and Robertson, and fig. 36.7 by kind permission of the Clarendon Press.

Virtually all of the thin sections illustrated in the igneous and metamorphic divisions of the book are available in the Harker Collection of Rock Slices in the Department of Mineralogy and Petrology, Cambridge University. We are grateful to the many contributors and to the successive curators who have built up and conserved this valuable collection.

S.R.N.
R.W.O'B.K.
G.A.C.

December 1976

1: Igneous rocks

I

Igneous rocks and their textures

Igneous rocks are formed from relatively high-temperature melts or *magmas*, as such melts are usually termed. When a magma cools it solidifies to give a rock that may consist of an aggregate of minerals, or of minerals and glass or, more rarely, of glass alone. The mutual relationship of the mineral constituents, including glass when present, defines the *texture* of the rock. The nature of the texture developed depends mainly on the rate of cooling of the magma, its viscosity, and the molecular concentrations of its components.

Some insight into the way in which at least some textures of igneous rocks are produced is provided by experimental work on simple melts (1). As a melt of a given substance cools down from its melting point, the number of centres of crystallisation (*nucleii*) formed in unit volume in unit time (the *nucleus number*) gradually increases. This increase is slow at first and then, at a given degree of undercooling, reaches a maximum value, thereafter sinking almost to zero.

The *viscosity* of the melt at first increases slowly with increasing undercooling until a *critical temperature* is reached at which the rate of increase rises rapidly until, finally, a glass may be formed. The marked increase in viscosity occurs in many cases just at that temperature where the nucleus number reaches its maximum. But with some substances the tendency towards spontaneous crystallisation is greatest when the melt has become rigid while, with other substances, the nucleus number attains its maximum value well above the point at which the great increase in viscosity becomes evident. The *initial* viscosity depends to a large degree on the composition of the melt. Thus, the viscosity increases with increasing SiO_2 content, the increase being rapid when the latter exceeds some 60 per cent. A large content of Al_2O_3 causes an increase, but the presence of much FeO or MnO causes a marked decrease: the substitution of some of the CaO in basic slags by MgO also decreases the viscosity of the melt, and the same effect is observed when MgO is substituted for CaO in soda-silicate glasses.

The *rate of growth* of the crystals forming in the melt rises to a maximum for a given degree of undercooling, and then decreases gradually. The maximum of the growth rate lies at a temperature above that at which the marked increase in viscosity sets in and, usually, above that defining the maximum of the nucleus number.

The facts outlined above may be applied most directly to the textures of lavas and very shallow-seated intrusions. Here, the volatile constituents of the magma have been largely lost, giving rise to essentially dry melt conditions, and cooling is relatively rapid.

If the cooling is very rapid, the magma quickly reaches the point where the marked increase in viscosity occurs, preventing the formation of any crystals, and causing the whole to solidify as a glass. The resulting rock is said to be *holohyaline*. Siliceous magmas are particularly prone to solidify as glass, once they have lost their volatile constituents; basic magmas much less so, owing to their lower initial viscosity. More often than not, incipient crystallisation of the magma has taken place so that *crystallites* (which do not react to polarised light), in the form of tiny globules, rods or hair-like bodies, or the somewhat larger *microlites*, having prismatic, acicular or arborescent forms, are scattered through the glass.

A glass is metastable and always has a tendency to crystallise, or *devitrify*. As a result, few natural glasses are known from before the Carboniferous era. Nevertheless, their former presence may frequently be proved, even after complete devitrification, by the presence of characteristic curved cracks, known as *perlitic* cracks (fig. 4.1A).

Spherulitic and *felsitic* textures are produced with a rather less rapid rate of cooling than that which provides a glass. Experimental work indicates that *spherulitic crystallisation*, involving the formation of acicular crystals radially arranged in the form of a sphere (spherulites) (fig. 4.2), takes place at a fairly moderate degree of undercooling when the nucleus number is still low but the growth rate is rather large. It also requires a certain degree of viscosity in the melt and is thus quite commonly found in siliceous lavas and siliceous furnace slags but is much rarer and less well developed in basic lavas and basic furnace slags, and practically unknown in highly fluid melts, such as those of metals. The less well-developed texture of this kind found in basic lavas is the *variolitic* texture in which irregular branching fibres of the mineral concerned (commonly plagioclase feldspar) have a crudely spherical disposition. In ultramafic lavas a different quench texture, the *spinifex* texture, is due to the rapid crystallisation of bladed skeletal olivine crystals (see p. 152).

Felsitic texture (fig. 4.1B), comprising an almost cryptocrystalline, more or less equigranular, aggregate of minerals, is formed at a greater degree of undercooling where the nucleus number is approaching its maximum, the growth rate has been reduced considerably, and the viscosity of the melt is high. It is thus possible to find spherulites embedded in a felsitic or a glassy groundmass (2).

The felsitic texture is found in siliceous lavas but basic lavas, cooling at approximately the same rate, do not normally show it because of their lower viscosity and because one of the main minerals to crystallise out, plagioclase feldspar, has a very strong tendency to develop a lath-like habit. Instead, either the *intergranular* texture, in which wedge-shaped areas between the plagioclase laths are filled with granules of the other minerals, or the *intersertal* texture, in which the interstices between the minerals are filled with glass (fig. 12.1), is developed.

A texture somewhat resembling the last is found in some lavas of intermediate composition. This is the *hyalopilitic* texture, in which a felt of crystals is set in a residuum of glass (fig. 9.1B). Again, certain lavas are composed of little feldspar laths aligned parallel to the direction of flow of the magma, giving rise to the *trachytic* texture (fig. 6.1).

Many lavas, and many of the rocks found in shallow-seated intrusions, carry larger crystals (*phenocrysts*) embedded in their fine-grained or glassy groundmasses. This is the *porphyritic* texture (fig. 4.3). One way in which this may develop is when the magma starts to cool slowly at greater depths and is then extruded, or intruded at a higher level, where, owing to the more rapid cooling, the loss of volatiles, and consequent increased viscosity, the remainder solidifies as a fine-grained aggregate, or as a glass.

But in some instances it is clear that the phenocrysts commenced to grow after the emplacement of the magma. During the normal cooling of such a magma there is, however, an early period some time before the marked increase in viscosity occurs and during which the nucleus number is small, while the growth rate may be relatively large. Phenocrysts might well be expected to form under these conditions, providing the magma remained long enough within the appropriate temperature interval (3). Phenocrysts may be grouped or clustered together to form *glomeroporphyritic aggregates*. This is one aspect of the phenomenon of *synneusis*, the process of drifting together and mutual attachment of crystals suspended in a melt (4).

The textures considered so far have been produced under conditions more or less comparable with those of artificial melts, but in magmas that solidify at an appreciable depth below the surface, the crystallised product is coarser in grain. Another factor enters here, namely, the presence of volatile constituents. These will lower the general viscosity and, more especially, prevent the marked increase of viscosity in the later stages as the volatiles become concentrated in the residual liquid. This, while important, should not be overemphasised. The dominant factor is the slower rate of cooling (5).

If cooling is slow enough, the characteristic texture produced is the *subhedral-granular* (or *granitic*) texture (fig. 3.4A) in which some of the constituents are anhedral while others have a more or less subhedral outline. The *microgranitic* texture (fig. 4.4A) is developed with more rapid rates of cooling, and there are all transitions from this to granitic texture on the one hand, and from this to felsitic on the other (6).

In some basic igneous rocks whose plagioclase feldspar has started to crystallise out early and then been joined by pyroxene, a *subophitic* texture may take the place of the subhedral-granular. In this, wedge-shaped areas between the subhedral crystals of plagioclase are occupied by plates of pyroxene that may be in optical continuity over a considerable area (fig. 11.1).

If the nucleii of some particular constituent are widely separated in comparison with those of the other constituents then the *poikilitic* texture, in which relatively large plates of one constituent enclose grains of the others, may result. *Micropoikilitic* texture is the same on a finer scale.

These textures, listed above as being formed by the relatively slow cooling and crystallisation of a magma, have their analogues in igneous rocks formed in a different way. These are the accumulative rocks (or *cumulates*), in which the minerals have been aggregated to a large extent by gravitational processes or by deposition during convective circulation in the magma chamber. Such consist, in their pristine state, of solid crystals and a variable amount of interstitial liquid that eventually solidifies. In their final state the rocks may have a more or less granular texture (fig. 11.2B), a subhedral-granular texture, a poikilitic texture (fig. 13.2) or a subophitic texture, depending on the way in which the interstitial liquid crystallises (7).

An entirely different class of textures embraces those due to the simultaneous crystallisation of two or more constituents under appropriate conditions. The first of these is the *graphic* texture, in which apparently isolated rods and wedges of one mineral are embedded in the other, the whole bearing a resemblance to ancient cuneiform writing. *Micrographic*, or *granophyric*, texture is the same on a finer scale (fig. 4.4B). The minerals most commonly involved in this texture are quartz and feldspars with the latter forming the host. But it may sometimes occur with other minerals, e.g. titanaugite and nepheline, and may be found with magnetite and fayalite at the eutectic composition in magnetite–fayalite slags.

That a particular rate of cooling, degree of supercooling (8), or other condition is necessary for the formation of textures of this kind is well shown in the case of certain granite-aplites whose quartz and feldspars have crystallised out simultaneously. It is possible to find two such aplites, of virtually the same composition, one of which shows a micrographic intergrowth of quartz with potash feldspar and with plagioclase, while the other shows a microgranular aggregate of the same constituents.

Though at first sight an unlikely-looking texture, the *ophitic* texture is also due to the simultaneous crystallisation of its constituents (9). As far as rocks are concerned, it consists of a plexus of plagioclase laths partially and wholly enclosed in plates of pyroxene (fig. 12.2). The spacing of the nucleii of the constituents was unsuitable for the development of graphic intergrowth, and a more or less granular texture was inhibited by the wide spacing of the nucleii of one of the constituents. A similar texture is observed in magnetite–fayalite slags when the magnetite and fayalite are crystallising together under conditions that preclude the formation of a graphic texture.

References and notes

(1) Tamman, G., 1926. *The states of aggregation.* London: Constable, especially chapter IX. Eitel, W., 1954. *The physical chemistry of silicates.* University of Chicago Press, 569–84.
For an early application of the experimental work to the texture of igneous rocks see Vogt, J. H. L., 1931. The physical chemistry of the magmatic differentiation of igneous rocks, III second half. *Skrift. der Norsk Vidensk.-Akad. Oslo. I Mat.-Nat.-Kl.* 1930, no. **3**, 177–208.

(2) This applies to primary spherulitic and felsitic textures. Secondary spherulitic, particularly microspherulitic, and felsitic textures may be produced by the devitrification of glass (see pp. 39, 50), and by the secondary crystallisation of ash-flow tuffs (see p. 214).

(3) Hawkes, L., 1930. On rock glass, and the solid and liquid states. *Geol. Mag.* **67**, 23–4.

(4) Vance, J. A., 1969. On synneusis. *Contr. Miner. Petrol.* **24**, 7–29.

(5) As, in general, the temperature of the country rock increases with depth, a deep-level intrusion would be expected to have a coarser texture than one of the same size intruded at a higher level. This is usually the case, but when a magma solidifies in the lower parts of a volcanic conduit where the walls have been heated to a considerable degree by the continued passage of lava, coarse-grained rocks may result at very shallow depths. These are the *subvolcanic* rocks of Washington.

(6) It is believed that these textures may be produced also, in some cases, by unmixing and recrystallisation as, for instance, in some subsolvus granites (p. 19). See Tuttle, O. F. & Bowen, N. L., 1958. Origin of granite in the light of experimental studies in the system $NaAlSi_3O_8$–$KAlSi_3O_8$–SiO_2–H_2O. *Mem. geol. Soc. Am.* **74**, 137–42.
(7) See Wager, L. R., Brown, G. M. & Wadsworth, W. J., 1960. Types of igneous cumulates, *J. Petrology* **1**, 73–85. Good illustrations of textures found in cumulate igneous rocks can be seen in Jackson, E. D., 1961. Primary textures and mineral associations in the ultramafic zone of the Stillwater complex, Montana. *Prof. Pap. U.S. geol. Surv.* **358**, 106 pp.
(8) Dunham, A. C., 1965. The nature and origin of the groundmass textures in felsites and granophyres from Rhum, Inverness-shire. *Geol. Mag.* **102**, 8–22.
(9) Bowen, N. L., 1928. *The evolution of the igneous rocks*. Princeton University Press, 68–9.

2

The classification of igneous rocks

The ideal classification of igneous rocks would be a *genetic* one, and tentative steps in this direction are now being made. The most important of these has been the recognition of igneous rock series whose members may range in composition from mafic (or ultramafic) to felsic but all of which have certain features in common. Such series can be grouped into three categories.

1. *Tholeiitic igneous rock series* are relatively rich in iron throughout and show a varying degree of absolute iron enrichment in the intermediate members. Typical tholeiitic rocks include tholeiitic basalts, tholeiitic andesites, ferrodacites, ferrorhyodacites and ferrorhyolites, many gabbros, some ultramafic rocks, ferrodiorites, ferrogranodiorites and ferrogranites.

2. *Calc-alkali igneous rock series*, relatively poor in iron throughout, show little or no absolute iron enrichment in the intermediate members and are generally richer in alumina than the rocks of tholeiitic series. Typical calc-alkali rocks include calc-alkali basalts, andesites, latite-andesites, dacites, rhyodacites and rhyolites, certain gabbros and ultramafic rocks, diorites, monzodiorites, tonalites, granodiorites and granites.

3. *Alkali igneous rock series* are very variable in character, some showing a rather feeble absolute enrichment in iron in intermediate members, others not, but all are characterised by a higher content of alkalies for a given silica percentage than tholeiitic or calc-alkali series. Typical alkali rocks include mildly alkaline types such as alkali basalts, hawaiites and mugearites, alkali gabbros, alkali monzodiorites and alkali monzonites; and more highly alkaline types such as basanites, essexites, teschenites and ultra-alkaline rocks. Felsic types include phonolites and nepheline syenites, peralkaline and alkali trachytes, peralkaline and alkali syenites, peralkaline and most alkali rhyolites, peralkaline and most alkali granites.

Much work remains to be done, however, in sorting out the various rock series within each of these three main groups before a true genetic classification can be erected.

At present it is necessary to be content with a *convenient* classification which, as far as possible, brings out the relationship of similar, or closely allied, rocks. Many and varied are the attempts which have been made to this end, and some of the factors involved will now be reviewed briefly.

In the first place, it is possible to make a very broad subdivision of igneous rocks, based on their mode of occurrence, into extrusive (or *volcanic*), shallow-seated intrusive (or *hypabyssal*), and deep-seated intrusive (or *plutonic*). This was essentially the method adopted to define the three main groups in the classification of the great nineteenth-century German petrologist, H. Rosenbusch, and some classifications in use today adopt the same plan.

Difficulties arise, however, if these distinctions are rigidly applied. Thus the chilled margin of a deep-seated mass intruded into cold country rocks may have all the characters of a hypabyssal rock; and the chilled margin, or even the whole, of a hypabyssal rock occurring, for instance, as a thin dyke may have all the characters of a volcanic rock. If, therefore, mode of occurrence is rigidly enforced as the prime classificatory characteristic, identical rocks may have to be placed in different subdivisions. Moreover, the exact mode of occurrence of some particular specimen may not be known and its degree of crystallinity must then be relied upon to place it in its appropriate group.

As a consequence, these terms, as used nowadays, have come more and more to indicate degree of crystallinity rather than mode of occurrence. Igneous rocks are said to be coarse grained if the average grain size of their constituents is greater than 5mm; medium grained if the grain size lies between 5mm and 1mm; fine grained if the grain size is between 1mm and 0.05mm; and very fine grained if the grain size is less than 0.05mm. On this basis, 'plutonic' rocks are those of medium to coarse grain, 'hypabyssal' rocks are fine-grained, and 'volcanic' rocks are very fine grained to glassy.

Some petrologists do not attach much importance to this three-fold division and are content to divide igneous rocks into *phanerocrystalline*, in which the constituents can be detected by the eye, and *aphanitic* in which the constituents are no longer visible to the eye.

In the second place, the mineralogical composition of the igneous rock may be used and it is on this that most of the classifications in general use depend. The simpler classifications are qualitative in nature, the various rock groups being determined simply by the presence or absence of certain essential minerals. The more sophisticated classifications are of a quantitative nature and correspondingly more complex, the actual amounts of the various minerals present (the *mode* of the rock) having to be determined. The most complete attempt along these lines is that of A. Johannsen (1). In all, a distinction is made between *essential* minerals, whose presence or absence affects the naming of the rock, and *accessory* minerals, which normally occur in small amounts and are not used for diagnostic purposes.

In some classifications that are predominantly mineralogical, a chemical factor is also introduced. The most ingenious of such classifications is that due to S. J. Shand (2) who introduces the concept of silica saturation. If a rock is oversaturated with silica, free quartz will appear; if undersaturated, the rock will contain minerals with a deficiency of silica, such as olivine and/or feldspathoids; if just saturated, quartz on the one hand and olivine or feldspathoids on the other will be lacking. Shand further applied the same principle to alumina and the case of most interest here is where a rock is undersaturated with alumina, leading to the formation of soda-pyroxenes and/or soda-amphiboles. He called such rocks *peralkaline*, and this term has come into general use.

In most quantitative or semiquantitative mineralogical classifications use is made directly, or indirectly, of the so-called *colour index* of the rock. This is defined as the percentage, by volume, of the minerals in the rock whose density is greater than 2.8 (i.e. just sink in bromoform) (3). This includes all the dark coloured constituents, hence the term, but it also includes a few constituents, such as apatite and muscovite, that are light coloured.

If the rock has a colour index of less than 5 it is said to be *hololeucocratic*; if it lies between 5 and 30 the rock is described as *leucocratic*; between 30 and 60 as *mesotype* (4); between 60 and 90 as *melanocratic*; and if the colour index is greater than 90, the rock is spoken of as *hypermelanic*.

In the third place, chemical composition may be used to classify igneous rocks, and a number of rather elaborate classifications have been founded on this basis (5). All suffer from the serious objections that no rock can be classified until it has been analysed chemically, while the mineralogy and the texture of the rock, both important features, are ignored. The main value of such classifications lies in the opportunity they afford of comparing chemical analyses of rocks with one another.

The simple classification used in this book is a semiquantitative mineralogical one (table 2.1). Three large divisions are erected on the basis of essential (> 10 volume per cent of the rock) quartz; essential

TABLE 2.1. *Classification of igneous rocks*

Other essential minerals	Alkali feldspars only	Essential lime-bearing plagioclase and potash feldspar Percentage plagioclase in total feldspar content				No essential feldspar (<10% of any feldspar present)
		10 to 40	40 to 60	60 to 90	>90	
Quartz (>10% of the rock)	**Alkali granite** **Peralkaline granite** *Alkali rhyolite* *Peralkaline rhyolite*		**Granite** *Rhyolite*	**Granodiorite** *Rhyodacite*	**Tonalite** *Dacite*	
	Alkali syenite **Peralkaline syenite** *Alkali trachyte* *Peralkaline trachyte*	**Syenite** *Trachyte*	**Monzonite** *Latite* **Alkali monzonite** *Tristanite*	**Monzodiorite** **Monzogabbro** *Latite–andesite* **Alkali monzodiorite** **Alkali monzogabbro** *Hawaiite*, etc. *Trachybasalt*, etc.	**Diorite** *Andesite* **Gabbro** *Basalt* **Alkali gabbro** *Alkali basalt*	**Ultramafic igneous rocks**
Feldspathoid (>10% of the rock)	**Nepheline syenite**, etc. *Phonolite*, etc.		**Nepheline monzonite**, etc. *Nepheline tristanite*, etc.	**Essexite** **Teschenite** *Nepheline hawaiite*, etc. *Nepheline trachybasalt*, etc.	**Theralite** *Tephrite*, etc.	**Ultra–alkaline plutonic rocks** *Nephelinite*, etc.

Names of coarse to medium grained (plutonic) igneous rocks given in heavy type.
Names of very fine grained (volcanic) igneous rocks given in italics.
Names of fine grained (hypabyssal) igneous rocks not shown.

feldspathoid; and neither quartz nor feldspathoid present in essential amounts. The other large divisions are based on the relative proportions of the different feldspars present or, in one case, on the absence of any essential feldspar. The classification is thus based primarily on the light constituents present in the rocks (6).

Within each rock family so formed, three textural subdivisions are recognised, based on the grain size. Thus, in the granite–rhyolite family there is granite (coarse to medium grained), microgranite (fine grained), and rhyolite (very fine grained to glassy). Any member can be defined further by prefixing the dark minerals present, e.g. biotite granite, and/or any necessary textural term, e.g. porphyritic biotite granite. The pigeon-holing of rocks in a classification such as this is arbitrary and purely a matter of convenience. In reality, there are all transitions from one group to the next. Thus, for instance, there are all transitions from diorite, through quartz-bearing diorite, to tonalite or from gabbro to feldspathic peridotite or pyroxenite or, again, from pyroxenite to peridotite. Equally, there are all transitions between the coarse-grained and fine-grained to glassy members of any family (7).

In any mineralogical classification of igneous rocks there are obvious difficulties to be faced when dealing with volcanic rocks. Many of these have a glassy, or partially glassy, groundmass or are so fine grained that the individual constituents, and particularly their relative proportions, cannot be determined with any precision. Here, the purely mineralogical classification breaks down and the chemical composition of the given volcanic rock must be compared with the chemical compositions of coarser igneous rocks whose mineralogy is known. In this way, the volcanic rock can be placed in its proper family.

Such comparison may be made directly or the chemical analysis may be recalculated in the manner required for one or other of the chemical classifications already mentioned. Two such methods of recalculation are in general use, one due to P. Niggli and the other due to four American petrologists, W. Cross, J. P. Iddings, L. V. Pirsson and H. W. Washington.

In the first, the so-called *Niggli values* are calculated from the chemical analysis: this method is much used on the Continent (8). The *C.I.P.W. method* of recalculating chemical analyses is used extensively in America and Britain, and this will be considered in a little more detail.

Briefly, the chemical analysis of the igneous rock is recast into a set of simple theoretical standard minerals. These are divided into two groups: *salic*, including quartz (*qz*), orthoclase (*or*), albite (*ab*), anorthite (*an*), nepheline (*ne*), leucite (*lc*), corundum (*C*); and *femic*, including diopside (*di*) (calculated as its components $CaSiO_3$, $MgSiO_3$, $FeSiO_3$), hypersthene (*hy*) (calculated as its components $MgSiO_3$, $FeSiO_3$), olivine (*ol*) (calculated as its components Mg_2SiO_4, Fe_2SiO_4), acmite (*ac*), magnetite (*mt*), ilmenite (*il*), apatite (*ap*), etc.

The mineral composition of the rock, as calculated from the chemical analysis in terms of these standard minerals according to a definite set of rules (9), is known as the *norm* and must be carefully distinguished from the *mode*, or actual composition of the rock. The norm and the mode may approach each other closely for rocks carrying pyroxenes and/or olivine. But where complex minerals like hornblende or micas become prevalent, the norm may depart widely from the mode.

Hornblende will be represented in the norm as diopside and hypersthene with a varying amount of anorthite, depending on its alumina content. *Muscovite* will be represented in the norm as orthoclase and corundum. *Biotite*, a mineral low in silica, appears in the norm of mafic rocks largely as leucite and olivine. But in more siliceous rocks where free silica is available, these normative components are replaced by normative orthoclase and hypersthene, owing to the way in which the norm is calculated. Many biotites occurring in the more siliceous rocks have more aluminium than in the ideal formula, and this appears in the norm as corundum. Thus, in siliceous rocks carrying micas alone, corundum appears in the norm, diopside is absent, and normative quartz is usually less than modal quartz. Peralkine rocks, with soda-pyroxenes and/or soda-amphiboles in the mode, have acmite in the norm.

The norm has uses other than that of comparing analyses of igneous rocks. First, it may be used to distinguish rocks whose chemical analyses, at first sight, resemble each other closely. Thus, in table 2.2, columns 1 and 2 give the chemical analyses of two black volcanic glasses (obsidians). Inspection shows the high silica, fairly high alkalies and very low lime characteristic of alkali rhyolites. But the norms show that whereas the first has a composition corresponding with that of an ordinary alkali rhyolite, the second corresponds with a peralkine rhyolite.

Columns 3 and 4 give chemical analyses of two basalts very rich in olivine, picrite-basalts as they are termed. The norm of the first has considerable hypersthene as well as some diopside, while the norm of the second has diopside with only a trace of hypersthene. These features are characteristic of tholeiitic basalts and alkali basalts respectively (see p. 129), and the first rock is a tholeiitic picrite-basalt, while the second is an alkali picrite-basalt.

TABLE 2.2

	1	2	3	4	5	6	7
SiO_2	74.22	74.01	45.61	43.82	57.48	67.30	68.10
TiO_2	0.11	0.11	1.77	2.07	0.81	0.50	0.51
Al_2O_3	12.98	13.08	9.06	9.59	19.27	16.65	15.87
Fe_2O_3	0.40	1.38	1.32	3.11	2.30	2.50	0.50
FeO	1.60	1.21	10.49	10.36	4.63	1.75	2.66
MnO	0.01	tr.	0.17	–	0.10	tr.	0.59
MgO	0.04	tr.	21.48	20.51	2.23	0.75	0.83
CaO	0.67	0.13	7.79	8.21	5.84	0.95	2.60
Na_2O	4.13	5.78	1.28	1.52	2.73	6.20	5.48
K_2O	4.97	4.31	0.30	0.62	2.49	2.60	2.88
H_2O^+	0.27	0.16	0.27	–	1.19	0.40	0.38
P_2O_5	0.14	tr.	0.16	0.26	0.13	0.15	0.19
F	–	–	–	–	–	–	0.19
qz	28.9	24.4	–	–	14.0	18.5	16.9
or	29.5	25.6	1.7	3.3	15.0	15.6	17.2
ab	35.1	43.0	11.0	12.6	23.1	52.4	46.6
an	2.0	–	18.1	17.8	28.1	3.9	10.0
C	–	–	–	–	1.7	2.1	–
di	0.5	0.8	15.7	16.6	–	–	0.9
hy	2.2	1.7	11.1	0.7	11.0	2.2	6.4
ol	–	–	36.2	39.9	–	–	–
ac	–	4.2	–	–	–	–	–
mt	0.7	–	1.9	4.4	3.3	3.7	0.7
il	0.2	0.2	3.5	4.0	1.5	0.9	0.9
ap	0.3	tr.	0.3	0.6	0.3	0.3	0.4
ns	–	0.4	–	–	–	–	–
fl	–	–	–	–	–	–	0.2

1. Chemical analysis and norm of an alkali rhyolite–obsidian.
2. Chemical analysis and norm of a peralkaline rhyolite–obsidian.
3. Chemical analysis and norm of a tholeiitic picrite–basalt.
4. Chemical analysis and norm of an alkali picrite–basalt.
5. Chemical analysis and norm of a rock described as a biotite-bearing hypersthene–amphibole andesite.
6. Chemical analysis and norm of a rock described as an arfvedsonite soda–granite.
7. Chemical analysis and norm of a rock described as biotite granodiorite with accessory muscovite.

Secondly, the norm may be used for checking the accuracy of a chemical analysis and/or the freshness of the rock analysed. Some illustrations of this are given in table 2.2, columns 5–7, all representing chemical analyses of igneous rocks whose norms were not determined at the time the analyses were published.

Columns 5 and 6 are chemical analyses of rocks described as 'biotite-bearing hypersthene-amphibole andesite' and 'arfvedsonite soda-granite' respectively. Corundum would not be expected in the norm of the first, still less so in that of the second where, indeed, normative acmite would cause no surprise. But determination of the norm gives quite large amounts of corundum in both cases. Clearly, the rocks are either somewhat altered or the chemical analyses are inaccurate. Incidentally, the norm of the first rock also shows that it has too much normative quartz and orthoclase to be an andesite.

Column 7 is a chemical analysis of a rock carrying biotite and accessory muscovite. Here, an appreciable amount of normative corundum would be expected. But the norm shows none at all, and even has a little diopside. In this case, the chemical analysis must be inaccurate.

These examples illustrate how important it is to determine the norms of chemical analyses of igneous rocks before using them, for instance, to obtain the average chemical composition of an igneous rock type.

Thirdly, the norm finds use in the concept of the *differentiation index* (D.I.). Introduced by Thornton and Tuttle (10), it is defined as the sum of the weight percentages of normative quartz + albite + nepheline + leucite + kalsilite, and indicates how far a rock has proceeded towards what Bowen has called petrogeny's 'residual' system, namely, the system $NaAlSiO_4$–$KAlSiO_4$–SiO_2 (11).

References and notes

(1) Johannsen, A., 1931. *A descriptive petrography of the igneous rocks.* University of Chicago Press. 4 vols.

(2) Shand, S. J., 1947. *Eruptive rocks* (3rd ed.). London: T. Murby & Co., chapter XIV.

(3) see Shand, *ibid.*, pp. 233–4.

(4) Shand (*ibid.*, p. 235) has pointed out that the term 'mesocratic', still used by some petrologists, is meaningless as the middle cannot predominate. He substituted the term 'mesotype'.

(5) The interested student will find all the more important chemical classifications of igneous rocks detailed in Johannsen, 1931, vol. I.

(6) A good review of various mineralogical classifications is given in Streckeisen, A. L., 1967. Classification and nomenclature of igneous rocks (final report of an inquiry). *Neues Jb. Miner. Abh.* **107**, 144–240.

(7) The transitional nature of igneous rocks has resulted in a very large number of different names being given to them, many of which are quite unnecessary. Definitions of most of these are in Johannsen, 1931, vol. I, 238–88 and in Holmes, A., 1928. *The nomenclature of petrology.* London: T. Murby & Co.

(8) For the method of calculating Niggli values see Johannsen, 1931, vol. I, 103–7.

(9) Rules for calculating the norm are detailed in Johannsen, 1931, vol. 1, 83–99; in Washington, H. S., 1917. Chemical analyses of igneous rocks. *Prof. Pap. U.S. geol. Surv.* **99**, 1162–5; or in Holmes, A., 1921. *Petrographic methods and calculations*. London: T. Murby & Co., 410–32.
(10) Thornton, C. P. & Tuttle, O. F., 1960. Chemistry of igneous rocks. I. Differentiation index. *Am. J. Sci.* **258**, 664–84.
(11) Bowen, N. L., 1937. Recent high-temperature research on silicates and its significance in igneous geology. *Am. J. Sci.* (5) **33**, 11–13.

3
Granitic rocks

Under this general heading are placed all those holocrystalline igneous rocks of medium to coarse grain that are composed essentially of quartz (> 10 per cent by volume) and one or more feldspars, usually with some ferromagnesian mineral in addition. They thus include *peralkaline granites* and *alkali granites* which have alkali feldspar only; *granites* which have alkali feldspar and lime-bearing plagioclase, the former making up more than 40 per cent by volume of the total feldspar content; *granodiorites* in which lime-bearing plagioclase is accompanied by subordinate alkali feldspar, the latter making up between 40 and 10 per cent of the total feldspar content; and *tonalites* in which alkali feldspar is either absent or makes up less than 10 per cent of the total feldspar content.

They are all normally leucocratic with an average colour index of about 10 for granites, 15 for granodiorites, and 20 for tonalites. Such granitic rocks are responsible for numerous bosses, stocks and ring-dykes while granites and, more especially, granodiorites make up the bulk of the world's great batholiths.

Mineralogy

Feldspars make up the greater part of a granitic rock. Two types of peralkaline and alkali granite may be distinguished according to the nature of the alkali feldspar present. Hypersolvus types have a single alkali feldspar, normally a microperthite, while subsolvus types have two alkali feldspars, microperthite (or orthoclase or microcline) and independent albite (1). In the hypersolvus types the microperthite may show a wide range of composition in different rocks, and is sometimes a soda-potash feldspar rather than a potash-soda feldspar. In subsolvus rocks the range of composition of the microperthite is more restricted.

In the remaining granitic rocks, the alkali feldspar present is a potash-soda or potash variety, microperthite or orthoclase or microcline. Microperthite, which may be either orthoclase-microperthite or microcline-microperthite, is the commonest of these, ranging in composition from about $Or_{60}Ab(+An)_{40}$ to $Or_{85}Ab(+An)_{15}$ and averaging about $Or_{75}Ab(+An)_{25}$. The alkali feldspar is usually more or less anhedral, except when it occurs as phenocrysts as is the case with certain granites and granodiorites.

The lime-bearing plagioclase present in granites, granodiorites and tonalites is, on the other hand, commonly subhedral and frequently zoned. Its composition falls within the oligoclase to andesine range.

The *quartz* is found filling up the spaces between the other constituents as small plates or aggregates of grains. It is frequently rich in minute fluid inclusions and sometimes in tiny hair-like crystals of rutile. Few granitic rocks have more than 40 per cent of quartz, and the average content for granites is some 30 per cent, for granodiorites and tonalites about 25 per cent.

The *micas* of granitic rocks include both biotite and muscovite in characteristic plates and flakes. When occurring together, they are not infrequently in parallel growth. Biotite, but not muscovite, may also be found in association with amphibole and pyroxene. In general, the biotite of the granitic rocks is moderately rich in iron, brown and strongly pleochroic in thin section. But in certain granites and granodiorites, and in many peralkaline and alkali granites, it is a very strongly pleochroic, very iron-rich variety. A feature of the biotite is the occurrence of dark pleochroic haloes round minute included crystals (fig. 3.3A), due to the radioactive properties of the latter (2).

Amphibole, when present, is a strongly pleochroic sodic variety in the peralkaline granites, commonly riebeckite or arfvedsonite; while ferrohastingsite may occur in alkali granites and in certain granites and granodiorites. In the majority of granites and granodiorites, and in tonalites, the amphibole is a somewhat aluminous hornblende, green or brownish green in thin section. The crystals are usually prismatic but in peralkaline granites they may be irregular in outline, wrapping round the feldspar and quartz.

Pyroxene is uncommon in granitic rocks, except in peralkaline granites, where aegirine or aegirine-augite may be present. Some alkali granites and certain granites and granodiorites have hedenbergite, and a few granodiorites and tonalites carry augite and orthopyroxene.

Olivine is normally absent, but some granitic rocks very low in magnesia have some fayalite.

Among the *accessory constituents*, apatite in little prisms and needles, small crystals of zircon, and grains of opaque ore are almost ubiquitous.

The main ore is magnetite but ilmenite is present in some rocks, and ilmenite–haematite exsolution intergrowths are not uncommon. Pyrite is found sometimes, also pyrrhotite and chalcopyrite, the latter only in very small amount.

Crystals of light brown pleochroic sphene are often seen and may be relatively abundant, but this mineral is usually absent in rocks having muscovite as a primary constituent. Sporadic crystals of orthite are widely distributed, and both monazite and xenotime may be found. Small reddish garnets (spessartite-almandite) occur in some muscovite-bearing rocks, as does tourmaline. Fluorite appears in some granites, and the peralkaline granites, in particular, have a number of rare accessory constituents such as astrophyllite and aenigmatite not found in the other granitic rocks.

Textures and structures

The texture of these rocks is normally subhedral-granular; sometimes there is some micrographic intergrowth of quartz and feldspar. A number of granites and granodiorites are porphyritic, commonly with phenocrysts of potash feldspar. A few granitic rocks develop the rapakivi texture, in which ovoids of potash feldspar are mantled by lime-bearing plagioclase (3).

With regard to the larger-scaled structures, some granites have a banded or layered structure in the field (4); a few granitic rocks show orbicular structure (5); while high-level granitic rocks may develop a drusy structure.

Main varieties with some examples

1. Peralkaline granites

The peralkaline granites have crystallised from a magma undersaturated with alumina, leading to the formation of sodic pyroxenes or amphiboles and to the occurrence of acmite in the norm (p. 14). The common varieties have riebeckite, riebeckite and aegirine, aegirine, or arfvedsonite as their mafic constituents.

Riebeckite-bearing granites in the Younger Granite province of northern Nigeria form the most extensive occurrence of these granites known in the world (6). A few are riebeckite granites, a few others are riebeckite–biotite granites, but the bulk are riebeckite–aegirine granites (fig. 3.1B). The riebeckite and aegirine in these may form subhedral crystals, may be intergrown, or may be found interstitial to the other constituents. The common feldspar is microcline-microperthite and in some rocks this is the only feldspar present (hypersolvus type). In others a variable amount of independent albite is present (subsolvus type) (fig. 3.2B), and in yet others there appears to have been extensive replacement of all the minerals by small laths of albite.

The well-known riebeckite–aegirine granite of Quincy, Massachusetts, (fig. 3.1A) also has microcline-microperthite as the main alkali feldspar. The latter is sheathed with fine-grained albite and the rock is believed to represent a stage transitional between hypersolvus and subsolvus granite (7).

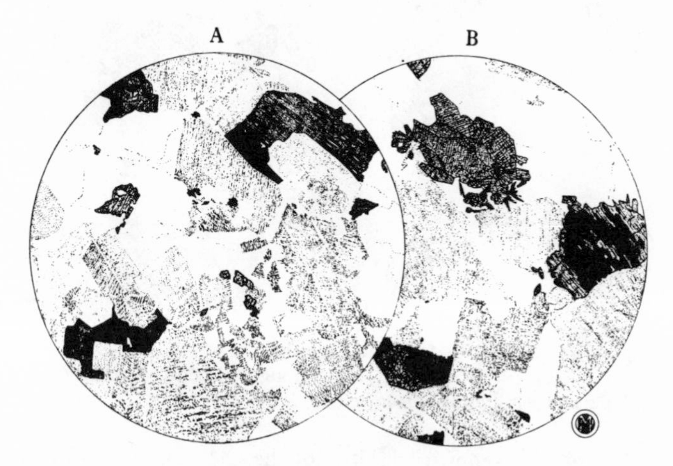

Fig. 3.1. Riebeckite–aegirine granite.

A. Quincy, Mass., U.S.A. Microperthite sheathed with albite (clear), quartz, aegirine and riebeckite (dark) are the essential constituents. A few grains of zircon are visible in the centre; × 18.

B. Kila Hills, N. Nigeria. A variety relatively rich in aegirine, accompanied by microperthite and quartz. Minor riebeckite replacing the pyroxene and, at the eastern edge of the figure, accessory aenigmatite surrounded by aegirine; × 24.

Peralkaline granites are very rare in Britain but *aegirine granite* occurs on the islet of Rockall in the Atlantic, some 320km west of the Outer Hebrides. This is a subsolvus granite with albite and microcline and with the aegirine partly as euhedral crystals, partly moulded upon feldspar. The green aegirine is accompanied by brownish acmite, mostly mantling the former (8).

Good examples of *arfvedsonite granite* (fig. 3.2A) are to be found at Narsak and in the Ilimausak complex, S. Greenland. That from Ilimausak has albite-antiperthite, quartz with abundant fluid inclusions, and arfvedsonite as irregular prisms and large anhedra. Aegirine and aenigmatite are present among the accessory constituents.

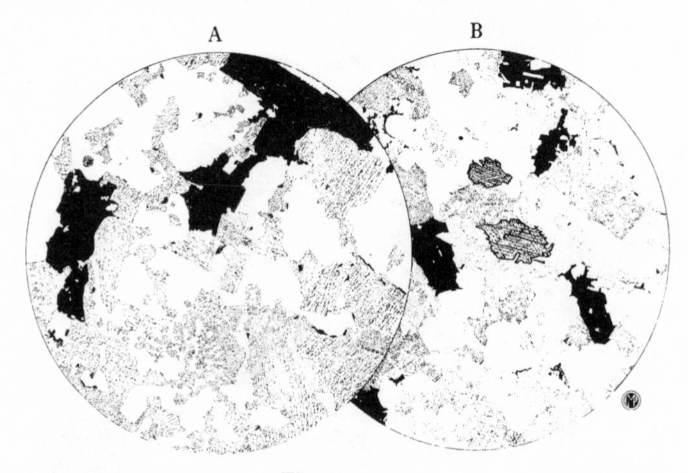

Fig. 3.2. × 22.

A. Arfvedsonite granite, N. Siorasuit, Tunugdliarfik, S. Greenland. Strongly coloured arfvedsonite in rather irregular crystals, quartz and microperthite. In places, arfvedsonite is rimmed by aegirine.
B. Riebeckite granite, Teria, N. Nigeria. Strongly pleochroic riebeckite, quartz, small laths of clear albite and somewhat turbid microcline.

Most of the well-known 'ekerites' of the Oslo district, Norway, are *arfvedsonite–aegirine granites*, and interesting rocks of this kind occur as ejected blocks in trachytic agglomerate and tuff on Ascension Island, remarkable for the abundance of fluid inclusions in both the quartz and the alkali feldspar. These ejected blocks are hypersolvus granites in which the feldspar is either a homogeneous soda-sanidine or a sanidine-perthite. Sanidine is a characteristic feldspar of volcanic rocks and these granites must be regarded as subvolcanic (9).

2. *Alkali granites*

The more important varieties of alkali granite are ferrohastingsite alkali granite, with or without biotite; biotite alkali granite; and muscovite–biotite alkali granite.

The *ferrohastingsite alkali granite* of the Percy ring-dyke complex in the northern part of New Hampshire, U.S.A., is a hypersolvus granite with microperthite as its only feldspar. It carries accessory hedenbergite and shows transitions through riebeckite–ferrohastingsite granite to riebeckite granite (10). By contrast, the Chocorua granite of New Hampshire is a subsolvus ferrohastingsite granite with both microperthite and albite, and with transitions through ferrohastingsite–biotite alkali granite to biotite alkali granite.

A number of ferrohastingsite alkali granites carry a little hedenbergite and fayalite (e.g. N. Nigeria) but in some alkali granites these become the main mafic constituents, usually accompanied by subordinate amphibole. Alkali granites of this kind are to be found among the drusy, shallow-seated, Tertiary granites of the Western Red Hills complex in the isle of Skye. These are typically hypersolvus granites with a coarse granophyric texture and phenocrysts of alkali feldspar (Beinn Dearg Mhòr and Loch Ainol granites) (11).

Biotite alkali granite, with fluorite as a noticeable accessory constituent, is widely distributed and forms some of the largest individual intrusions in the Younger Granite province of northern Nigeria, already mentioned. Both hypersolvus and subsolvus types appear to be present. Further good examples are to be found in the different phases of the Conway biotite alkali granite, forming part of the White Mountain batholith, New Hampshire, and in the younger members of the Tertiary Mourne Mountains granites, N. Ireland (12).

The ferrohastingsite granites and the biotite alkali granites are found associated with peralkaline granites or with ferrogranites (p. 29), and their biotite is a variety rich in iron, either deep brown, deep green, or brownish green in thin section. The *muscovite–biotite alkali granites*, on the other hand, are associated with biotite granite or muscovite–biotite granite and their biotite is not noticeably rich in iron. Moreover, they are all subsolvus granites.

There are a number of muscovite–biotite alkali granites among the granites of New England, associated with biotite granites and muscovite–biotite granites. They have microcline and albite as their feldspars and muscovite is present in excess of biotite. In the Chelmsford, Concord, and Waldoboro granites the albite has an average composition of about An_6, while in the North Jay–Hallowell granite it averages about An_9 (13).

The granites of Lundy Island are British examples. Here, an Upper Granite, a relatively coarse muscovite–biotite alkali granite with phenocrysts of microperthite orthoclase, is intruded by the Lower Granite of similar type but somewhat more siliceous, and with phenocrysts of both feldspar and quartz (14).

3. Granites, granodiorites and tonalites

The main varieties of these three rock types carry muscovite and biotite, or biotite, or hornblende and biotite. Many of the granites here have about equal amounts of potash feldspar and lime-bearing plagioclase and these are named 'quartz-monzonite' or 'adamellite' by some petrologists. The plagioclase has an average composition of about An_{22} in granites, An_{28} in granodiorites and An_{32} in tonalites. The average composition of the plagioclase varies in each rock group with the character of the ferromagnesian minerals present as is shown by the following table:

	Muscovite–biotite	Biotite	Hornblende–biotite
Granite	An_{19}(27)*	An_{21}(70)	An_{25}(29)
Granodiorite	An_{22}(22)	An_{27}(47)	An_{31}(54)
Tonalite	An_{27}(7)	An_{27}(13)	An_{36}(29)

(*The figure in brackets gives the number of determinations averaged.)

Excellent examples of *muscovite–biotite granites* are provided by the Armorican intrusions of Devon and Cornwall (15). All are composite, with two or more important phases. The earliest granite in each case has large phenocrysts of orthoclase-microperthite set in a coarse matrix of quartz, microperthite and subordinate oligoclase, together with flakes of biotite and muscovite. Sometimes, as at Dartmoor, the latter becomes very subordinate. The later intrusions in each case are more siliceous and finer grained. They have less biotite and more muscovite and, in some, a lithium mica makes its appearance. In a number the plagioclase is still oligoclase, but in others it is albite, these latter being muscovite–biotite alkali granites.

Other British examples are of Caledonian age, including the important Grey Granites (known commercially as 'Aberdeen granite') of Aberdeen, with microcline as their alkali feldspar (16); the porphyritic muscovite–biotite granite of Moy, in Nairnshire; the central parts of the Cairnsmore of Fleet granite and of the Beinn Dearc intrusion at Glen Tilt, Perthshire; and part of the Leinster granite, Eire (fig. 3.3A).

Muscovite–biotite granodiorites are less common than the corresponding granites but have been recorded, for instance, from the West Dummerston (17) and Derby (18) districts of Vermont, U.S.A. The West Dummerston rock has muscovite in excess of biotite and carries accessory garnet while, in those from Derby, biotite exceeds muscovite

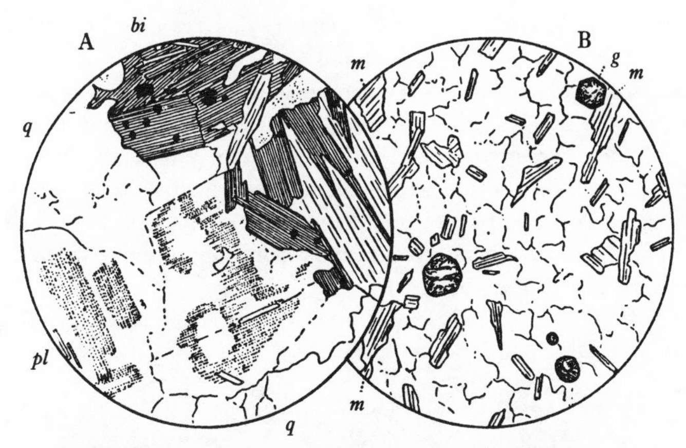

Fig. 3.3. Granitic rocks, Killiney, near Dublin, Eire; × 20.

A. Muscovite–biotite granite. Two micas (partly intergrown), two feldspars and clear quartz (*q*). Biotite (*bi*) shows strongly pleochroic haloes, plagioclase (*pl*) is often turbid in the interior, owing to alteration.

B. Granite-aplite, veining the preceding rock. A fine-grained aggregate of quartz, plagioclase and microcline, with abundant flakes of muscovite (*m*) and some crystals of garnet (*g*).

or else the two micas occur in roughly equal proportions. These rocks are non-porphyritic, but the Härno 'granite' of Sweden is a muscovite–biotite granodiorite in which potash feldspar sometimes builds rather large phenocrysts.

Muscovite–biotite tonalites are less common still, but some of the 'trondhjemites' of Norway belong here. That of Skavlien is very coarse grained and sometimes feebly porphyritic. The feldspar is oligoclase, potash feldspar occurring only as an accessory constituent. The rock at Frenstad is similar but finer grained, while that from Mastravarde in the Stavanger district is richer in biotite and has a little garnet. The muscovite–biotite tonalites from the Tatra region of Poland have more biotite and less quartz while their plagioclase is more anorthitic and lies close to An_{30}.

Biotite granite is a very common type and one of the best known is the beautiful rock from Shap, Westmorland (fig. 3.4A). This has some 20 per cent of potash feldspar phenocrysts, from three to five centimetres long, in a matrix of zoned oligoclase, quartz, potash feldspar and flakes of biotite, and with sphene as a prominent accessory constituent. A slightly earlier phase has fewer phenocrysts and rather more biotite. This is of Caledonian age and biotite granites of the same age are found as individual intrusions or as a late phase in a number of the so-called

'Newer Granite' complexes of Scotland (19). Here may be mentioned the even-grained Main Granite and the Porphyritic Granite of the Cairngorm pluton, both with microperthitic microcline as their alkali feldspar; the outer portions of the Beinn Dearc intrusion at Glen Tilt and of the Cairnsmore of Fleet granite; the granite of the Foyers complex, with accessory hornblende; and the porphyritic biotite granite making up the main part of the Skene igneous complex, west of Aberdeen, in which the phenocrysts of microperthite are sometimes accompanied by phenocrysts of quartz.

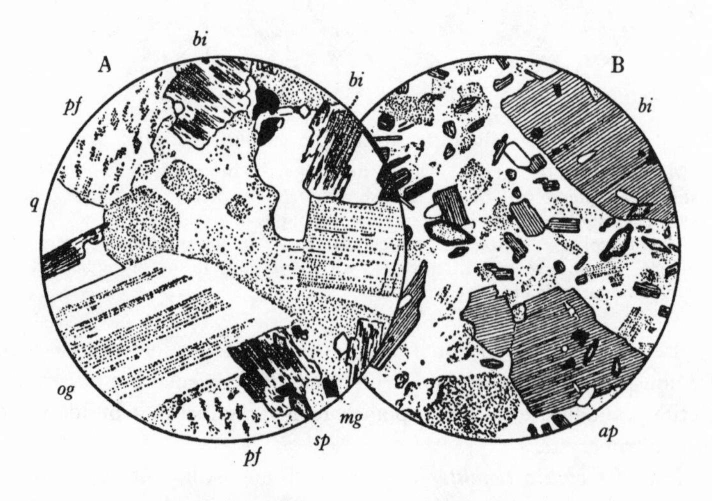

Fig. 3.4. Biotite granite, Shap Fell, Westmorland; × 20.

A. The normal rock. Biotite (*bi*) partly chloritised, oligoclase (*og*), potash feldspar (*pf*) and quartz (*q*), with accessory sphene (*sp*), magnetite (*mg*) and apatite (*ap*). The rock has large phenocrysts of potash feldspar in addition.

B. Dark mafic inclusion in the foregoing. This is of finer texture and richer in biotite, sphene and apatite.

Biotite granodiorite is also a common type. That of the Strontian complex has phenocrysts of potash feldspar and some of oligoclase in a rather fine-grained groundmass with scanty biotite, while that comprising the normal type of 'Inner Granite' at Ben Nevis is non-porphyritic (20).

Portions of the great Idaho batholith, U.S.A., are composed of biotite granodiorite with more or less euhedral oligoclase and subordinate microperthitic microcline, quartz, and varying amounts of biotite (21). The Granite Creek granodiorite of the Pend Oreille district of northern Idaho is a similar rock but the plagioclase is markedly zoned, averaging andesine in composition and both sphene and orthite are relatively abundant. Further examples are the biotite granodiorites of the central Österbotten district, Finland, with large phenocrysts of microcline-microperthite; the well-known Andlau 'granite' of the Vosges, carrying accessory phenacite; and some of the granodiorites of central and middle Bohemia.

Biotite tonalites are less common. Portions of the 'Old Granite' of the Vredefort region, South Africa, are made up of a rock of this kind with oligoclase (An_{20} to An_{27}) as the feldspar together with accessory micro-perthite. A similar, but more siliceous, tonalite forms the Heany 'granite' stock, also of Precambrian age, in Southern Rhodesia, while the Rewnick tonalite of New Hampshire and the almost holoeucocratic tonalites of the Derby district, Vermont, are further examples. Certain of the Norwegian 'trondhjemites' belong here, also, such as those of Malletuen and Dragaasen, the latter with accessory hornblende and augite.

Hornblende–biotite granite is not as common as biotite granite, and most of the rocks that fall here are of the 'quartz-monzonite' type. British examples are to be found among the 'Newer Granites' of Scotland, such as the porphyritic and non-porphyritic phases of the Starav granite in the Etive complex; the medium-grained hornblende–biotite granite of the Cairnsmore of Carsphairn; and that forming the inner part of the Portencorkrie complex of Wigtownshire. All these have biotite in excess of the green hornblende.

Similar rocks occur as relatively late phases in a number of the great composite batholiths such as the Coast Range batholith in south-eastern Alaska, and the Sierra Nevada batholith in California (22).

Hornblende–biotite granodiorite is a very common type and most of the great composite batholiths such as those of the Coast Range and the Sierra Nevada (fig. 3.5B) just mentioned, and that of southern California (23), are composed predominantly of this. There are usually a number of different plutons of this type in such batholiths, the earlier ones being less silicic and richer in hornblende than the later ones and, not infrequently, having cores of pyroxene in some of the hornblende crystals.

Both porphyritic (with phenocrysts of potash feldspar) and non-porphyritic varieties may be present.

Numerous examples, on a smaller scale, are to be found in Britain among the 'Newer Granites' of Scotland as, for instance, the Cruachan 'granite' of the Etive complex, the porphyritic outer 'granite' and the

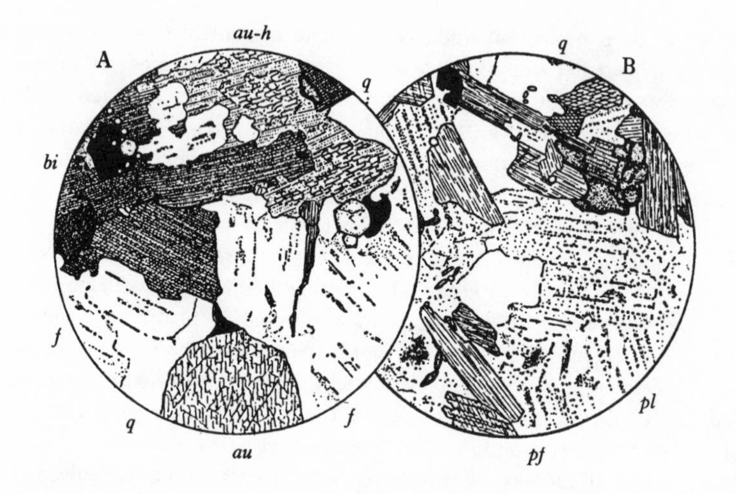

Fig. 3.5. × 20.

A. Hornblende–biotite tonalite with some clinopyroxene, Beinn Cruachan, Argyllshire, Scotland. Biotite (*bi*), green hornblende with cores of augite (*au-h*), augite (*au*), feldspar (*f*), averaging oligoclase but strongly zoned between crossed polars, and quartz (*q*).

B. Hornblende–biotite granodiorite, near Grouse Lake, Sierra Nevada, U.S.A. Green hornblende with some biotite, plagioclase (*pl*), potash feldspar (*pf*) and quartz (*q*), with accessory sphene, iron-ore and apatite.

basic type of inner 'granite' at Ben Nevis, the Strontian hornblende–biotite granodiorite, the porphyritic granodiorite of the Foyers complex, the granodiorites of the Criffel–Dalbeattie mass and those of the Garabal Hill–Glen Fyne complex. In the latter, a medium-grained granodiorite with a colour index of 19 and with hornblende nearly as abundant as biotite is followed by a porphyritic rock with phenocrysts of microcline-microperthite, a colour index of 10, and with biotite strongly dominant over hornblende. This grades in the central portions of its outcrop into a fine-grained granodiorite with a colour index of 7 and with hornblende reduced almost to the status of an accessory constituent.

Hornblende–biotite tonalite is, again, the most widespread variety of tonalite and occurs frequently as a relatively early phase in composite granitic batholiths and plutons, as in the Strontian complex and the Foyers complex among the 'Newer Granites' of Scotland. Tonalites of this kind are to be found in the western parts of the Coast Range batholith in Alaska. The feldspar is andesine with accessory potash feldspar and the average colour index is 25. The tonalite of Baranof

Island, west of the main batholith, is somewhat richer in quartz and has a colour index of 10. Other excellent examples are provided by the tonalites of the southern California batholith. Some carry variable amounts of clinopyroxene (fig. 3.5A).

Pyroxene granodiorites and tonalites are relatively rare and usually have biotite as an additional ferromagnesian constituent. Here fall the rocks from south Norway called 'opdalite'. These are medium-grained *biotite–pyroxene granodiorites* with strongly zoned plagioclase often mantled by the microperthitic microcline, some 15 per cent of interstitial quartz, biotite, hypersthene and augite, the latter sometimes partly replaced by brownish green hornblende. Rocks resembling these are to be found elsewhere, as in the Peruvian Andes, and in the Boulder batholith of Montana (the Unionville granodiorite) (24).

Pyroxene tonalites seem to be more variable in character. Thus, *biotite–pyroxene tonalite* with two pyroxenes (Austberg, Norway), *biotite–augite tonalite* (east of Ornas, Radmanso, Sweden), and *biotite–hypersthene tonalite* (Vasaukari, Finland) have all been recorded.

The biotites and amphiboles in the granites and granodiorites dealt with so far have only a moderate content of iron. There are, however, varieties in which the biotite is a very iron-rich type and the amphibole, when present, is a ferrohastingsite. These may be distinguished as *ferrogranites* and *ferrogranodiorites*.

Probably the best-known examples are to be found among the rapakivi granites of Fennoscandia and, in particular, those of the great Wiborg massif in Finland. This is made up of a number of different rock types. One of these is coarse grained and composed essentially of microcline-microperthite, minor oligoclase, quartz, and some iron-rich biotite; another shows the rapakivi texture, the mantled ovoids of potash feldspar being set in a medium-grained matrix of potash feldspar, subordinate oligoclase, and quartz with iron-rich biotite and minor ferrohastingsite. The somewhat earlier 'Lappee granite' ranges from granite to granodiorite, has ferrohastingsite normally in excess of biotite and may carry some hedenbergite and a little fayalite.

Rocks similar to these are to be found elsewhere. They occur among the Tertiary intrusions of western Scotland and Northern Ireland as, for instance, in the Mourne Mountains where an early ferrohastingsite–biotite granite is followed by biotite granite and this, in turn, by a biotite alkali granite (25). Both ferrogranodiorites and ferrogranites are present in the upper portions of the great Bushveld igneous complex in the Transvaal.

Some intrusions of granitic rocks carry an abundance of dark *mafic inclusions*, also termed 'basic segregations' or 'autoliths'. They are normally finer grained than the rock in which they occur and their minerals are similar to those of the enclosing rock but they are much enriched in the ferromagnesian minerals and also in plagioclase. At one time believed to be clots of early formed minerals (hence the name 'basic segregation'), they are now generally thought to be fragments of country rock, or of an earlier basic intrusive phase, modified by reaction with the magma in which they were caught up (26). Detailed descriptions of such inclusions have been given, for example, for those present in the Bonsall tonalite of the southern California batholith, and for those in the Shap granite (fig. 3.4B) (27).

Chemistry

Average chemical compositions for peralkaline granite, alkali granite, granite, granodiorite and tonalite are given in table 3.1. All have high or fairly high contents of silica, with decrease in the amount of this oxide on passing from granites through granodiorite to tonalite. This decrease is coupled with an increase in alumina and lime, and a decrease in potash, leading to an increase in normative anorthite and a decrease in normative orthoclase. Normative quartz remains fairly high throughout.

Note the low alumina content, the excess of ferric over ferrous oxide and the higher content of soda in average peralkaline granite. These features are reflected in the norm by the absence of anorthite and the presence of diopside and acmite. Note also the low lime content here and in average alkali granite.

Average alkali granite and average granite both show appreciable normative corundum owing to the fact that micaceous varieties of these rocks are dominant in the average. With the increasing importance of hornblende-bearing rocks in average granodiorite and average tonalite, the former shows neither corundum nor diopside in the norm, while in the latter a little normative diopside is present.

The last two columns of table 3.1 give the compositions of average ferrogranite and ferrogranodiorite, and show well the marked differences between these and average granite and granodiorite. Characteristic are the relatively low contents of alumina, and the high iron contents coupled with low magnesia.

Note the high (> 80) values of the differentiation index for peralkaline granite, alkali granite, and granite, and the lower values for granodiorite and tonalite.

TABLE 3.1. *Average chemical compositions of granitic rocks*

	1	2	3	4	5	6	7
SiO_2	72.72	73.86	70.41	66.88	66.15	71.00	64.90
TiO_2	0.32	0.20	0.45	0.57	0.62	0.50	0.89
Al_2O_3	11.51	13.75	14.38	15.66	15.56	12.93	13.33
Fe_2O_3	2.72	0.78	1.04	1.33	1.36	1.40	2.17
FeO	1.76	1.13	1.93	2.59	3.42	3.16	6.29
MnO	0.10	0.05	0.06	0.07	0.08	0.08	0.14
MgO	0.18	0.26	0.81	1.57	1.94	0.35	0.52
CaO	0.73	0.72	1.97	3.56	4.65	1.88	3.52
Na_2O	4.84	3.51	3.23	3.84	3.90	3.18	3.74
K_2O	4.66	5.13	4.95	3.07	1.42	4.91	3.38
H_2O^+	0.38	0.47	0.55	0.65	0.69	0.51	0.78
P_2O_5	0.06	0.14	0.20	0.21	0.21	0.09	0.23
qz	26.3	32.2	27.2	22.0	24.2	28.0	20.2
or	27.8	30.0	28.9	18.4	8.3	28.9	20.0
ab	33.0	29.3	27.3	32.5	33.0	27.3	31.4
an	–	2.8	8.3	16.4	20.9	6.1	9.5
C	–	1.4	0.7	–	–	–	–
di *wo*	1.3	–	–	–	0.4	1.2	2.8
di *en*	0.2	–	–	–	0.2	0.2	0.4
di *fs*	1.2	–	–	–	0.1	1.1	2.6
hy *en*	0.2	0.6	2.0	3.9	4.6	0.7	0.9
hy *fs*	1.5	1.1	2.2	2.9	4.0	2.9	5.9
ac	6.9	–	–	–	–	–	–
mt	0.5	1.2	1.4	1.9	2.1	2.1	3.3
il	0.6	0.5	0.8	1.1	1.2	0.9	1.7
ap	0.1	0.3	0.5	0.5	0.5	0.2	0.6
DI	87.2	91.6	83.4	72.8	65.5	84.1	71.7

1. Peralkaline granite (average of 30).
2. Alkali granite (average of 48) (*Bull. Geol. Soc. Am.* **65**, 1012, table 1, no. III).
3. Granite (average of 166).
4. Granodiorite (average of 137) (*ibid.*, p. 1014, table 2, no. III).
5. Tonalite (average of 58) (*ibid.*, p. 1015, table 2, no. V).
6. Ferrogranite (average of 26).
7. Ferrogranodiorite (average of 9).

Later alteration

Few granitic rocks have remained entirely unchanged after their solidification. They have usually been modified, though to a very varying extent, by the action of residual fluids rich in water and have undergone some degree of *hydrothermal* alteration. The common results of this process are:

1. Chloritisation of biotite with excretion of iron-ore, needles of rutile, or granules of sphene. Sometimes lenses of epidote or prehnite are developed in the cleavage planes of the chlorite.
2. Potash feldspar may be sericitised, i.e. altered to an aggregate of finely divided white mica.
3. Lime-bearing plagioclase may be altered to a mixture of zoisite (or epidote) and secondary white mica, lying in an albitic base. In some cases lime is leached away during the alteration and only the white mica and albite appear. In general, lime-bearing plagioclase is less resistant to alteration than is potash feldspar.

Quartz and white mica are unaffected by the hydrothermal fluids; hornblende, if present, usually remains unchanged unless alteration is severe when it may be represented by chlorite and epidote, or chlorite and calcite. Peralkaline granites seem to show little change of this kind but the alteration of soda amphiboles and aegirine to magnetite or haematite may, however, be ascribed to hydrothermal activity. The albitisation of potash feldspar which is found in some granites also belongs to this period, and certain albite granites have been produced from more normal granites in this manner. Alternatively, plagioclase may be replaced to a varying degree by potash feldspar, leading to the formation of potash-rich granites.

While most granitic rocks show at least traces of such hydrothermal alteration, some have been affected considerably. Thus, the Eskdale granite in Cumberland shows a marked sericitisation of feldspar and chloritisation of biotite throughout.

More striking effects are produced when the residual fluids contained boron or fluorine. This occurs, more particularly, in connection with certain muscovite–biotite granites and biotite granites, and the granites of Devon and Cornwall may be considered in this connection.

With boron present in some quantity, *tourmaline granites* appear as modifications of the more normal rocks, the tourmaline apparently taking the place of the mica. As a further modification, the feldspars may be replaced partly or wholly by tourmaline and quartz, leading eventually to a *tourmaline–quartz* rock (fig. 3.6A).

When fluorine is important, the rock known as *greisen* is produced from the granite. The Cornish greisens consist of quartz and a fluorine- and lithium-bearing mica accompanied by topaz and a varying amount of tourmaline (fig. 3.6B). The micas of such topaz-bearing greisens are typically lithium-iron micas falling largely within the zinnwaldite range, as in the greisens of the Erzgebirge (28), more rarely siderophyllite, as in the greisens of the Mourne Mountains, N. Ireland (29), and in those associated with the Younger Granites of northern Nigeria.

In other examples, however, greisen has formed from granite through the agency of heated water only, as is the case with that found in association with the Skiddaw granite of Cumberland (30), where the mica is muscovite and both topaz and tourmaline are absent.

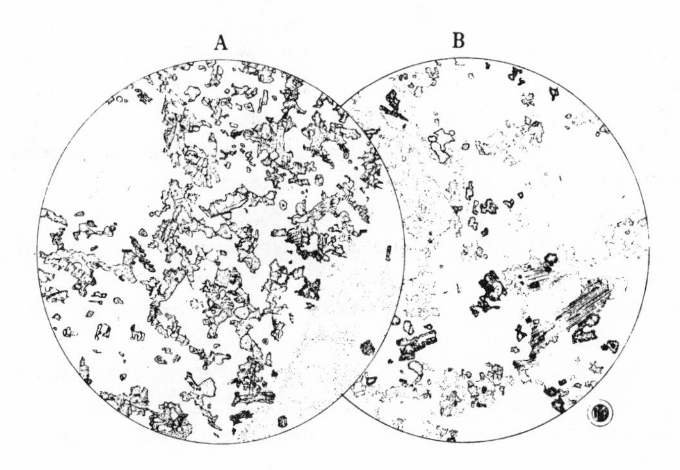

Fig. 3.6. × 22.

A. Tourmalinised granite, Trevalgan, St. Ives, Cornwall. Most of the granite has been replaced by an aggregate of quartz and tourmaline grains, the latter having brown cores and blue margins in thin section. Remnants of a large feldspar crystal can be seen at the south-east edge of the figure, and the clear area of quartz in the north-west may represent original material.

B. Greisen, Cligga Head, Cornwall. Quartz, pale mica, topaz, and tourmaline (both brown and blue in thin section).

Another type of alteration that affects some of the Cornish granites is ***kaolinisation***, best seen in the St. Austell granite mass (31). The feldspars are replaced to a varying degree by kaolinite, the plagioclase feldspar being most strongly affected while the potash feldspar phenocrysts are more resistant.

Granite-aplites and granite-pegmatites

Many granitic intrusions and their immediately surrounding country rocks are traversed by dykes and veins of more silicic character. These may be fine grained (aplite), or they may have a very coarse grain (pegmatite). Though usually occurring separately, they are sometimes to be found occupying the same vein (32). Most commonly these rocks have the mineralogical composition of a granite and may be referred to as granite-aplites and granite-pegmatites.

Granite-aplites (fig. 3.3B) have the typical equigranular, sugary, aplitic texture and are usually composed of alkali feldspars and quartz with muscovite and, sometimes, a little biotite. The minor constituents are commonly garnet and minerals such as tourmaline and topaz.

The coarse *granite-pegmatites* may have a zonal arrangement of their main constituents, frequently have a portion of their potash feldspar and quartz in graphic intergrowth, and may grade into massive quartz veins. They sometimes contain a large variety of rare minerals not found in the parent granitic rock. Local concentrations of these and the large, or even immense, size of the crystals of common minerals such as feldspar and mica, give many of these pegmatites considerable economic importance. Pegmatites in which no appreciable amount of hydrothermal replacement has taken place are said to be *simple*; those that owe their present character partly or wholly to such replacement are referred to as *complex*.

Simple granite-pegmatites are commonly made up essentially of quartz and potash feldspar (normally microcline-perthite) with or without minor oligoclase, muscovite, biotite, spessartite-almandite garnet, and tourmaline that is black in hand specimen.

Complex granite-pegmatites started as simple pegmatites but were subjected to hydrothermal replacement processes, often in stages. All show replacement of potash feldspar by albite and one of the commonest types of complex pegmatite shows also an introduction of lithium, fluorine, manganese, and often phosphorous as well, leading to the formation of new minerals such as lepidolite mica, red and green lithium tourmalines, spodumene, and various iron and manganese phosphates (33).

Although the aplites and pegmatites associated with granitic rocks usually have the mineral composition of a granite, this is not always the case and they may have a granodioritic or even tonalitic nature. *Granodiorite-aplites and pegmatites* and *tonalite-aplites and pegmatites* occur in addition to granite-aplites and pegmatites associated with the granitic rocks of the Coast Range batholith in Alaska; the pegmatites of the Sprucepine district, N. Carolina, range from tonalite-pegmatite through granodiorite-pegmatite to granite-pegmatite; and tonalite-aplites and tonalite-pegmatites ('trondhjemite'-aplites and pegmatites) occur abundantly in connection with the 'trondhjemite' intrusions of southern Norway.

These less silicous pegmatites do not seem to show much evidence of any hydrothermal replacement and appear to be simple.

References and notes

(1) Tuttle, O. F. & Bowen, N. L., 1958. Origin of granite in the light of experimental studies in the system $NaAlSi_3O_8$–$KAlSi_3O_8$–SiO_2–H_2O. *Mem. geol. Soc. Am.* **74**, 129–30.
(2) These minute crystals are commonly identified as zircon but may, in fact, also be monazite, xenotime or apatite (see, for instance, Smetsinger, K. G., 1967. Nuclei of pleochroic halos in biotites of some Sierra Nevada granitic rocks. *Am. Miner.* **52**, 1901–3).
(3) *E.g.*, Elders, W. A., 1968. Mantled feldspars from the granites of Wisconsin. *J. Geol.* **76**, 37–49.
(4) Emeleus, C. H., 1963. Structural and petrographic observations on layered granites from southern Greenland. *Spec. Pap. Min. Soc. Am.* **1**, 22–9. Claxton, C. W., 1968. Mineral layering in the Galway Granite, Connemara, Eire. *Geol. Mag.* **105**, 149–59.
For layering caused by differences in texture see Gates, R. M. & Scheerer, P. E., 1963. The petrology of the Nonewaug Granite, Connecticut. *Am. Miner.* **48**, 1040–69.
(5) Sederholm, J. J., 1928. On orbicular granites, spotted and nodular granites etc., and on the rapakivi texture. *Bull. Comm. géol. Finlande*, no. **83**.
(6) Jacobson, R. R. E., MacLeod, W. N. & Black, R., 1958. Ring complexes in the Younger Granite province of northern Nigeria. *Mem. geol. Soc. Lond.* **1**, 16–19.
(7) Tuttle & Bowen, 1958.
(8) Sabine, P. A., 1960. The geology of Rockall, North Atlantic. *Bull. geol. Surv. U.K.* **16**, 156–78.
(9) Roedder, E. & Coombs, D. S., 1967. Immiscibility in granitic melts, indicated by fluid inclusions in ejected granitic blocks from Ascension Island. *J. Petrology* **8**, 417–51.
(10) Chapman, R. W., 1935. Percy ring-dike complex. *Am. J. Sci.* (5) **30**, 426–8.
(11) Wager, L. R., Vincent, E. A., Brown, G. M. & Bell, J. D., 1965. Marscoite and related rocks of the Western Red Hills complex, Isle of Skye. *Phil. Trans. Roy. Soc. Lond.*, Ser. A., **257**, 279.
(12) Brown, P. E., 1956. The Mourne Mountains granites – a further study. *Geol. Mag.* **93**, 72–84.
(13) Chayes, F., 1952. The finer-grained calcalkaline granites of New England. *J. Geol.* **60**, 207–54.
(14) Dollar, A. T. J., 1941. The Lundy complex: its petrology and tectonics. *Q. Jl. geol. Soc. Lond.* **97**, 39–77.
(15) Edmonds, E. A., McKeown, M. C. & Williams, M., 1969. South-West England. *Brit. Regional Geol.* (3rd ed.), chapter 5, and references to papers on individual intrusions given there.
(16) Bisset, C. B., 1932. A contribution to the study of some granites near Aberdeen. *Trans. geol. Soc. Edinb.* **13**, 72–88.
(17) Church, M. S., 1937. A quantitative petrographic study of the Black Mountain leucogranodiorite at West Dummerston, Vermont. *J. Geol.* **45**, 763–74.

(18) Grantham, D. R., 1928. The petrology of the Shap Granite. *Proc. geol. Ass. Lond.* **39**, 299–331.

(19) Accounts of most of the 'Newer Granites' of Scotland mentioned here and on subsequent pages will be found in: Pringle, J., 1948. The South of Scotland. *Brit. Regional Geol.* (2nd edit.), chapter 4. Phemister, J., 1960. The Northern Highlands. *Brit. Regional Geol.* (3rd edit.), chapter 8. Johnstone, G. S., 1966. The Grampian Highlands. *Brit. Regional Geol.* (3rd edit.), chapter 8, and references to papers on individual intrusions given in these.

(20) Haslam, H. W., 1968. The crystallisation of intermediate and acid magmas at Ben Nevis, Scotland. *J. Petrology* **9**, 84–104.

(21) Langton, C. M., 1935. Geology of the north-eastern part of the Idaho batholith and adjacent region in Montana. *J. Geol.* **43**, 37–8.

(22) Bateman, P. C., Clark, L. D., Huber, N. K., Moore, J. G. & Rinehart, C. D., 1963. The Sierra Nevada batholith: a synthesis of recent work across the central part. *Prof. Pap. U.S. geol. Surv.* **414**-D, 1-46.

(23) Larsen, E. S., 1948. Batholith and associated rocks of Corona, Elsinore, and San Luis Rey Quadrangles, Southern California. *Mem. geol. Soc. Am.* **29**, 182 pp.

(24) Knopf, A., 1957. The Boulder Bathylith of Montana. *Am. J. Sci.* **255**, 81–103.

(25) Brown, 1956.

(26) It has been suggested recently that such mafic inclusions could, in some instances, be the refractory residue remaining after partial melting of crustal material to give the magma in which they were suspended. See Piwinskii, A. J., 1968. Studies of batholithic feldspars: Sierra Nevada, California. *Contr. Miner. Petrol.* **17**, 218–21.

(27) Hurlbut, C. S., 1935. Dark inclusions in a tonalite of southern California. *Am. Miner.* **20**, 609–30. Grantham, 1928.

(28) Rieder, M., 1970. Chemical composition and physical properties of lithium-iron micas from the Krusněhory Mts. (Erzgebirge): part A. Chemical composition. *Contr. Miner. Petrol.* **27**, 131–58.

(29) Nockolds, S. R. & Richey, J. E., 1939. Replacement veins in the Mourne Mountains Granite, N. Ireland. *Am. J. Sci.* **237**, 27–47.

(30) Hitchen, C. S., 1934. The Skiddaw Granite and its residual products. *Q. Jl. geol. Soc. Lond.* **90**, 158–99.

(31) Exley, C. S., 1959. Magmatic differentiation and alteration in the St. Austell Granite. *Ibid.* **114**, 197–230.

(32) See, for example, Heinrich, E. W., 1965. Composite aplite-pegmatites of the Franklin-Sylva district, North Carolina. *Am. Miner.* **50**, 1681–97.

(33) Good accounts of particular complex granite-pegmatites will be found in:

Landes, K. K., 1925. The paragenesis of the granite pegmatites of Central Maine. *Ibid.* **10**, 355–411.

Quensel, P., 1952. The paragenesis of the Varuträsk Pegmatite. *Geol. Mag.* **89**, 49–60.

4

Silicic volcanic and hypabyssal rocks

The silicic volcanic rocks comprise the very fine-grained to glassy equivalents of the granitic rocks. The volcanic representatives of granites are *rhyolites* and *rhyolite-obsidians*; of granodiorite, *rhyodacite* and *rhyodacite-obsidian*; and of tonalite, *dacite* and *dacite-obsidian*.

These siliceous volcanics occur as flows and as volcanic domes, and most of the great ash-flows of the world, also, have been produced by magmas of this type (see p. 212). As many of these rocks are crypto-crystalline or have a partially, or wholly, glassy groundmass it is often necessary to have a chemical analysis before they can be assigned to their proper group.

1. Rhyolites and rhyolite-obsidians

If the cooling of a magma corresponding in composition with a granite is very rapid, it will solidify as a glass, with or without phenocrysts, and the resulting rock is termed a *rhyolite-obsidian*. These are commonly black, or sometimes green in hand specimen, with a high lustre and break with a conchoidal fracture. In thin section they are very pale coloured or colourless, frequently showing perlitic cracks (fig. 4.1A) and numerous crystallites or microlites. Many are vesicular which, in the extreme, produces *pumice*.

Recorded refractive indices for the glass of fresh rhyolite-obsidians range from 1.483 to 1.523 and there is some evidence that the refractive index tends to be higher for peralkaline rhyolite-obsidians than for the others. The average of seven is 1.503 for peralkaline rhyolite-obsidians (range 1.488 to 1.523), the average of nine is 1.489 for the other rhyolite-obsidians (range 1.483 to 1.498).

Rhyolite-pitchstones resemble obsidians but are more commonly red or dark green in hand specimen and have a dull, pitch-like lustre. They are volcanic glasses, also, and may be indistinguishable from obsidian in thin section. They differ from the latter in containing relatively large amounts of water, up to a maximum of some 10 per cent, but averaging about 4 per cent. In many cases, at least, this water is secondary and has been absorbed by the glass. Such absorption is, not infrequently, accompanied by a leaching of alkalies, especially sodium.

In the *rhyolites* proper, the groundmass is more or less crystalline and composed essentially of feldspar and one or more silica minerals. Its texture may be cryptocrystalline; finely microcrystalline (fig. 4.1B); micropoikilitic; or microspherulitic (fig. 4.1C). The feldspars present are

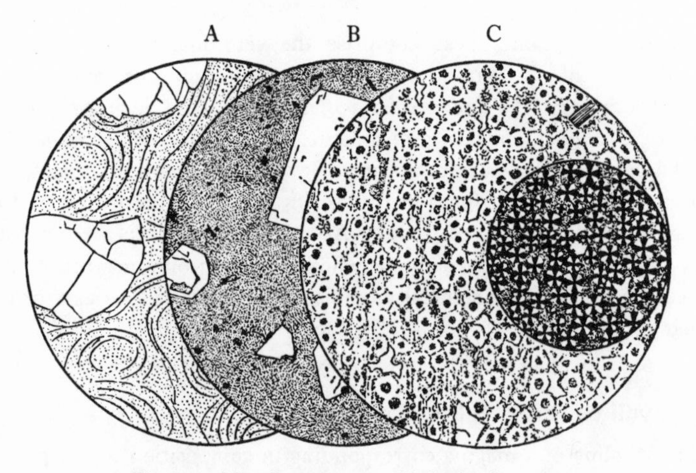

Fig. 4.1. Rhyolites, Antrim, N. Ireland; × 20.

A. Rhyolite-obsidian, Sandy Braes. Phenocrysts of quartz set in pale brownish glass with perlitic cracks.
B. Rhyolite, Kirkinriola. Phenocrysts of sanidine and quartz, a few scattered flakes of biotite and little crystals of iron-ore, lying in a finely microcrystalline groundmass of feldspar and quartz.
C. Microspherulitic rhyolite, Cloughwater. Composed essentially of closely packed little spherulites, each giving a black cross between crossed polars, as shown in the small inset circle.

sanidine in little laths or grains with or without oligoclase, depending on the type of rhyolite. The silica minerals present in the groundmass of recent rhyolites may be tridymite, cristobalite, or quartz. Cristobalite tends to form when crystallisation was very rapid or retarded and may be the silica mineral intergrown with the feldspar in the microspherulites. Tridymite and quartz form under conditions of less rapid crystallisation. Both cristobalite and tridymite have crystallised metastabily and in older rhyolites the silica mineral of the groundmass is always quartz due, no doubt, to the inversion of original cristobalite and tridymite.

The textures of the groundmass have been listed individually but many rhyolites show several of them associated, sometimes within the limits of a thin section. Thus successive bands may have different textures and such differences may produce a marked fluxion structure in the rock, sometimes further enhanced by differences in the colour of the bands in hand specimen.

These textures are not always primary. Glasses have a tendency to crystallise, or devitrify, and many do so in the course of time. As a result of this, rhyolite-obsidians are virtually unknown in pre-Permian times. Sometimes, however, remnants of perlitic cracks can still be seen although the rock is now completely crystalline, proving its original glassy nature.

Many, perhaps most, rhyolites and rhyolite-obsidians are *porphyritic*. The commonest phenocrysts are those of feldspar and, of these, *sanidine* is the most important. It is usually a soda-sanidine corresponding in composition with the microperthite of the granites; in some peralkaline rhyolites its place is taken by *anorthoclase*. Some phenocrysts of *lime-bearing plagioclase* may be present in subalkali rhyolites and obsidians; *albite* is sometimes found in alkali and peralkaline types.

Phenocrystal quartz, if present, has the bipyramidal habit characteristic of the (originally) high-temperature form, often much embayed. Phenocrysts of *micropegmatite* are found in certain rocks.

In many rhyolites, ferromagnesian constituents are practically confined to the phenocryst generation. The commonest is probably *biotite* in small euhedral tablets; *augite*, more or less colourless in thin section, is not uncommon; *hypersthene* is found in some; but *hornblende* is rarer and may show signs of resorption. Certain subalkali rhyolites and pitchstones have phenocrysts of *ferroaugite* and *fayalite*. In a few instances, spessartite-almandite *garnet* takes the place of the more usual ferromagnesian constituents. A noticeable feature is the absence of any primary muscovite.

Peralkaline rhyolites have *aenigmatite* as their most constant ferromagnesian phenocryst associated with sodic *hedenbergite*, sometimes accompanied by *fayalite*, or with a *soda-amphibole*. In this type of rhyolite, ferromagnesian constituents also appear to be more prominent in the groundmass and include aenigmatite, *aegirine*, *arfvedsonite* and *riebeckite*.

Accessory *iron-ore*, *apatite* and *zircon* may be observed at times in a phenocryst generation, though some part of these is doubtless hidden away in the groundmass.

In addition to the tiny microspherulites already mentioned as sometimes making up the groundmass, many rhyolites and some rhyolite-obsidians have *larger spherulites* (fig. 4.2), ranging in diameter from a fraction of a centimetre to several centimetres in different occurrences. These are more or less isolated growths, and are made up of sanidine fibres alone. They are frequently of complex structure; the radiating fibres may start from numerous centres and they often bifurcate with the production of a tufted or plumose arrangement. The interstices between the fibres are filled with a silica mineral or with glass. Large spherulites may have a cavity at the centre (hollow spherulites), or may have within a number of roughly concentric spaces (lithophysae). The hollow spaces in both may contain aggregates of tridymite or cristobalite and small crystals of minerals such as topaz, beryl, garnet and pseudobrookite or little tablets of fayalite.

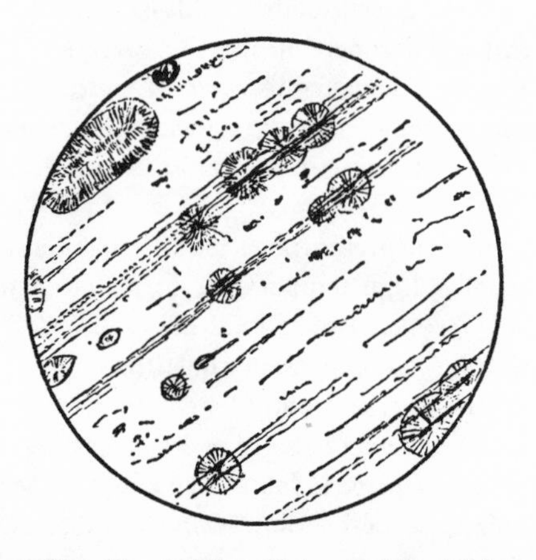

Fig. 4.2. Rhyolite-obsidian, Vulcano, Lipari Isles; × 20.

The glassy matrix encloses isolated spherulites which have some tendency to coalesce in bands following the direction of flow. Note that the flow-lines pass uninterruptedly through the spherulites.

Some of the large spherulites have formed before the lava came to rest, as crystallites or microlites can be seen curving round them (1). But many are of late formation as they lie across lines of flow marked out by crystallites (fig. 4.2). They may also be found in some obsidians as incomplete growths on the sides of cracks, and must then certainly be due to devitrification.

Main varieties, with some examples

Rhyolites and rhyolite-obsidians are known corresponding with peralkaline granite, alkali granite and granite, and the same kind of nomenclature may be used.

Peralkaline rhyolites ('pantellerites' and 'comendites') and peralkaline rhyolite-obsidians are well displayed in the islands of Sardinia and Pantelleria (2) in the Mediterranean. Those of Pantelleria usually have phenocrysts of anorthoclase, sodic hedenbergite, aenigmatite, and sometimes fayalite. Some have quartz in addition. The groundmass is variable consisting in some of a very fine-grained intergrowth of alkali feldspar and a silica mineral together with numerous little crystals of aegirine and aenigmatite; in others minute laths of the feldspar are enclosed micropoikilitically by small plates of quartz, or the feldspar and quartz form a minutely granular base for the little crystals of the ferromagnesian minerals, accompanied by more or less glass. The latter form a kind of transition to the obsidians, where the groundmass is of glass. While many of these obsidians are porphyritic, some are virtually devoid of phenocrysts.

Similar rocks have been described from parts of Africa, including Kenya where they are associated with others carrying riebeckite as their only, or dominant, ferromagnesian mineral; from Malagasy (Madagascar); from Mayor Island, New Zealand (3); from the Canary Islands; from Ascension Island, and elsewhere.

Among the Malagasian rocks there are some with phenocrysts of quartz and alkali feldspar in a spherulitic and partly vitreous groundmass. Chemical analysis shows these to be typical peralkaline rhyolites although no individualised ferromagnesian constituents are present, indicating the relatively late period of crystallisation of the latter.

The examples given above are all of Tertiary or a more recent age. Peralkaline rhyolites seem to be less conspicuous in earlier geological periods, perhaps because alteration of the ferromagnesian minerals renders the rocks difficult to identify. That they can still be found comparatively well-preserved under favourable conditions is shown by the peralkaline rhyolites of middle Devonian age in the Lahn district, Germany, where they occur among the so-called keratophyres of the area. Here, the freshest rocks may still carry some riebeckite and aegirine or aegirine-augite, though in many chlorite occurs instead and there is a development of carbonate.

Alkali rhyolites are highly leucocratic rocks, commonly with phenocrysts of sanidine and quartz, or of sanidine alone. In some, small amounts of lime-bearing plagioclase may appear among the phenocrysts, but this mineral is always absent from the groundmass and is never more than an accessory constituent.

Biotite, more rarely amphibole, may be present in small amount in the phenocryst generation; alternatively, pyroxene, often a ferroaugite, with or without fayalitic olivine may occur.

The Recent alkali rhyolites and rhyolite-obsidians forming a chain of plug-domes south of Mono Lake, California (4), have sparse phenocrysts of sanidine, quartz and plagioclase, sometimes with amphibole and fayalite in addition. The groundmass, which varies from glassy to microcrystalline, is frequently riddled with small laths of feldspar and tiny crystals of amphibole, pyroxene and iron-ore, and spherulitic patches of intergrown feldspar and quartz may be common.

The alkali rhyolite of Sugarloaf Hill, San Franciscan volcanic field, Arizona (5), has phenocrysts of alkali feldspar with a few of quartz, oligoclase and biotite in a groundmass made up of about equal parts of glass and of finely microcrystalline alkali feldspar and quartz. That of Capanna, in Sardinia, has small phenocrysts of sanidine and somewhat altered biotite in a groundmass with small spherulites and micropoikilitic patches of quartz and alkali feldspar with minute needles, apparently of pyroxene. That of Phira, Antiparos, has phenocrysts of quartz and sanidine in a partly crystalline groundmass containing a little biotite.

On the other hand, the alkali rhyolites of Lipari (6) have small phenocrysts of pyroxene and feldspar in a very fine-grained holocrystalline or partly glassy groundmass, and the alkali rhyolite found as a plug at Thingmuli, Iceland, has rare phenocrysts of ferroaugite and fayalitic olivine in a finely crystalline groundmass of alkali feldspar and quartz.

Many of the *alkali rhyolite-obsidians* are virtually non-porphyritic. This is the case with the well-known obsidian of Lipari and with that of the Obsidian Cliff flow, Yellowstone National Park. Both have spherulites and lithophysae embedded in the glass.

Here also belong some of the pitchstones of Arran and elsewhere in the Brito-Icelandic Tertiary igneous province (7) with sparse phenocrysts of ferroaugite with or without orthopyroxene and fayalitic olivine, a few of lime-bearing plagioclase, and microphenocrysts of iron-ore, set in a glass that often contains abundant microlites.

Lime-bearing plagioclase, occurring, if at all, only as an accessory constituent in alkali rhyolites, becomes an essential constituent in the ***rhyolites.***

The rhyolites of the Potosi series, San Juan region, S.W. Colorado (8), are somewhat feebly porphyritic with phenocrysts of sanidine common but not abundant, some plagioclase ranging in composition from oligoclase to calcic andesine in different cases, and a few of biotite. Phenocrysts of quartz are rare.

The groundmass may be in large part cryptocrystalline with small fibrous patches, or it may be microspherulitic. X-ray investigation of such groundmasses indicates that they are made up of alkali feldspar and cristobalite. A number of the rocks have more coarsely crystalline and porous streaks in their groundmass, usually composed of tridymite and alkali feldspar. More rarely, the place of tridymite is taken by quartz. At the base of the flows the groundmass is commonly of glass, giving rise to ***rhyolite-obsidian.***

There are all gradations from these rocks to others termed 'rhyolitic quartz-latites'. These have more abundant phenocrysts of which the chief is zoned plagioclase, averaging oligoclase or andesine in different examples, together with biotite, and sometimes augite or hornblende, or both, in addition. Some also have phenocrysts of sanidine and quartz. The groundmass of these rocks is similar to that found in the rhyolites but tends to be less siliceous. While some of these are true rhyolites, others would be more correctly termed rhyodacites and they can be regarded as a transitional group.

Tertiary rhyolites from Lörinci in the Matra district of Hungary have sparing phenocrysts of An_{35}, feebly zoned, and a few pseudomorphs probably after biotite in a cryptocrystalline to finely microcrystalline groundmass. When microcrystalline, it can be seen to be made up of little laths of sanidine and oligoclase cemented by quartz.

The rhyolites of the second period (?Cretaceous) in the Mount Somers district, Canterbury, New Zealand, have phenocrysts of embayed quartz, sanidine and oligoclase-andesine in about equal amounts, occasional biotite more or less replaced by iron-ore, and very occasional rounded grains of almandite garnet, set in a cryptocrystalline to finely microcrystalline groundmass of feldspar and of quartz, some of which appears to be secondary. *Pitchstones* with the same phenocrysts but with a groundmass of glass are also present.

The coarsely porphyritic vesicular rhyolite of Black Range, New Mexico (9), is of interest as it is one of a number of occurrences in which the vesicles contain cassiterite as well as minerals such as garnet, topaz, pseudobrookite, specularite, etc.

The Recent *rhyolite-obsidian* south of Borax Lake, Clear Lake area, California, is a colourless glass in section with bands of crystallites and a few microlites. It has numerous small xenoliths, probably cognate, composed essentially of acicular green-brown hornblende and andesine with some interstitial glass. The same minerals are scattered through the obsidian as little clots or as isolated individuals. A *rhyolitic pumice* from pumice breccia in the same area has phenocrysts of sanidine, oligoclase, quartz and biotite.

Flows of rhyolitic lava are not normally of any very great volume, but a remarkable flow composed of three main flow units apparently erupted from the same vent and having a volume estimated at no less than 24km^3 occurs in the Andes of northern Chile (10). The rock is a somewhat silica-poor porphyritic rhyolite-obsidian, white and crumbly in hand specimen, corresponding in composition with certain 'quartz monozonites'. Phenocrysts of poorly zoned andesine, fractured and fragmented, of highly fractured quartz, of biotite and some of euhedral green hornblende make up about 30 per cent of the whole, the remainder being a colourless glass with perlitic cracks.

2. Rhyodacites, dacites and their corresponding obsidians

These lavas are not always easy to tell apart in the absence of a chemical analysis as their groundmass is often partly or wholly glassy or else cryptocrystalline. Moreover, some rhyodacites closely resemble rhyolites, while dacites and other rhyodacites may look very much like certain andesites.

They are usually porphyritic and may have euhedral phenocrysts of quartz (often much embayed, as in the rhyolites), and this serves to distinguish them from andesites. Rhyodacites may have phenocrysts of *sanidine* and can then be distinguished from dacites. But frequently, both the quartz and alkali feldspar are occult, hidden in the glass of the groundmass. Both rhyodacites and dacites have phenocrysts of *lime-bearing plagioclase*, usually zoned, and sometimes riddled with glass inclusions.

Ferromagnesian constituents occurring as phenocrysts may be *biotite, oxybiotite, hornblende* (green or brown in thin section), *oxy-hornblende, augite* and *hypthersthene*. A few of the least siliceous dacites have a little *olivine*. Certain rhyodacites and dacites have phenocrysts of *ferroaugite* and *fayalite*.

Green hornblende seems to be more or less confined to rocks with a glassy or semiglassy groundmass, while the brown hornblende, probably derived from the green variety by partial oxidation, is more characteristic of rocks with a crystalline or feebly glassy groundmass. The red-brown oxyhornblende, with virtually straight extinction, occurs in rocks that have been strongly oxidised. The copper-red oxybiotite, with very strong pleochroism, definitely biaxial character and strong dispersion, is found in a similar environment.

Both the mica and amphibole phenocrysts are often much resorbed and replaced by a very fine-grained aggregate, frequently appearing virtually opaque, of iron-ore and pyroxene. When somewhat coarser, both ortho- and clinopyroxene can be identified, and also some plagioclase. Sometimes almost the whole of these phenocrysts is replaced in this way and the products can be seen being strewn about in the groundmass of the rock.

The *groundmass*, when crystalline, consists of feldspar and a silica mineral together with granules of pyroxene and iron-ore. Both sanidine and a lime-bearing plagioclase are present in the groundmass of rhyodacites, only the latter in that of dacites. In both, the groundmass plagioclase is richer in the albite molecule than the phenocrystal plagioclase. Tridymite and cristobalite are frequently present in the groundmass of fresh, young rocks but only quartz is found in the older lavas.

Rhyodacites may show any of the groundmass textures listed under rhyolite, but many dacites, and also some rhyodacites, have a groundmass with hyalopilitic texture in which a felt of plagioclase laths, granules of pyroxene and little grains of iron-ore are embedded in a matrix of glass.

Some lavas falling here are composed wholly of glass, with or without phenocrysts. These are the *rhyodacite-* and *dacite-obsidians*. Normally black in hand specimen and colourless in thin section like the rhyolite-obsidians, they can be distinguished from the latter only by chemical analysis. These glasses may also absorb water, lose their bright lustre, and form pitchstones.

Few determinations of refractive index seem to have been made on the glasses of these obsidians but an average of five rhyodacite-obsidians gives 1.493, with a range of 1.490 to 1.499. A dacite-obsidian gave a value of 1.515.

Some examples

Rhyodacites are very common lavas, much commoner, in fact, than a perusal of the literature would suggest. The reason for this is that petrographers tended to classify any glassy siliceous volcanic rock as a 'rhyolite', while in rocks with a hyalopilitic groundmass the nature of the glass was ignored and they were classed as 'andesite'. Sometimes, again, rhyodacites were lumped together with true dacites as 'dacites'.

Many of the 'dacites' of the Lassen Peak Volcanic National Park, California, have appreciable contents of potash and are really rhyodacites (11) (fig. 4.1A). Nearly all have prominent phenocrysts of somewhat zoned plagioclase, commonly rich in glass inclusions. In the commonest

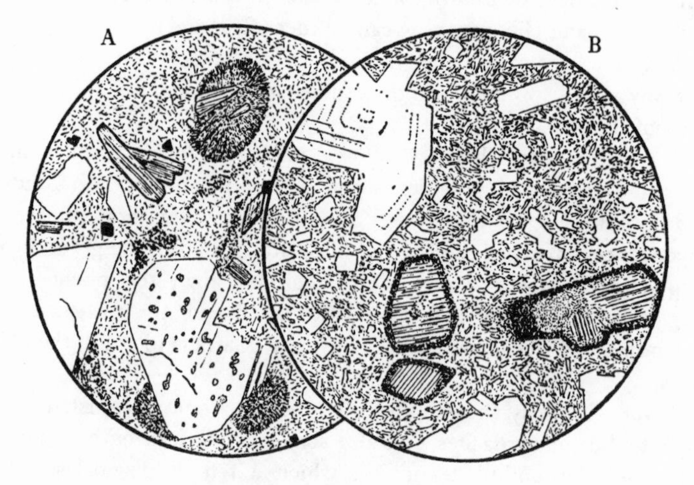

Fig. 4.3. × 22.

A. Rhyodacite (pre-Lassen flow), Lassen Peak, California, U.S.A. Phenocrysts of andesine, brownish green hornblende, biotite, iron-ore and, elsewhere in the section, sparing orthopyroxene and quartz. These are set in a groundmass of colourless glass carrying microlites of plagioclase and pyroxene, and crowded with acicular crystallites. There are, in addition, some imperfect spherulitic growths.

B. Hornblende dacite, Black Butte, Mt. Shasta, California, U.S.A. The phenocrysts are of hornblende (with resorption border) and zoned bytownite-labradorite. The groundmass has numerous microlites of andesine, some orthopyroxene and iron-ore set in an almost colourless glass.

type (Raker Peak, Chaos Crags, etc.) the phenocrysts are of oligoclase-andesine and there are also phenocrysts of green or brownish green hornblende and of biotite. Both may be oxidised and may show partial resorption. In some there are a few phenocrysts of quartz. These are set in a groundmass consisting of little prisms of granules of pyroxene (usually both ortho- and clinopyroxene), laths of plagioclase and tiny grains of iron-ore, all embedded in a colourless, often pumiceous glass that may be clear or crowded with minute laths of alkali feldspar. Zircon and apatite are common accessory constituents.

Less siliceous rocks (partly described as 'andesite', but having from 15 to 20 per cent of normative quartz) may have phenocrysts of plagioclase, averaging medium andesine, commonly accompanied by some of augite and orthopyroxene. The groundmass has microlites of andesine; grains and small prisms of augite and orthopyroxene; and granules of iron-ore, with a varying amount of pale grey to brownish glass. Some of these rocks have accessory olivine (e.g. 'Juniper Andesites'), or biotite with or without olivine (1915 eruption, Lassen Peak).

Some of the rhyodacites of the Clear Lake area, Coast Range, California, have a residuum of cryptocrystalline material, rather than glass. Some also have scattered phenocrysts of sanidine in addition to plagioclase and pyroxene.

Among the rhyodacites ('dacites') of the San Franciscan volcanic field, Arizona (12), are some that have the appearance of rhyolites. They range from holohyaline obsidians to cryptocrystalline rocks with recognisable minute scattered crystals of biotite and magnetite and may be aphanitic or have phenocrysts of plagioclase (An_{37}) and biotite, with or without a little hornblende and small crystals of orthopyroxene.

Other rhyodacites here have phenocrysts of plagioclase and green or brown hornblende, with or without a few of quartz.

Rhyodacites and rhyodacite-obsidians similar to the above may be found at numerous other localities, not only in America, but in other parts of the world such as Mexico, the Andes Mountains of South America, the East Indies, Hungary, the Aegean Islands, the Rotorua-Taupo subdivision in the North Island of New Zealand, etc., in connection with orogenic zones or island arcs of Tertiary to Recent age, and associated with andesites and latite-andesites.

They are also common, though usually less fresh, in older geological epochs as among the Old Red Sandstone lavas of Scotland (13), and among the Silurian volcanic rocks of the eastern Mendip Hills, Somerset (14).

A number of rhyodacite flows and domes carry *cognate xenoliths* of more crystalline character and coarser texture, ranging from sharply defined subangular nodules to ill-defined 'clots' visible only under the microscope. A common type consists of abundant brown more or less oxidised hornblende, and plagioclase richer in the anorthite molecule than that of the enclosing rock. In another type, the place of amphibole is taken by ortho- and clinopyroxenes. The first occurs largely in hornblende-bearing rhyodacites, the second in those with pyroxene.

Dacites are less abundant than rhyodacites but are found in the same kinds of environment. They are well developed, for instance, in the volcanic island arc of the Lesser Antilles. The dacites of Montserrat (15) have zoned phenocrysts of plagioclase, averaging about An_{65}, hypersthene with some augite, and varying amounts of hornblende, normally brown to red-brown (oxyhornblende) in colour. Quartz appears as an additional phenocrystal constituent in the more siliceous rocks. There are microphenocrysts of iron-ore and apatite while some rocks have microphenocrysts of plagioclase, pyroxene and hornblende in addition.

The groundmass may be holocrystalline and very fine-grained with plagioclase, granules of pyroxene and iron-ore, and cristobalite or tridymite or both; but in some there is a rather sparse residuum of glass.

Similar dacites are present in other islands such as Martinique, St. Kitts (16), Nevis and Guadeloupe.

Unlike the dacites just described, those of Mount Shasta, California (17), are usually rich in glass. Typically they have phenocrysts of zoned labradorite with some of orthopyroxene and augite, lying in a matrix of abundant glass, dusty with iron-ore and carrying varying amounts of andesine or oligoclase microlites, granules of augite, tiny prisms of orthopyroxene and accessory apatite.

The normal lava of the Black Butte dome (fig. 4.3B) differs in having large phenocrysts of brown, rarely green, hornblende partly or completely resorbed, and smaller ones of plagioclase zoned from bytownite to labradorite. Microlites of andesine, rather scanty orthopyroxene, accessory iron-ore and apatite lie in a base of pale yellow to colourless glass.

A more siliceous type from the main cone has plagioclase, zoned from labradorite to andesine, prisms of oxyhornblende, flakes of biotite and a few rounded crystals of quartz set in a dusty glass base.

Just as with granitic rocks, so with their corresponding lavas, there are certain types rich in iron and poor in alumina carrying iron-rich ferromagnesian minerals such as fayalitic olivine, ferroaugite and iron-rich pigeonite. These *ferrorhyolites*, *ferrorhyodacites* and *ferrodacites* are found associated with tholeiitic basalts and tholeiitic andesites as, for instance, in the Tertiary to Recent Brito-Icelandic igneous province (18) (ferrorhyolites and ferrorhyodacites) and in the pigeonite series of the Hakone volcano and surrounding area, Japan (19) (ferrodacites).

Chemistry of the silicic volcanic rocks

Average chemical compositions of silicic volcanic rocks are given in table 4.1, columns 1–7. Note the high Na_2O and low Al_2O_3 in peralkaline rhyolite, leading to the absence of *an* and presence of *ac* in the norm, and

TABLE 4.1. *Average chemical compositions of silicic volcanic rocks*

		1	2	3	4	5	6	7	8	9
SiO_2		72.18	74.57	71.10	66.27	62.68	68.40	64.73	75.04	71.23
TiO_2		0.39	0.17	0.37	0.66	0.57	0.85	0.83	0.10	0.47
Al_2O_3		10.59	12.58	14.15	15.39	17.07	13.18	15.02	13.39	13.74
Fe_2O_3		2.77	1.30	1.56	2.14	2.31	2.59	2.13	1.61	0.45
FeO		2.80	1.02	1.33	2.23	3.01	3.44	4.37	0.37	1.18
MnO		0.15	0.05	0.05	0.07	0.12	0.11	0.16	0.05	–
MgO		0.15	0.11	0.55	1.57	2.44	0.77	1.91	0.18	0.27
CaO		0.59	0.61	1.87	3.68	6.14	3.17	5.63	0.40	0.58
Na_2O		5.47	4.13	3.47	4.13	3.82	3.96	3.50	6.36	0.53
K_2O		4.45	4.73	4.73	3.01	1.21	2.51	1.03	0.83	10.31
H_2O^+		0.41	0.66	0.71	0.68	0.46	0.78	0.46	1.07	0.58
P_2O_5		0.03	0.07	0.11	0.17	0.16	0.22	0.21	0.08	0.13
qz		26.6	31.1	28.3	20.8	19.0	27.8	24.8	33.9	26.6
or		26.1	27.8	27.8	17.8	7.2	15.0	6.1	5.0	61.2
ab		29.9	35.1	29.3	35.1	32.0	33.5	29.3	54.0	4.7
an		–	2.0	8.3	14.5	26.1	10.6	22.2	1.4	2.0
C		–	–	0.2	–	–	–	–	1.4	0.9
di	*wo*	1.2	0.1	–	1.3	1.5	1.5	1.9	–	–
	en	0.1	–	–	0.9	1.0	0.6	0.9	–	–
	fs	1.2	0.1	–	0.3	0.4	0.9	0.9	–	–
hy	*en*	0.3	0.3	1.4	3.0	5.1	1.3	3.9	0.4	1.1
	fs	3.6	0.5	0.5	1.0	2.5	2.1	4.4	–	0.7
ac		8.3	–	–	–	–	–	–	–	–
mt		–	1.9	2.3	3.0	3.3	3.7	3.0	1.2	0.7
il		0.8	0.3	0.8	1.4	1.1	1.5	1.5	0.2	0.9
ap		0.1	0.2	0.3	0.4	0.3	0.5	0.5	0.2	0.3
ns		1.6	–	–	–	–	–	–	0.8*	–
DI		84.2	94.0	85.4	73.7	58.2	76.3	60.2	–	–

1. Peralkaline rhyolite and rhyolite-obsidian (average of 55).
2. Alkali rhyolite and rhyolite-obsidian (average of 21) (*Bull. Geol. Soc. Am.* **65**, 1012, no. IV).
3. Rhyolite and rhyolite-obsidian (average of 80).
4. Rhyodacite and rhyodacite-obsidian (average of 115) (*Ibid.*, p. 1014, no. IV).
5. Dacite and dacite-obsidian (average of 32).
6. Ferrorhyodacite (average of 9).
7. Ferrodacite (average of 18).
8. Soda quartz-keratophyre, Baker quadrangle, E. Oregon (Gilluly, *Am. J. Sci.* (5), **29**, p. 228).
9. Spherulitic potash keratophyre with relict perlitic structure, S.S.W. of Brodie's Creek, N.Z. (Battey, *Geol. Mag.* **92**, 116, anal. 5299).

**hm*

the very low normative *an* in alkali rhyolite. Rhyolite, rhyodacite and dacite form a series in which decreases in SiO_2 and K_2O are matched by increases in Al_2O_3, MgO, iron and CaO and a lowering of the differentiation index. Ferrorhyodacite and ferrodacite (columns 6, 7) differ essentially from rhyodacite and dacite (columns 4, 5) by having lower MgO and higher iron, reflected in the norm by lower *en* and higher *fs* in normative *di* and *hy*, and lower Al_2O_3. The same differences are seen between ferrorhyolite and rhyolite but enough good analyses of the former to give an average composition could not be found.

There is a fairly close correspondence, on the whole, in chemical composition between these volcanic rocks and their equivalent plutonic types (table 3.1), and it is difficult to know whether such differences as do appear are real or are due to sampling. It is perhaps significant that where large and similar numbers of analyses can be averaged, as is the case with rhyodacite and its equivalent, granodiorite, the two averages are very much alike. One real difference is the higher ratio of Fe_2O_3 to FeO in the volcanic representative, and the virtual absence of normative *C* in fresh alkali rhyolite and rhyolite (contrasting with its presence in the norms of the corresponding granites), due to the absence of muscovites in these volcanic rocks.

Alteration of silicic volcanic rocks

Most of the examples chosen to illustrate the petrography of silicic lavas have been of Tertiary or more recent date. Such lavas erupted in earlier times have usually suffered a greater or less degree of alteration and glassy representatives are particularly susceptible to change. Some changes, such as the absorption of water by obsidian to give rise to pitchstone, and the devitrification of silicic glasses (20), have been noted already, but alteration here and in more crystalline rocks may go further with pronounced changes in the chemical composition.

A common type of alteration is one where original feldspar is replaced by albite, giving rise to a *soda keratophyre* (or soda quartz-keratophyre) (21) in which there is marked enrichment in soda at the expense of potash (table 4.1, column 8). Albite may, in its turn, be replaced by orthoclase, and glass may absorb potash or be replaced by microspherulitic orthoclase, producing a *potash keratophyre* (or potash quartz-keratophyre) with enrichment in potash at the expense of soda (table 4.1, column 9). Original ferromagnesian minerals are usually represented by scattered chlorite, and epidote, calcite and sericite may be present as further alteration products.

Beautiful examples of both are found in some keratophyres from the North Island of New Zealand (22), where they possess relict textures such as characterise normal fresh rhyolitic lavas. There is no evidence here of outside introduction of alkalies and the rocks are regarded as rhyolites that have been buried to depths where the temperature was high enough to induce recrystallisation and limited mobility of alkalies and silica in pore fluids.

Another common type of alteration is *silicification*, which may range from the appearance of small amounts of secondary quartz in the groundmass to a virtually complete replacement of the whole rock. Spherulites are very prone to replacement by silica, giving rise to the so-called *nodular rhyolites* well seen among the Ordovician lavas of Caernarvonshire and elsewhere. The Devonian silicic volcanic rocks of the Cape Colville peninsula, New Zealand, afford striking examples of extensive silicification. Relics of phenocrysts and traces of spherulitic texture can still be seen although almost the whole rock has been replaced.

Hypabyssal equivalents of siliceous lavas and granitic rocks

These are the *microgranites*, *microgranodiorites*, and *microtonalites*. They occur typically in small intrusions such as dykes and sills or at the margins of a larger intrusion intruded into cold country rocks. Porphyritic varieties of these are often referred to as 'quartz porphyry', 'granite porphyry' or 'porphyrite'.

Mineralogically they are more similar to their slowly cooled equivalents than to the lavas. Thus quartz is usually the only silica mineral present. Quartz paramorphs after tridymite have been found in rare instances, but there is no evidence that cristobalite has ever been formed. The potash feldspar may be sanidine but is more commonly orthoclase or orthoclase-microperthite, though microcline and microcline-microperthite are relatively rare. The lime-bearing plagioclase is free from glass inclusions. Biotite and hornblende show no signs of resorption, oxybiotite and oxyhornblende do not occur, muscovite may appear as a primary constituent, pyroxene (except in peralkaline microgranites) is rare.

Texturally, on the other hand, these rocks are frequently porphyritic with phenocrysts of ferromagnesian constituents as well as light constituents, and in this they resemble their volcanic equivalents. But the groundmass is coarser and free of glass, and subhedral-granular on a fine scale. Some, however, have a groundmass in which the light constituents are intergrown to give a granophyric texture and these are distinguished separately as *granophyres*.

Some examples

The 'grorudites' of the Oslo district, Norway, afford good examples of *peralkaline microgranites*. They are essentially *aegirine microgranites*, with little needles and grains of aegirine, microperthite or soda-orthoclase with some microcline and albite, and quartz, with a little soda-amphibole in some. They may have phenocrysts of microperthite and aegirine in varying amount. Rocks like this are found in many areas as dykes and sills associated with major intrusions of alkaline character and, in this connection, those of the Assynt district of Scotland may be mentioned (23) (fig. 4.4A). Another common type is *riebeckite micro-*

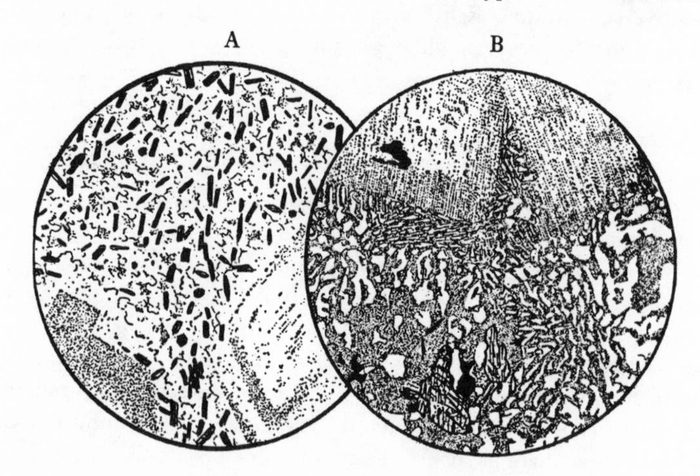

Fig. 4.4. × 20.

A. Aegirine microgranite, Inchnadamph, Assynt, Scotland. Abundant little crystals of aegirine scattered through the microcrystalline groundmass of alkali feldspar and quartz.

B. Pyroxene granophyre, Meall Dearg, Glen Sligachan, Skye. Showing the granophyric intergrowth of turbid feldspar and clear quartz.

granite with the riebeckite occurring as moss-like patches between the alkali feldspar and quartz, such as occurs at Ailsa Craig, in the Firth of Clyde (24). The *riebeckite–acmite microgranite* of Mynydd Mawr, Caernarvonshire, has microlites of acmite as well as spongy riebeckite.

Porphyritic alkali microgranite with biotite is well represented by the striking rock at Llano, central Texas, which has phenocrysts of quartz and soda microcline in a groundmass, partly microgranular, partly granophyric, of quartz and alkali feldspar with some biotite and accessory fluorite, magnetite, apatite and zircon (25), and an example of a rather coarse *muscovite–biotite alkali microgranite* with accessory topaz is provided by that intruding the alkali granites of Lundy Island (26).

The well-known 'quartz porphyry' or 'elvan' dykes associated with the granites of Devon and Cornwall are *porphyritic microgranites* with phenocrysts of potash feldspar and some oligoclase often turbid through alteration, pyramidal or embayed quartz, and usually both muscovite and biotite, in a microcrystalline groundmass of the same constituents with numerous flakes of secondary white mica.

Porphyritic microgranites occur as large laccoliths at Thunder Mountain and Big Baldy Mountain, Little Belt Mountains, Montana (27). These are light coloured rocks with numerous phenocrysts of oligoclase or andesine, sometimes zoned, and soda-orthoclase or microperthite together with some of biotite and hornblende, olive-green at Thunder Mountain, pale brown at Big Baldy Mountain. The groundmass, whose texture varies from microgranitic to micropoikilitic, is essentially made up of alkali feldspar, and quartz with a little oligoclase.

Microgranodiorites, usually porphyritic and usually described as 'porphyrites', are very common as dykes and more irregular intrusions in connection with major intrusions of granodiorite as, for instance, those associated with the Newer Granites of Scotland (Criffell–Dalbeattie (28), Etive, Garabal Hill, etc.). These normally have phenocrysts of oligoclase or andesine, which may be zoned, biotite and green or brownish green hornblende, and sometimes quartz, more rarely potash feldspar. The groundmass has laths of plagioclase, flakes of biotite and grains of iron-ore in a residuum of quartz and alkali feldspar which sometimes show micrographic intergrowth.

Less siliceous varieties may have augite and rhombic pyroxene as phenocrysts with brown hornblende, calcic andesine, and some biotite.

Comparable microgranodiorites from other regions may have hornblende as well as biotite in the groundmass, while highly siliceous types may have biotite as the only ferromagnesian constituent in both the phenocryst and groundmass generation (fig. 4.5).

Descriptions of *microtonalites* appear to be rare but they have been recorded from near Bergalia, Moruya district, New South Wales (29). One type, rather altered, has phenocrysts of plagioclase (An_{35}) and greenish brown hornblende, largely pseudomorphed by epidote, carbonate, etc. The groundmass is mainly of plagioclase and quartz with scattered alteration products such as chlorite, epidote, calcite and

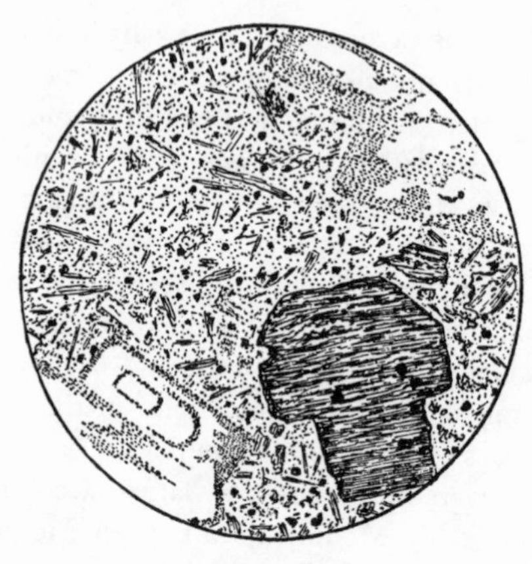

Fig. 4.5. Porphyritic microgranodiorite, Colvend, near Dalbeattie, Scotland; × 20.
With phenocrysts of zoned plagioclase and partially altered biotite.

iron-ore. Another type has phenocrysts of oligoclase, quartz and biotite in a groundmass of the same with accessory potash feldspar, iron-ore and apatite. Here, too, belong the 'trondhjemite-porphyrites' of Norway, hypabyssal representatives of the 'trondhjemites' (p. 27).

The *granophyres*, characterised by having their groundmass feldspar and quartz in micrographic (granophyric) intergrowth, show a certain range of composition. Some have the mineralogy of microgranites, others of microgranodiorites. The Buttermere and Ennerdale intrusion of the Lake District, England, with biotite or augite as ferromagnesian constituents is an example of the first, while the laccolith intruding Tertiary basalts at Deer Creek, Park County, Wyoming (30), is an example of the second. This has a micrographic groundmass of plagioclase, potash feldspar and quartz making up just over half the rock, the remainder being phenocrysts of zoned andesine with subordinate phenocrysts of biotite and green hornblende, and a few of quartz. The rock is interesting because it contains numerous xenoliths, derived from the underlying pre-Cambrian basement, which must have been carried up a minimum distance of 1950 metres by the magma.

The commonest granophyres, however, appear to be *ferrogranophyres*, corresponding in composition with ferrogranites and ferrogranodiorites. Occurring typically as late-stage differentiates in differentiated sills of quartz dolerite (31), they are also found as independent minor intrusions and as chilled margins to ferrogranite intrusions as, for instance, in the Tertiary to Recent Brito-Icelandic province (fig. 4.4B). The ferromagnesian constituents are rich in iron and are usually ferrohedenbergite with or without some fayalite. Hypersthene, amphibole and biotite are less common. A rock of this kind from Coire Uaigneich, Skye, is unusual in that it has quartz paramorphs after tridymite (32).

References and notes

(1) Colony, R. J. & Howard, A. D., 1934. Observations on spherulites. *Am. Miner.* **19**, 515–18.
(2) Washington, H. S., 1913. The volcanoes and rocks of Pantelleria: part II. Petrography. *J. Geol.* **21**, 691–708. Carmichael, I. S. E., 1962. Pantelleritic liquids and their phenocrysts. *Miner. Mag.* **33**, 86–113.
(3) Ewart, A., Taylor, S. R. & Capp, A. C., 1968. Geochemistry of the pantellerites of Mayor Island, New Zealand. *Contr. Miner. Petrol.* **17**, 116–40.
(4) Carmichael, I. S. E., 1967. The iron-titanium oxides of salic volcanic rocks and their associated ferromagnesian silicates. *Ibid.* **14**, 36.
(5) Robinson, H. H., 1913. The San Franciscan volcanic field, Arizona. *Prof. Pap. U.S. geol. Surv.* **76**, 103–5.
(6) Washington, H. S., 1920. The rhyolites of Lipari. *Am. J. Sci.* (4) **50**, 446–53.
(7) Carmichael, I. S. E., 1960. The pyroxenes and olivines from some Tertiary acid glasses. *J. Petrology* **1**, 309–36 (see, especially, chemical analyses and norms in tables 6 and 7).
(8) Larsen, E. S. & Cross, W., 1956. Geology and petrology of the San Juan Region, South-western Colorado. *Prof. Pap. U.S. geol. Surv.* **258**, 162–6.
(9) Fries, C., Schaller, W. T. & Glass, J. J., 1942. Bixbyite and pseudobrookite from the tin-bearing rhyolite of the Black Range, New Mexico. *Am. Miner.* **27**, 305–22.
(10) Guest, J. E. & Sanchez, R. J., 1970. A large dacitic lava flow in Northern Chile. *Bull. Volc.* **33**, 778–89.
(11) Williams, H., 1932. Geology of the Lassen Volcanic National Park, California. *Bull. Dept. geol. Sci. Univ. Calif. Publ.* **21**, 195–385.
(12) Robinson, 1913, pp. 114–36.
(13) *E.g.* Francis, E. H., *et al.*, 1970. The geology of the Stirling district. *Mem. Geol. Surv. U.K.* 45–53 (Ochil Hills). Haslam, H. W., 1968. The crystallization of intermediate and acid magmas at Ben Nevis, Scotland. *J. Petrology* **9**, 93–4.
(14) Van de Kamp, P. C., 1969. The Silurian volcanic rocks of the Mendip Hills, Somerset; and the Tertworth area, Gloucestershire, England. *Geol. Mag.* **106**, 543–8.
(15) MacGregor, A. G., 1938. The volcanic history and petrology of Montserrat, with observations on Mt Pelé, in Martinique. *Phil. Trans. Roy. Soc.* Ser. B, **229**, 1–90.
(16) Baker, P. E., 1968. Petrology of Mt Misery Volcano, St Kitts, West Indies. *Lithos* **1**, 124–50.
(17) Williams, H., 1934. Mount Shasta, California. *Zeit. Vulk.* **15**, 225–53.
(18) Carmichael, I. S. E., 1964. The petrology of Thingmuli, a Tertiary volcano in eastern Iceland. *J. Petrology* **5**, 441–4 (ferrorhyodacites, described as 'andesites (icelandites)'). Walker, G. P. L., 1966. Acid volcanic rocks in Iceland. *Bull. Volc.* Ser. II, **29**, 375–402. King, B. C.,

1955. The Ard Bheinn area of the central igneous complex of Arran. *Q. Jl. geol. Soc. Lond.*, **110**, 335–6, table 1, anal. 5 p. 328 (ferrorhyodacite).

(19) Kuno, H., 1950. Petrology of Hakone Volcano and the adjacent areas, Japan. *Bull. geol. Soc. Am.* **61**, 957–1020 (ferrodacites).

(20) Good descriptions and illustrations of textures formed by the devitrification of ancient siliceous lavas can be found in Bascom, F., 1896. The ancient volcanic rocks of South Mountain, Pennsylvania. *Bull. U.S. geol. Surv.* **136**, especially pp. 42–63.
See also, Lofgren, G., 1971. Experimentally produced devitrification textures in natural rhyolitic glass. *Bull. geol. Soc. Am.* **82**, 111–23.

(21) Following Battey (see next reference) and others, the term quartz–keratophyre would be used when the rocks have phenocrysts of quartz.

(22) Battey, M. H., 1955. Alkali metasomatism and the petrology of some keratophyres. *Geol. Mag.* **92**, 104–26 (numerous references to other localities are given here).

(23) Sabine, P. A., 1953. The petrography and geological significance of the post-Cambrian minor intrusions of Assynt and the adjoining districts of North-West Scotland. *Q. Jl. geol. Soc. Lond.* **109**, 140–2.

(24) Teall, J. J. H., 1891. On a micro-granite containing riebeckite from Ailsa Craig. *Miner. Mag.* **9**, 219–21.

(25) Iddings, J. P., 1904. Quartz–feldspar–porphyry (graniphyro liparese–alaskose) from Llano, Texas. *J. Geol.* **12**, 225–31.
Goldich, S. S., 1941. Evolution of the Central Texas granites. *Ibid.* **49**, 706–8.

(26) Dollar, A. T. J., 1941. The Lundy complex: its petrology and tectonics. *Q. Jl. geol. Soc. Lond.* **97**, 51–2.

(27) Pirsson, L. V., 1900. *In* Geology of the Little Belt Mountains, Montana. *20th Ann. Rept. U.S. geol. Surv.*, Part III, 506–12.

(28) MacGregor, M., 1937. The western part of the Criffell–Dalbeattie igneous complex. *Q. Jl. geol. Soc. Lond.* **93**, 465–7.

(29) Brown, I. A., 1928. The geology of the south coast of New South Wales. Part I. The Palaeozoic geology of the Moruya district. *Proc. Linn. Soc. N.S.W.* **53**, 176–8.

(30) Rouse, J. T., 1933. The structure, inclusions and alteration of the Deer Creek intrusive, Wyoming. *Am. J. Sci.* (5), **26**, 139–46.

(31) *E.g.* McDougall, I., 1962. Differentiation of the Tasmanian dolerites. Red Hill dolerite–granophyre association. *Bull. geol. Soc. Am.* **73**, 279–316 (granophyre with pyroxene and fayalite). Hawkes, D. D., 1966. Differentiation of the Tumatumari–Kopinaug dolerite intrusion, British Guiana, *Ibid.* **77**, 1131–58 (granophyre with iron-rich amphibole).

(32) Wager, L. R., Weedon, D. S. & Vincent, E. A., 1953. A granophyre from Coire Uaigneich, Isle of Skye, containing quartz paramorphs after tridymite. *Miner. Mag.* **30**, 263–75.

5
Syenites

Syenites are medium to coarse grained igneous rocks composed essentially of alkali feldspar with or without subordinate lime-bearing plagioclase (<40 per cent of the total feldspar content), normally with one or more ferromagnesian constituents in addition. A distinction is made between *peralkaline* and *alkali syenite* that have alkali feldspar only, and *syenite* that has both alkali feldspar and lime-bearing plagioclase.

These rocks are much less widely distributed and less abundant than the granitic rocks, occurring as relatively small independent intrusions or as a local facies of some larger intrusion of a different nature.

General mineralogy

Alkali feldspars that may be present include orthoclase, microcline, microperthite and cryptoperthite, with the addition of albite in peralkaline and alkali syenites. Of these, the microperthites, both orthoclase-microperthite and microcline-microperthite, are particularly common and often beautifully developed (fig. 5.1). *Lime-bearing plagioclase* in syenite is usually oligoclase or andesine, with a range of composition from about An_{20} to An_{40}.

The feldspars tend to be subhedral and sometimes, especially the alkali feldspars, tabular.

The *mica* of syenite is a brown biotite, similar to that in the granites. Peralkaline and alkali syenites may have a very deep brown or green iron-rich variety. Muscovite is not found in these rocks as a primary constituent.

Green hornblende, with a notable content of alumina, is the common amphibole of syenite, building subhedral prisms. Soda- and iron-rich amphiboles such as *arfvedsonite* and *barkevikite* may occur in peralkaline syenite and *ferrohastingsite* in alkali syenite.

Clinopyroxene, when present in syenite, is a diopsidic or somewhat aluminous augite. It is commoner in peralkaline syenites as *aegirine* or *aegirine-augite*, or, more rarely, as pale lilac to green titaniferous augite, while *ferrohedenbergite* may occur in alkali syenites.

Olivine is found in some rocks. Certain rather mafic syenites have a variety fairly rich in magnesia, but the olivine found in peralkaline and alkali syenites is rich in the fayalite molecule.

The usual accessory constituents of granites, *apatite*, *zircon*, and *iron-ore*, are present also in most syenites and the first two, especially, are frequently more conspicuous. Iron oxides are variable in amount. The most important is magnetite, but titaniferous magnetite with exsolution lamellae of ilmenite is common. Ilmenite and ilmenite-haematite intergrowths occur in a few rocks. *Sulphides* are sometimes present in small amount, the commonest being pyrrhotite, chalcopyrite, and pyrite.

Sphene is prominent in many syenites, usually euhedral and sometimes twinned on {001}, a feature particularly noticeable in some peralkaline syenites. *Orthite* is common but not abundant, and a number of additional accessory constituents, such as deep brown *melanite garnet*, may be present in peralkaline syenites.

Many syenites have a small but appreciable (<10 per cent by volume) content of *quartz*, marking a transition to granite. Others have small amounts of *nepheline* and are transitional to nepheline syenite.

Texture and structure

The texture of syenites is commonly subhedral-granular, but when the feldspar crystals are tabular they may tend to have a subparallel disposition, giving rise to a trachytoidal texture. Syenites may be porphyritic, but this is not so frequent as in the granites. A number have the drusy structure.

Main types with examples

1. Peralkaline syenite

The syenites of the Singida district, Tanzania (1), are *aegirine syenites* with melanite garnet. That of Tumuli Tor is fairly coarse grained and the crystals of microcline-microperthite have a subparallel arrangement. The rock from Itunda Hill is also trachytoidal with abundant aegirine and melanite garnet, the latter having crystallised late as is often the case with this mineral. Calcite occurs as large plates enclosed by feldspar, aegirine and garnet, while the accessory constituents are sphene, iron oxide, orthite and apatite. Other specimens have green mica, fluorite, tourmaline, sodalite and nepheline in addition.

The 'pulaskite' of Cnoc nan Cuilean, Ben Loyal igneous complex, Sutherland (2), is an *aegirine-augite syenite*, made up of microperthite and a little albite, subhedral prisms of zoned aegirine-augite, with abundant accessory sphene and some orthite. In the Loch Ailsh mass (3) there is an almost hololeucocratic syenite of this kind, associated with more normal aegirine-augite syenite and aegirine syenite with melanite garnet.

Some syenites of this type carry fayalite. Thus the border phase of the syenite in the Agamenticus complex, York County, Maine (4), is a leucocratic aegirine-augite syenite in which fayalite is as abundant as pyroxene.

Many have soda amphiboles as an essential constituent. The drusy syenite on the mainland opposite the island of Cabo Frio, near Rio de Janeiro, is essentially an *arfvedsonite syenite* with barkevikite forming cores to the larger crystals of arfvedsonite. The amphibole is poikilitic with respect to the microperthite and the rock also contains a little iron-rich mica and pyroxene.

The 'nordmarkites' of the Oslo district, Norway, are usually highly leucocratic. Medium to rather coarse grained, they consist essentially of microperthite, a little interstitial quartz, and a small proportion of dark minerals, including arfvedsonite and aegirine, aegirine-augite, and iron-rich mica (fig. 5.1). These rocks are often drusy.

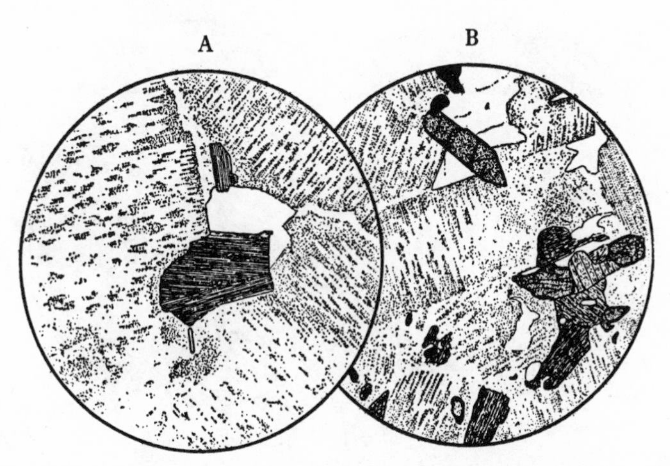

Fig. 5.1. Hypersolvus quartz-bearing syenites, Nordmarken, Norway; × 20. Composed essentially of microperthite and a little interstitial quartz with A a little biotite and B arfvedsonite and conspicuous accessory sphene (upper part of figure).

All the peralkaline syenites cited have been hypersolvus types in which independent albite has been more or less lacking. But subsolvus types do occur as in the Malagasy Republic where the feldspars are microcline and albite, and in the Kerguelen Islands where they are orthoclase-microperthite and albite.

2. *Alkali syenite*

Ferrohastingsite syenites belonging to the White Mountain magma series have a wide distribution in New Hampshire, as at Red Hill (5), Belknap Mountains, Chandler Mountain, and elsewhere. They are composed largely of microperthite with varying amounts of ferrohastingsite; most are quartz-bearing and are sometimes transitional to granite. A little biotite is not infrequently present, and there may be some hedenbergite commonly mantled by the ferrohastingsite. Accessory constituents include sphene, zircon, apatite, iron-ore, orthite and fluorite.

Augite alkali syenite is found at Marblehead, Massachusetts (fig. 5.2A), and among the syenites occurring at Mount Caribou, Bonneville Co., S.E. Idaho (6), there is a 'diopside' alkali syenite with pronounced trachytoidal texture and with unusually abundant accessory magnetite, a little biotite and some apatite accompanying the clinopyroxene.

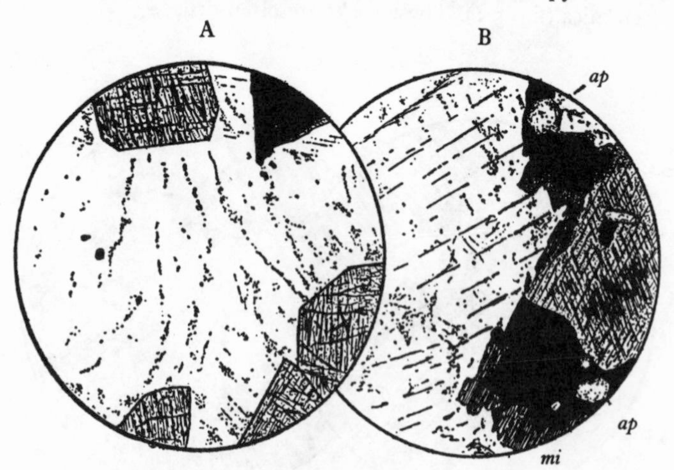

Fig. 5.2: × 20.

A. Augite alkali syenite, Marblehead, Mass., U.S.A. Micro- to cryptoperthitic alkali feldspar, pale euhedral augite and accessory iron-ore.

B. Alkali monzonite ('Larvikite'), Larvik, S. Norway. Pale lilac titaniferous augite with schiller structure, fringed with brown alkali amphibole and deep brown mica (*mi*); large crystals of feldspar; accessory apatite (*ap*) and iron-ore.

The small bodies of syenite in the Cebolla basin, Uncompahgre quadrangle, Colorado (7), may be taken as examples of *augite–biotite alkali syenites*. These are rather fine-grained rocks that may have abundant phenocrysts of augite. The subordinate biotite is in poikilitic plates and orthopyroxene was seen in a few thin sections.

Biotite alkali syenites seem to be rare but some of the alkali syenites associated with nepheline syenite in the Serra de Monchique intrusion, Portugal, have biotite as their only ferromagnesian constituent.

Shonkinite is the name given to mesotype and melanocratic alkali syenites. The type rock from Square Butte, Highwood Mountains, Montana (8), has a colour index of 70. Zoned diopsidic augite makes up about half the rock and is accompanied by biotite and a little olivine. The light constituents are potash feldspar, a little nepheline, and zeolites derived from these two minerals. Other shonkinites from the Highwood Mountains and from the Bearpaw Mountains, Montana, may have a little plagioclase (An_{60}, but zoned); in some the diopsidic pyroxene is mantled by aegirine-augite, and in some the content of nepheline, and zeolites derived from nepheline, is great enough for the rock to be classed among the nepheline syenites.

3. *Syenite*

The type syenite, sometimes showing good trachytoidal texture, is the *hornblende syenite* of Plauen'scher Grund, near Dresden (fig. 5.3A). It is composed of soda-orthoclase with subordinate oligoclase and green hornblende, with or without a little biotite, minor quartz, and sphene, apatite and magnetite as noticeable accessories. Unfortunately, although it is possible to find specimens answering this description, the rock is very variable and most of it is a quartz-bearing monzonite. The same is true of the other famous hornblende syenite, that of Biella, Italy.

The hornblende syenite of the Ratagan complex, Glenelg, Inverness-shire (9) (fig. 5.3B), is a British representative. There are darker and lighter varieties, including one almost purely feldspathic.

These syenites are sometimes porphyritic, as is the case with one of the prevailing types found in the Carmacks district, Yukon. This hornblende syenite has large phenocrysts of potash feldspar, making up about a quarter of the rock.

A *hornblende–biotite syenite*, the Coryell syenite, occurs in the Rossland district, British Columbia. It is medium to coarse grained, has a

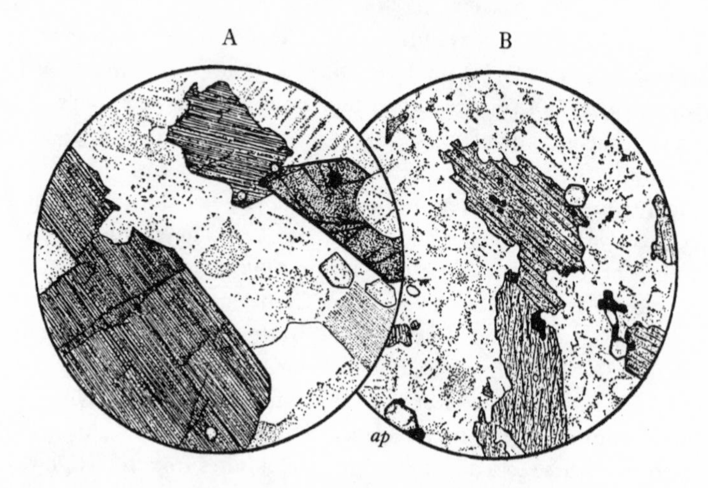

Fig. 5.3. Hornblende syenite; × 20.

A. Plauen 'scher Grund, Dresden. Green hornblende, large crystals of sphene and apatite (right side of figure), potash feldspar, some oligoclase, and interstitial quartz.

B. Ratagan, Glenelg, Scotland. A finer-grained rock with conspicuous apatite.

small amount of interstitial quartz, and grades in places into hornblende syenite.

The syenite associated with monzonite and granodiorite in the composite Deboullie stock, N. Maine (10), is a rather variable rock but the average mode is that of a *hornblende–biotite–augite syenite*, transitional to monzonite, with hornblende dominant over the other ferromagnesian constituents. The main feldspar is orthoclase crypto- to microperthite, the subordinate plagioclase ranges in composition from An_{21} to An_{28}, and most specimens have a little quartz.

All these examples are of leucocratic rocks, but the *pyroxene-biotite syenites*, with or without olivine, are frequently mesotype or even melanocratic. A rock of this kind at Seeconnell, at the eastern end of the Newry complex, N. Ireland (11), has about equal amounts of biotite and pale green augite, replaced to a small extent by green hornblende, soda-orthoclase and subordinate andesine, and a little quartz.

The mesotype syenite from Christina Lake, Rossland district, British Columbia (12), is an augite–olivine–biotite syenite of medium to rather coarse grain in which the biotite is present, in part, as very large thin foils giving a porphyritic aspect to the rock. The subordinate plagioclase occurring in association with the soda-orthoclase is a labradorite.

Chemistry

Average chemical compositions of peralkaline syenite, alkali syenite and syenite are given in table 5.1. Note the moderate content of SiO_2 and high alkalies, with higher Al_2O_3, CaO, MgO, total iron, and TiO_2 than in the corresponding average granites.

TABLE 5.1. *Average chemical composition of syenites and trachytes*

		1	2	3	4	5	6
SiO_2		61.65	61.86	59.41	62.16	61.95	58.31
TiO_2		0.52	0.58	0.83	0.40	0.73	0.66
Al_2O_3		14.73	16.91	17.12	16.56	18.03	18.05
Fe_2O_3		4.56	2.32	2.19	3.55	2.33	2.54
FeO		3.68	2.63	2.83	1.90	1.51	2.02
MnO		0.20	0.11	0.08	0.21	0.13	0.14
MgO		0.70	0.96	2.02	0.33	0.63	2.07
CaO		1.87	2.54	4.06	1.39	1.89	4.25
Na_2O		6.69	5.46	3.92	7.44	6.55	3.85
K_2O		4.65	5.91	6.53	5.20	5.53	7.38
H_2O^+		0.58	0.53	0.63	0.65	0.54	0.53
P_2O_5		0.17	0.19	0.38	0.20	0.18	0.20
qz		1.9	1.7	2.0	–	–	–
or		27.2	35.0	38.4	30.6	32.8	43.9
ab		49.8	46.1	33.0	52.4	54.0	28.8
an		–	4.2	10.0	–	3.3	9.7
ne		–	–	–	2.3	0.6	2.0
di	*wo*	3.5	3.0	3.1	2.3	2.1	4.2
	en	1.0	1.6	2.0	0.7	1.6	3.2
	fs	2.6	1.3	0.9	1.7	–	0.5
hy	*en*	0.7	0.8	3.0	–	–	–
	fs	1.6	0.8	1.2	–	–	–
ol	*fo*	–	–	–	0.1	–	1.4
	fa	–	–	–	0.1	–	0.2
ac		6.0	–	–	5.5	–	–
mt		3.7	3.3	3.3	2.3	3.3.	3.7
il		0.9	1.2	1.5	0.8	1.4	1.2
ap		0.4	0.4	0.9	0.5	0.4	0.5
DI		78.9	82.9	73.4	85.3	87.4	74.7

1. Peralkaline syenite (average of 47) (*Bull. Geol. Soc. Am.* **65**, p. 1016, no. VI).
2. Alkali syenite (average of 25) (*Ibid.*, p. 1016, no. IV).
3. Syenite (average of 18) (*Ibid.*, p. 1016, no. I).
4. Peralkaline trachyte (average of 12).
5. Alkali trachyte (average of 15) (*Ibid.*, p. 1016, no. V).
6. Trachyte (average of 24) (*Ibid.*, p. 1016, no. III).

Average alkali and peralkaline syenite differ from syenite in their markedly lower MgO and CaO, higher Na_2O and lower K_2O, features that become most pronounced in peralkaline syenite. Note, also, the relatively lower Al_2O_3 leading to the appearance of acmite in the norm, and high iron with excess of Fe_2O_3 over FeO, in the latter.

Syenite-aplites and syenite-pegmatites

Syenite-aplites and syenite-pegmatites mostly occur in connection with intrusions of alkali and peralkaline syenite.

The *syenite-aplites* have the typical aplitic texture and are composed of alkali feldspars, including cryptoperthite, orthoclase- or microline-microperthite, microcline, and albite, in different cases; some have a little quartz. The sparing ferromagnesian constituents may be mica, aegirine, aegirine-augite and various soda amphiboles, while the common accessory constituents are sphene and apatite.

Such aplites have been described from the Oslo district, Norway; from Malagasy; from Alter Pedroso, Portugal; and from Essex Co., Massachusetts, where they cut the Beverley syenite (13).

Typical *alkali syenite-pegmatites* in the Ilmen State Mineralogical Reservation, Ilmen Mountains, U.S.S.R., are composed essentially of alkali feldspar and iron-rich mica; their typical accessory mineral, zircon, is found on the walls of druses as well as in the body of the rock. *Syenite-pegmatites* with orthite are a closely related group from the same area.

A similar syenite-pegmatite, with the mica largely altered to chlorite, forms the core of the Copper Mountain stock, British Columbia, and occurs also as numerous dykes in, and adjacent to, the stock (14). Apatite in large crystals is an abundant accessory but the most striking feature is the presence of blebs of primary chalcopyrite, bornite, and some chalcocite, freely scattered through the rock.

Alkali syenite-pegmatite of a different kind is found around the north-east edge of the Watson Ledge quartz-bearing syenite, Red Hills, New Hampshire (15). It consists of microperthite, a little quartz, and large bladed crystals of ferrohastingsite.

Peralkaline syenite-pegmatites of simple character occur at a number of localities. These may have soda amphibole, with or without aegirine (Alter Pedroso, Portugal; Mariupol, Sea of Azov, etc.) or they may have a more varied mineralogy, as in the quartz-bearing examples from the Wausau district, north-central Wisconsin, whose main constituents are microperthite, riebeckite, acmite, and iron-rich mica.

The well-known pegmatites of Fredriksvärn in the Oslo district of Norway, and those of Narsarsuk, S. Greenland, are examples of complex peralkaline syenite-pegmatites. The latter have microcline and aegirine as the main constituents with some iron-rich mica, quartz, and a little eudialyte and astrophyllite. Drusy cavities are sometimes present and it is in these that the rare minerals for which this locality is famous are found. It is clear in both these examples that the pegmatites have had a long cooling history with a succession of hydrothermal stages.

References and notes

(1) Eades, N. W. & Reeve, W. M., 1938. Explanation of the geology of degree sheet no. 29 (Singida). *Bull. geol. Surv. Tanganyika* **11**, 34–6.
(2) King, B. C., 1943. The Cnoc nan Cuilean area of the Ben Loyal igneous complex. *Q. Jl. geol. Soc. Lond.* **98**, 157–9.
(3) Parsons, I., 1965. The feldspathic syenites of the Loch Ailsh intrusion, Assynt, Scotland. *J. Petrology* **6**, 365–94.
(4) Wandke, A., 1922. Intrusive rocks of the Portsmouth Basin, Maine and New Hampshire. *Am. J. Sci.* (5) **4**, 153.
(5) Quinn, A., 1937. Petrology of the alkaline rocks at Red Hill, New Hampshire. *Bull. geol. Soc. Am.* **48**, 373–402.
(6) Anderson, A. L. & Kirkham, V. R. D., 1931. Alkaline rocks of the Highwood type in South-eastern Idaho. *Am. J. Sci.* (5) **22**, 61–2.
(7) Hunter, J. F., 1925. Pre-Cambrian rocks of Gunnison River, Colorado. *Bull. U.S. geol. Surv.* **777**, 64–9.
(8) Hurlbut, C. S., 1939. Igneous rocks of the Highwood Mountains, Montana. Part 1. The laccoliths. *Bull. geol. Soc. Am.* **50**, 1072–5.
(9) Nicholls, G. D., 1951. The Glenelg–Ratagan igneous complex. *Q. Jl. geol. Soc. Lond.* **106**, 319–20.
(10) Boone, G. M., 1962. Potassic feldspar enrichment in magma. Origin of syenite in Deboullie District, Northern Maine. *Bull. geol. Soc. Am.* **73**, 1451–76.
(11) Reynolds, D. L., 1936. The two monzonitic series of the Newry complex. *Geol. Mag.* **73**, 342–5.
(12) Daly, R. A., 1912. Geology of the North American cordillera at the Forty-Ninth Parallel. *Mem. geol. Surv. Canada* **38**, Part 1, 356–8.
(13) Clapp, C. H., 1921. Geology of the igneous rocks of Essex County, Massachusetts. *Bull. U.S. geol. Surv.* **704**, 85–8.
(14) Dolmage, V., 1934. Geology and ore deposits of Copper Mountain, British Columbia. *Mem. geol. Surv. Canada* **171**, 12–16.
(15) Quinn, 1937, p. 388.

6

Trachytes and microsyenites

Trachytes are the volcanic equivalents of syenites, occurring typically as lava flows and very commonly as volcanic domes or plugs. Usually pale coloured and porphyritic, they tend to have an open porous texture that gives them a rough feel in hand specimen.

The feldspar of the *phenocrysts* is commonly *sanidine* in columnar or tabular crystals, often of relatively large size, but some alkali and peralkaline trachytes have phenocrysts of *anorthoclase*. Smaller and subordinate phenocrysts of *lime-bearing plagioclase* may be present in some trachytic rocks.

A characteristic ferromagnesian phenocrystal constituent is *augite*, pale green to colourless in thin section. It is represented by *aegirine* or *aegirine-augite* in peralkaline trachytes. Any *rhombic pyroxene* is rare, but has been recorded in a few trachytes.

Biotite in little euhedral tablets is another common phenocrystal constituent; *hornblende* is less frequent, and both normally show some magmatic resorption. Peralkaline trachytes often carry *soda-amphiboles* such as riebeckite, arfvedsonite and kataphorite, but these are mostly found in the groundmass of the rock, as is *aenigmatite*. *Fayalitic olivine* occurs as phenocrysts in some trachytes, while small euhedral crystals of *sphene*, of *iron-ore*, of *apatite* and of *zircon* are frequently to be found in a phenocryst generation.

The groundmass is usually holocrystalline and made up largely of little laths of alkali feldspar with minor amounts of ferromagnesian minerals and iron-ore. The feldspar laths commonly have a more or less parallel arrangement as the result of magmatic flow, giving the typical *trachytic texture*. A little interstitial quartz is present in the groundmass of many trachytes; also tridymite aggregates, especially in small druses. On the other hand, peralkaline trachytes, in particular, may have a little nepheline, sodalite or analcite.

The groundmass of some trachytes has a certain amount of interstitial glass but holohyaline rocks, *trachyte-obsidians*, are rare. These are black in hand specimen, colourless or light brown in thin section, and cannot be distinguished from other obsidians except by chemical analysis.

Varieties with some examples

Three groups may be distinguished, namely, peralkaline trachytes, alkali trachytes, and trachytes. Many peralkaline and alkali trachytes have minor amounts of nepheline, analcite or a member of the sodalite group in the groundmass and are transitional to phonolite. These are frequently referred to as phonolitic trachytes (fig. 15.1A).

Peralkaline trachytes carry soda-rich pyroxene or amphiboles and have acmite in their norms when fresh. Some of the Italian trachytes (isle of Ischia, Phlegrean Fields, etc.) are of this type and have phenocrysts of soda-sanidine in a trachytic groundmass of the same with aegirine-augite, sometimes zoned, a little nepheline, sodalite or hauyne and accessory iron-ore, sphene and apatite.

The somewhat altered peralkaline trachytes of Mauritius (1) include both aphyric and porphyritic types. The aphyric rocks are composed essentially of little laths of anorthoclase with minor albite, little prisms of aegirine-augite and accessory iron-ore, apatite and, sometimes, aenigmatite. Most have a little nepheline, sodalite or analcite and in certain rocks, transitional to phonolite, these become noticeable. Porphyritic types have phenocrysts of anorthoclase and a few of pyroxene with, more rarely, one or two of a brown alkali amphibole surrounded by a reaction rim of aegirine-augite. A very fine-grained aphyric trachyte of this kind with minute grains of aegirine, aenigmatite and fayalite in a matrix of anorthoclase with a little nepheline and sodalite is found on the west flank of Mount Kenya (2).

Some of the Kenya peralkaline trachytes, like some of those from the Azores, have soda-amphiboles as essential constituents of the groundmass. This is the case at Mount Suswa (3) where arfvedsonite accompanies aenigmatite and a feebly sodic augite; with that east of Lake Naivasha (4) which has interstitial kataphorite and subordinate aegirine; and with some from Turkana (5) that have a blue alkali amphibole allied to riebeckite.

A similar blue amphibole is about as abundant as aegirine-augite in the groundmass of the trachyte of the Breigh a 'Choire Mhoir plug, Mull (6), and was possibly present in that of Ardnacross where original amphibole is represented by pseudomorphs in chlorite. These are of Tertiary age, but in the neighbourhood of Melrose, Roxburghshire (7), there are dykes of riebeckite trachyte of Carboniferous age. These have a little interstitial quartz.

Some of the peralkaline trachytes of Mount Suswa have a groundmass in which laths of alkali feldspar, rare crystals of nepheline and some iron-ore are embedded in abundant glass, and thus approach peralkaline trachyte-obsidian. True *peralkaline trachyte-obsidians* are rare, but that of Puu, Waawaa, Hawaii (8), is an example. This is crudely banded with black glass, forming a true obsidian and grey, partly crystalline, portions with grains of alkali feldspar and colourless microlites of acmite in a matrix of colourless glass. Peralkaline trachyte-pumice also occurs here.

Alkali trachytes do not have ferromagnesian minerals rich enough in soda to give acmite in the norm. They are distinguished from trachytes by the virtual absence of lime-bearing plagioclase.

Many alkali trachytes have a greenish clinopyroxene, commonly a somewhat sodic hedenbergite tending towards aegirine-augite, with or without fayalitic olivine. That forming a plug of Carboniferous age east of Townhead of Grange, Lugton, Ayrshire, has phenocrysts of alkali feldspar with a few of clinopyroxene, olivine and amphibole in a groundmass of alkali feldspar, greenish clinopyroxene, iron-ore and a little nepheline. Similar rocks of Carboniferous age, sometimes with appreciable nepheline or analcite in the groundmass, are not uncommon elsewhere in Scotland (fig. 6.1B).

Alkali trachytes of this kind are to be found in Réunion; some of the Society Islands, where they may contain a little sodalite; Malagasy; the island of Ischia, where biotite may accompany pyroxene and where porphyritic *alkali trachyte-obsidians* are present; Gough Island (9); St. Helena Island (10), etc.

Other alkali trachytes have biotite as their main ferromagnesian constituent. A good example comes from Cripple Creek, Colorado (11), where relatively abundant phenocrysts of biotite, partly replaced by augite and iron-ore, and a few of augite accompany the numerous phenocrysts of alkali feldspar.

Here, too, fall some of the 'domites' of the Auvergne district, France, for instance those of the Puy de Dôme. The main type has phenocrysts of alkali feldspar and biotite with a very few of andesine, small phenocrysts of sphene, iron-ore, apatite and zircon, and a groundmass essentially made up of little laths of alkali feldspar and grains of magnetite with some tridymite. Another type, found as volcanic bombs and blocks, has phenocrysts of brown amphibole in addition to those of biotite.

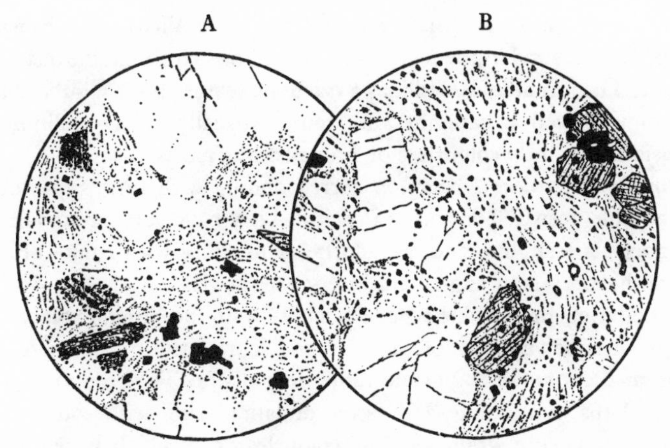

Fig. 6.1. Trachyte; × 20.

A. Trachyte, Drachenfels, Siebengebirge, Germany. Large phenocrysts of sanidine with smaller ones of oligoclase, partially resorbed biotite and iron-ore, lying in a fine-grained trachytic groundmass.

B. Alkali trachyte, Peppercraig, Haddington, Scotland. Phenocrysts of sanidine, pale green sodic hedenbergite and iron-ore; groundmass largely of small sanidine laths showing good flow structure.

It will have been noted that these 'domites' carry accessory lime-bearing plagioclase. This is the case elsewhere, as on Gough Island (12) where porphyritic alkali trachytes have phenocrysts of sanidine and a few of zoned plagioclase together with some of biotite, augite, olivine and iron-ore.

In the *trachytes* lime-bearing plagioclase becomes an *essential* but subordinate feldspathic constituent.

The famous trachyte of the Drachenfels, Rhine (fig. 6.1A), has large tabular and columnar phenocrysts of sanidine up to 5cm long, abundant smaller ones of oligoclase with some of augite and of biotite in all stages of resorption. There are also a few microphenocrysts of sphene, apatite and iron-ore. The trachytic groundmass is made up of alkali feldspar, a little plagioclase and some interstitial quartz. This rock has appreciable excess silica and is transitional to rhyolite.

Trachytes are well displayed in what has been termed the Roman comagmatic region (13). One type, found at Bolsena and in the Phlegrean Fields, has phenocrysts of sanidine and subordinate calcic plagioclase with fewer of augite, partially resorbed biotite and magnetite, making up some 20 per cent of the rock. The remainder is a trachytic groundmass with little prisms of augite, grains of magnetite, small laths of sanidine, and sometimes a little pale brownish amphibole.

Another slightly more mafic type, found at Mte. Vico and on the island of Ischia, has similar phenocrysts but may have a little olivine instead of biotite. The groundmass still has a trachytic texture but has subordinate andesine as well as sanidine and, though usually holocrystalline, has occasionally a little colourless or brownish interstitial glass.

Porphyritic ***trachyte-obsidians*** corresponding with these types occur in the Phlegrean Fields and at Procida. They have varying amounts of phenocrysts (sanidine and labradorite, sometimes with augite or with microphenocrysts of augite and biotite) set in glass.

A definitely more mafic type of trachyte from the Cimino volcano has about an equal number of sanidine and labradorite phenocrysts with some smaller ones of augite and rather less olivine. The groundmass is a felt of laths of sanidine and fewer of labradorite, with some augite prisms, magnetite grains and a colourless cement, believed to be sanidine.

Chemistry

Average peralkaline trachyte, alkali trachyte and trachyte resemble the corresponding average syenites in general character (compare table 5.1, columns 4, 5, 6 with columns 1, 2, 3), but are somewhat richer in Al_2O_3 and alkalies throughout, so that they have higher differentiation indices than the corresponding syenites. This is shown most markedly by average peralkaline and alkali trachyte, both of which also have less iron, MgO and CaO than the corresponding syenites.

Like the average syenitic rocks, the average trachytic rocks are just about saturated with SiO_2 but, whereas the former have a very little normative quartz, the latter have a little normative nepheline.

Alteration

A number of trachytes show varying degrees of alteration. The Tertiary trachyte of Ardnacross, Mull, with its chloritic pseudomorphs after amphibole has been mentioned already. Apart from this and similar pseudomorphing of much of the aegirine-augite, it is relatively fresh but in places is more decomposed with development of carbonate, chlorite and zeolites. In a trachyte of the same age from northern Skye original phenocrysts of alkali feldspar have been largely replaced by stilbite.

Some older trachytes, however, have been altered more radically as, for instance, peralkaline trachytes of Devonian age from the Lahn district, Germany. Here, original riebeckite and aegirine are often replaced by chlorite, original feldspar phenocrysts partly replaced, and most of the original groundmass feldspar entirely replaced, by albite. Such rocks show a high content of soda with little or no potash.

Equally remarkable are occurrences of trachytes with high contents of potash and little soda. Good examples have been described from the Kaiserstuhl, Germany (14), where trachytes have been metasomatised in such a way that nearly all the original potash-soda feldspar has been replaced by orthoclase.

Hypabyssal equivalents of syenites and trachytes

These are the *microsyenites*, intermediate between them with respect to grain size. In general, their mineralogy is more like that of the syenites than that of the trachytes. For instance, biotite and amphibole, when present, show no signs of resorption; microperthite is common and microcline may be present.

Many are porphyritic with phenocrysts of feldspar and one or more dark minerals, while the fine-grained groundmass is made up largely of stumpy crystals, or of laths, of alkali feldspar, giving in the former case the orthophyric texture and in the latter case a trachytic texture on a scale coarser than that found in trachytes.

They have received various names such as 'porphyry', 'syenite-porphyry' and 'orthopyre', quite apart from special varietal names.

The 'solvsbergites' of the Oslo district, Norway, and other areas may be taken as representatives of *peralkaline microsyenites*. They are fine-grained rocks, commonly aphyric, and often with trachytic texture. They are made up essentially of alkali feldspar which, in the rocks of the Oslo district, includes microcline-microperthite, microcline and albite. There is relatively abundant aegirine as acicular needles and short prisms, and there may be a little interstitial quartz or nepheline. Some have accessory flakes of green-brown mica, and a number have a little sodic amphibole. Occasionally the latter becomes the major ferromagnesian constituent and occasionally, also, the rocks have a few tabular phenocrysts of microcline-microperthite.

Elsewhere the alkali feldspar present has been recorded as anorthoclase (Mozambique, Tibesti, Abyssinia) or soda-orthoclase (Malagasy).

The 'nordmarkite porphyry' of the Oslo district is the hypabyssal equivalent of the 'nordmarkite' already mentioned. This is a *porphyritic peralkaline microsyenite* with some tabular phenocrysts of microperthite in a more or less trachytic matrix of microperthite with a little biotite and blue sodic amphibole and a little interstitial quartz.

A peralkaline microsyenite with orthophyric texture and with riebeckite interstitial to the alkali feldspar has been described from Middle Eildon Hills, Melrose, Scotland (fig. 6.2B). A similar rock with some aegirine and quartz occurs on Holy Island, near Arran (15), and is transitional to peralkaline microgranite.

The Bass Rock, Haddingtonshire, Scotland, an intrusion of Carboniferous age, affords an example of an *alkali microsyenite* with seriate porphyritic texture (16). The feldspar is mostly a soda-sanidine which is sometimes microperthitic and the larger crystals of which may hold a core of albite. The green clinopyroxene, probably a sodic hedenbergite, has an occasional core of titaniferous augite and is accompanied by subordinate fayalite. There is a little euhedral, altered nepheline, and analcite, sometimes associated with chlorite, fills small drusy cavities. A similar rock occurs at north Berwick Law (fig. 6.2A).

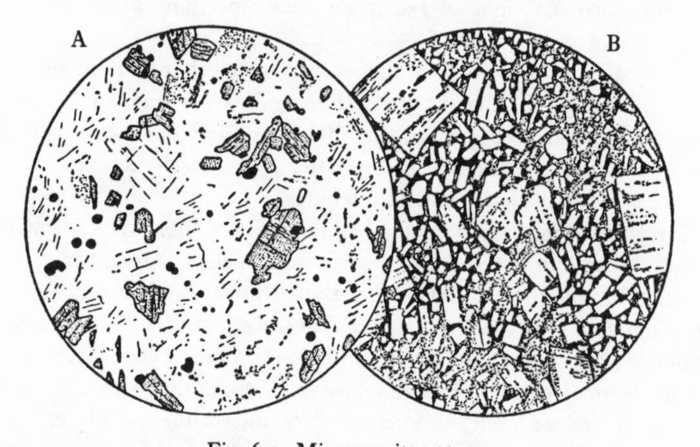

Fig. 6.2. Microsyenite; × 20.

A. Alkali microsyenite, North Berwick Law, Scotland. Alkali feldspar, green clinopyroxene and accessory iron-ore.

B. Peralkaline microsyenite, Middle Eildon Hills, Melrose, Scotland. Alkali feldspar and interstitial deep blue riebeckite.

Rocks with phenocrysts of rhomb-shaped anorthoclase to cryptoperthite, augite, altered fayalite, and iron-ore in an orthophyric groundmass are found among the alkali microsyenite dykes of Tugtutôq Island, S. Greenland (17).

Recorded examples of *microsyenite* seem to be very rare, but a rock from Pleasant Mountain, Maine (18), perhaps falls here. It has resorbed and zoned phenocrysts of oligoclase and of zoned euhedral augite, in a matrix of dominant potash feldspar, small grains of augite and ragged grains of green hornblende with a trace of biotite and interstitial quartz.

References and notes

(1) Walker, F. & Nicolaysen, L. O., 1954. The petrology of Mauritius. *Col. Geol. Miner. Res.* **4**, 12–14.

(2) Smith, W. Campbell, 1931. A classification of some rhyolites, trachytes, and phonolites from part of Kenya Colony, with a note on some associated basaltic rocks. *Q. Jl. geol. Soc. Lond.* **87**, 233.

(3) Nash, W. P., Carmichael, I. S. E. & Johnson, R. W., 1969. The mineralogy and petrology of Mount Suswa, Kenya. *J. Petrology* **10**, 309–39.

(4) Smith, 1931, *op. cit.*, 225.

(5) Smith, W. Campbell, 1938. Petrographic description of volcanic rocks from Turkana, Kenya Colony, with notes on their field occurrence from the manuscript of Mr A. M. Champion, *Q. Jl. geol. Soc. Lond.* **94**, 519–20.

(6) Bailey, E. B. *et al.*, 1924. Tertiary and post-Tertiary geology of Mull, Loch Aline, and Oban. *Mem. geol. Surv. U.K.* 191–3.

(7) McRobert, R. W., 1914. On acid and intermediate intrusions and associated ash-necks in the neighbourhood of Melrose, Roxburghshire. *Q. Jl. geol. Soc. Lond.* **70**, 303–14.

(8) Macdonald, G. A., 1949. Petrography of the Island of Hawaii. *Prof. Pap. U.S. geol. Surv.* **214**-D, 77.

(9) Le Maitre, R. W., 1962. Petrology of volcanic rocks, Gough Island, South Atlantic. *Bull. geol. Soc. Am.* **73**, 1315–16.

(10) Baker, I., 1969. Petrology of the volcanic rocks of Saint Helena Island. *Ibid.* **80**, 1283–1309.

(11) Lindgren, W. & Ransome, F. L., 1906. Geology and gold deposits of the Cripple Creek district, Colorado. *Prof. Pap. U.S. geol. Surv.* **54**, 77–8.

(12) Le Maitre, 1962, pp. 1314–15.

(13) Washington, H. S., 1906. The Roman comagmatic region. *Carnegie Inst. Wash. Publ.* **57**, 30–3, 58–66.

(14) Sutherland, D. S., 1967. A note on the occurrence of potassium-rich trachytes in the Kaiserstuhl carbonatite complex, West Germany. *Miner. Mag.* **36**, 334–41.

(15) Tyrrell, G. W., 1928. The geology of Arran. *Mem. geol. Surv. U.K.* 222–3, 236–7.

(16) Campbell, R. & Stenhouse, A. G., 1933. The occurrence of nepheline and fayalite in the phonolitic trachyte of the Bass Rock. *Trans. geol. Soc. Edinb.* **13**, 126–132.

(17) Upton, B. G. J., 1964. The geology of Tugtutôq and neighbouring islands, South Greenland: part II. Nordmarkitic syenites and related alkaline rocks. *Medd. om Gronland* **169**, no. 2, 12–16.

(18) Jenks, W. F., 1934. Petrology of the alkaline stock at Pleasant Mountain, Maine. *Am. J. Sci.* (5) **28**, 329.

7

Monzonites, monzodiorites, monzogabbros

These rocks differ from syenite in having greater amounts of lime-bearing plagioclase and, like syenite, they do not form intrusions of any great size. In monzonite, the alkali feldspar and lime-bearing plagioclase are present in roughly equal amounts, the plagioclase making up between 40 and 60 per cent of the total feldspar content (1). In monzodiorite and monzogabbro, plagioclase is dominant, making up between 60 and 90 per cent of the total feldspar content. It varies from oligoclase to andesine in composition in monzodiorite, whereas it is labradorite in monzogabbro.

Their minerals have much the same character as those of syenite but pyroxenes and, sometimes, olivine are more important here. They range from leucocratic to mesotype, with an average colour index of about 35. They may have a subhedral-granular texture but frequently develop the 'monzonitic' texture in which shapeless plates of alkali feldspar tend to enclose poikilitically the other constituents.

Varieties, with some examples

The intrusive rocks of Monzoni, S. Tyrol, provide examples of all three groups, and it is from here that monzonite is named. The monzonite is mainly an *augite–biotite monzonite* with subhedral zoned plagioclase averaging andesine, poikilitic alkali feldspar (about $Or_{70} Ab_{30}$), and often a little quartz (fig. 7.1A). In some, green hornblende and, more rarely, a little orthopyroxene are additional constituents. Iron-ore, zircon, sphene and apatite are the accessory minerals. The bulk of the intrusion, however, is made up of monzodiorite and monzogabbro with minor diorite, gabbro and pyroxenite, and the famous monzonite is only a local variant of the monzodiorite (2).

The *monzodiorite* here is similar to the monzonite in mineralogy but has more plagioclase, while the monzogabbro is an *augite–biotite–olivine monzogabbro* with dominant clinopyroxene, minor biotite and olivine, labradorite (An_{55-60}), and subordinate potash feldspar.

A number of monzonites and monzodiorites have orthopyroxene as an essential mineral in addition to augite, usually with a little biotite and sometimes a little hornblende as well. The core of the Adamant pluton, British Columbia (3), is composed of such a *pyroxene monzonite*,

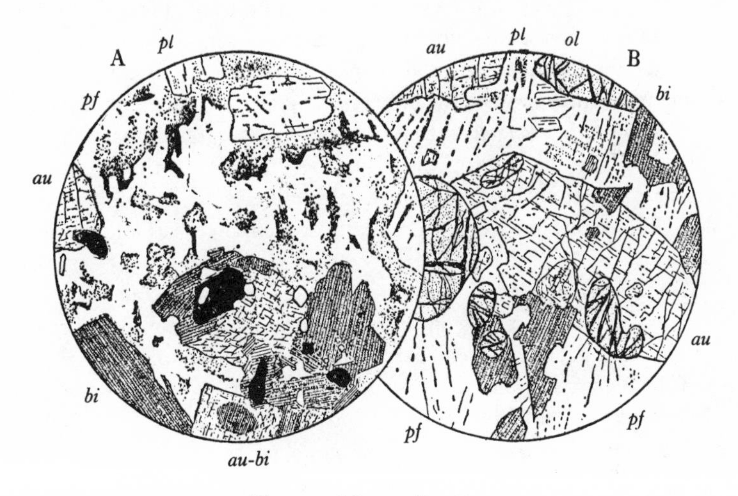

Fig. 7.1. Monzonite; × 20.

A. Monzonite, Predazzo, Tyrol. Colourless augite (*au*) and brown biotite (*bi*), plagioclase (*pl*) and potash feldspar (*pf*) with accessory iron-ore and apatite.
B. Olivine monzonite, Kentallen, near Ballachulish, Scotland. A more mafic type with noticeable olivine (*ol*).

medium-grained with subhedral-granular texture, and with a colour index of about 40. It is composed of a somewhat aluminous pale green augite, subordinate pleochroic hypersthene, plagioclase zoned from An_{45} to An_{39} and equant grains of orthoclase-microperthite. Apatite, magnetite, and haematite-ilmenite intergrowth are the accessory constituents. A very little hornblende is present as thin mantles to some of the pyroxene crystals, and clusters of radiating biotite flakes occur at junctions between hypersthene and alkali feldspar. Pyroxene monzonites of this kind are also found, for instance, at Slievegarron at the east end of the Newry complex, N. Ireland (4), and in the Glen Falloch area, north of Loch Lomond, Scotland (5).

The latter area also has small intrusions of *pyroxene monzodiorite* with appreciable biotite and a little olivine. Similar monzodiorites with two pyroxenes are the 'quartz–orthoclase gabbro' from Park Co., Montana

(6), with a small amount of biotite and a little interstitial quartz; and the 'Jotun-norite' of the Jotunheim district, Norway, in which variable, but usually sparing, biotite and sometimes some brown hornblende accompany the pyroxenes.

Pyroxene–biotite monzogabbro is present among the rocks of the Brocken massif, Harz Mountains, while some of the 'norite' of the Cortlandt series, Peekskill, New York, has appreciable potash feldspar and varies from pyroxene–biotite monzodiorite to monzogabbro.

In other monzonites and monzodiorites, hornblende becomes more prominent and may even be the only essential ferromagnesian mineral as in the *hornblende monzonite* that makes up the greater part of the Plauens'cher Grund outcrop, near Dresden, and as in the *hornblende monzodiorites* that occur among the satellitic intrusions of the Prince of Wales–Chichagof belt, west of the main Coast Range batholith in S.E. Alaska (7).

Olivine monzonite, in which olivine is associated in some quantity with pyroxene and biotite, makes up the whole, or part, of a number of small intrusions of Caledonian age in Scotland and is also found in the eastern part of the Newry complex, N. Ireland. Among the Scottish occurrences is that described as 'kentallenite' from Kentallen, near Ballachulish (fig. 7.1B). This is mesotype with rather large, fairly well-formed crystals of olivine and augite, each making up about 20 per cent of the rock, and semipoikilitic biotite making up some 10 per cent. The andesine is in lath-shaped crystals, potash feldspar is interstitial, and apatite is a notable accessory mineral.

Other Scottish 'kentallenites' have essential orthopyroxene in addition to augite, and among these are examples of *olivine monzodiorite*, as at A'Chrois, Argyllshire (8), and in the Glen Falloch area (9), in which andesine is dominant over potash feldspar.

Some monzonites, monzodiorites and monzogabbros have a more pronounced alkaline character, indicated by the presence of sodic pyroxenes or titaniferous augite, and sodic amphiboles. These will be distinguished as alkali monzonites, monzodiorites and monzogabbros. They occur associated with other alkali rocks and may sometimes form relatively large intrusions.

The most famous of the *alkali monzonites* is the rock known as 'larvikite' from the Oslo district, Norway. Originally regarded as a syenite, it is, in its typical development, now known to be a monzonite (10). It is relatively coarse grained with potash feldspar and oligoclase, often intimately intergrown. Ferromagnesian constituents make up about 15 per cent of the rock and comprise a very pale violet or greenish titaniferous augite with subordinate brown sodic amphibole, deep brown iron-rich mica and, usually, a little ferriferous olivine (fig. 5.2B). A little quartz or a little nepheline are normally present as accessory constituents in addition to titaniferous magnetite, zircon, and large and abundant crystals of apatite.

The 'kjelsåsites' of the same district are rather variable, but many of them are *alkali monzodiorites* with andesine dominant over potash feldspar. The main dark mineral is a feebly titaniferous augite with or without brown sodic amphibole, minor iron-rich biotite and sometimes a little olivine (about Fa_{50}). Apatite is, again, noticeable among the accessory constituents, and these rocks tend to have a somewhat higher colour index than the larvikites. There is commonly a little interstitial quartz.

The 'kauaiites' or 'Oslo-essexites' of the Oslo district are also alkali monzodiorites, but of a more mafic character (11). The plagioclase is zoned from labradorite to oligoclase but averages An_{34} to An_{44}, and may grade into anorthoclase marginally. The colour index ranges from 30 to 60, with titaniferous augite as the dominant dark mineral, minor biotite and, commonly, some olivine (about Fa_{30}).

The small stock at Rongstock, Bohemia, consists of rocks having titaniferous augite, brown sodic amphibole and biotite as their ferromagnesian constituents, zoned plagioclase averaging andesine, and interstitial potash feldspar. The proportions of the last two are variable and, while the bulk of the rocks are alkali monzodiorites, the border phase has a lower colour index and potash feldspar in excess of plagioclase. The accessory constituents include cancrinite and zeolites, replacing original nepheline, and also relatively abundant apatite and sphene.

Alkali monzonites with pale green clinopyroxene and biotite and with barium orthoclase as their potash feldspar occur in Montana (Highwood Peak, Highwood Mountains (12); Beaver Creek stock, Bearpaw Mountains), while *alkali monzogabbro* carrying titaniferous augite, biotite, and interstitial accessory analcite and nepheline, is found at Huerfano Butte, Spanish Peaks region, Colorado (13).

Subvolcanic, porous and granular *alkali monzonites* with aegirine-augite, sometimes some sodic amphibole, zoned plagioclase, relatively abundant sphene and some apatite, all enclosed by soda sanidine, occur among the ejected blocks found in the volcanic breccias and tuffs of Procida, Italy.

TABLE 7.1. *Average chemical compositions of monzonite, monzodiorite, latite, latite-andesite and allied rocks.*

	1	2	3	4	5	6	7	8	9	10	11	12
SiO_2	55.36	54.66	54.02	56.00	56.17	50.00	56.23	49.60	46.49	58.15	51.75	48.87
TiO_2	1.12	1.09	1.18	1.29	1.19	2.29	1.49	2.50	2.96	1.28	2.07	2.77
Al_2O_3	16.58	16.98	17.22	16.81	18.20	16.31	18.70	17.14	16.28	17.71	17.50	17.14
Fe_2O_3	2.57	3.26	3.83	3.74	2.12	3.99	2.81	3.24	3.38	2.37	3.72	3.86
FeO	4.58	5.38	3.98	4.36	3.72	6.26	3.09	6.65	7.82	4.58	6.49	7.48
MnO	0.13	0.14	0.12	0.13	0.10	0.20	0.12	0.15	0.18	0.18	0.21	0.18
MgO	3.67	3.95	3.87	3.39	2.31	4.46	1.66	5.08	7.02	1.34	2.69	4.44
CaO	6.76	6.99	6.76	6.87	5.90	8.33	4.82	7.97	9.96	3.82	6.06	7.98
Na_2O	3.51	3.76	3.32	3.56	4.72	4.27	5.04	3.87	2.93	6.23	5.56	4.47
K_2O	4.68	2.76	4.43	2.60	4.50	2.43	4.74	2.75	1.75	3.25	2.41	1.71
H_2O^+	0.60	0.60	0.78	0.92	0.49	0.83	0.89	0.48	0.75	0.59	0.82	0.49
P_2O_5	0.44	0.43	0.49	0.33	0.54	0.63	0.41	0.57	0.48	0.50	0.72	0.61
qz	–	2.0	0.4	7.4	–	–	–	–	–	–	–	–
or	27.8	16.7	26.1	15.6	26.7	14.4	27.8	16.7	10.6	19.5	14.5	10.0
ab	29.3	31.9	27.8	29.9	38.0	29.9	40.9	27.3	21.0	52.9	40.4	32.5
an	15.8	21.1	19.2	22.2	15.0	18.1	14.5	21.1	26.1	10.6	15.6	21.7
ne	–	–	–	–	1.0	3.4	0.9	2.8	2.0	–	3.7	2.8
di *wo*	6.3	4.5	4.6	3.9	4.4	8.2	2.8	6.1	8.2	2.0	4.1	6.0
di *en*	3.9	2.7	3.4	2.7	2.7	5.3	2.1	3.9	5.4	0.8	2.1	3.6
di *fs*	2.0	1.5	0.8	0.9	1.5	2.4	0.4	1.9	2.2	1.2	1.8	2.1

		1	2	3	4	5	6	7	8	9	10	11	12
hy	*en*	4.0	7.2	6.3	5.8	–	–	–	–	–	1.6	–	–
	fs	2.1	3.9	1.6	2.2	–	–	–	–	–	2.1	–	–
ol	*fo*	0.8	–	–	–	2.2	4.0	1.4	6.2	8.4	0.7	3.2	5.3
	fa	0.4	–	–	–	1.3	1.8	0.4	3.1	3.7	1.0	3.0	3.3
mt		3.7	4.9	5.6	5.3	3.0	5.8	4.2	4.6	4.9	3.5	5.3	5.6
il		2.1	2.1	2.3	2.4	2.3	4.4	2.9	4.7	5.6	2.4	4.0	5.3
ap		1.0	1.0	1.2	0.8	1.3	1.5	1.0	1.3	1.1	1.2	1.7	1.4
DI		57.1	50.6	54.3	52.9	65.7	47.7	69.6	46.8	33.6	72.4	58.6	45.3

1. Monzonite (average of 46) (*Bull. Geol. Soc. Am.* **65**, 1017, no. I).
2. Monzodiorite (and monzogabbro) (average of 56) (*Ibid.*, p. 1018, no. I).
3. Latite (average of 42) (*Ibid.*, p. 1017, no. II).
4. Latite-andesite (average of 38) (*Ibid.*, p. 1017, no. II).
5. Alkali monzonite (average of 17).
6. Alkali monzodiorite (and monzogabbro) (average of 53) (*Ibid.*, p. 1018, no. III).
7. Tristanite (average of 16).
8. Trachyandesite (average of 26).
9. Trachybasalt (average of 21).
10. Benmoreite (average of 22).
11. Mugearite (average of 17).
12. Hawaiite (average of 47).

Chemistry

Average chemical compositions of monzonite, nonzodiorite and monzogabbro, alkali monzonite and alkali monzodiorite + monzogabbro, are given in table 7.1.

Note the moderate SiO_2 content throughout. Monzonite and monzodiorite + monzogabbro have hypersthene in the norm and are more or less silica saturated. Alkali monzonite and monzodiorite + monzogabbro, on the other hand, are richer in Na_2O and have both olivine and a little nepheline in the norm, but no hypersthene. Note, too, the tendency for the alkali representatives to be richer in TiO_2 and P_2O_5.

References and notes

(1) In a recent classification of plutonic rocks approved by the I.U.G.S. (see Streckeisen, A. L., 1973. *Geol. Newsl.* **2**, 110, or Sabine, P. A., 1974. *Geol. Mag.* **111**, 165) the limits are taken as 35 to 65 instead of 40 to 60. These are too wide for rocks which, by definition, are supposed to have approximately equal amounts of the two feldspars.

(2) Del Monte, M., Paganelli, L. & Simboli, G., 1967. The Monzoni intrusive rocks. A modal and chemical study. *Miner. Petrogr. Acta* **13**, 75–118.

(3) Fox, P. E., 1969. Adamant pluton, British Columbia. *Pap. geol. Surv. Canada* **67–61**, 21–3.

(4) Reynolds, D. L., 1934. The eastern end of the Newry igneous complex. *Q. Jl. geol. Soc. Lond.* **90**, 619–21.

(5) Anderson, J. G. C., 1937. Intrusions of the Glen Falloch area. *Geol. Mag.* **74**, 464.

(6) Emmons, W. H., 1908. Geology of the Haystack stock, Cowles, Park County, Montana. *J. Geol.* **16**, 213–17.

(7) Buddington, A. F., 1927. Coast Range intrusives of South eastern Alaska. *Ibid.* **35**, 232–33, and table II, no. 2, p. 236.

(8) Anderson, J. G. C., 1935. The Arrochar intrusive complex. *Geol. Mag.* **72**, 279–80.

(9) Anderson, 1937, p. 463.

(10) Oftedal, I. W., Bergstøl, S. & Svinndal, S., 1960. Guide of excursions no. A 12 and no. C 8. *XXI Inter. geol. Congr. Norden*, p. 6.

(11) Barth, T. F. W., 1945. Studies on the igneous rock complex of the Oslo region. II. Systematic petrography of the plutonic rocks. *Skrift. Det Norske Vidensk.-Akad. Oslo* I. *Mat. Nat. Kl. 1944*, no. **9**, 31–40.

(12) Burgess, C. H., 1941. Igneous rocks of the Highwood Mountains, Montana: part IV. The Stocks. *Bull. geol. Soc. Am.* **52**, 1825.

(13) Knopf, A., 1936. Igneous geology of the Spanish Peaks region, Colorado. *Ibid.* **47**, 1776–8.

8

Latites, latite-andesites, allied volcanics and their hypabyssal equivalents

Latites are the volcanic equivalents of the monzonites, *latite-andesites* of the monzodiorites. Both are commonly porphyritic, with a very fine-grained holocrystalline or hypocrystalline groundmass.

The main *phenocrystal* mineral is *lime-bearing plagioclase* in euhedral crystals, often zoned and sometimes riddled with glass inclusions. This is accompanied by one or more mafic minerals and, of these, *clinopyroxene* is the commonest. *Orthopyroxene* is found in some, and *olivine* is not infrequent. *Biotite* is quite common but *hornblende* is, on the whole, rare and both these minerals may show varying degrees of magmatic resorption.

In some latites, *sanidine* is present in the phenocryst generation, either as independent crystals or as margins to the plagioclase phenocrysts. It is not normally found in the phenocryst generation of latite-andesites.

The texture of the groundmass is usually hyalopilitic or else more or less trachytic. The trachytic groundmass of latites is made up largely of little laths of sanidine and some of plagioclase, together with grains of iron-ore and some mafic constituent, commonly clinopyroxene. The corresponding groundmass of latite-andesites is composed dominantly of little laths of plagioclase with subordinate sanidine. The latter may fringe the plagioclase as a narrow rim and can easily be overlooked unless careful search is made.

When the groundmass is hyalopilitic, a part, or the whole, of the potential potash feldspar may be occult in the glassy residuum. Rocks belonging here with a groundmass of this kind may be indistinguishable from each other or from andesites unless a chemical analysis is available.

Some examples

True *latites*, like monzonites, are relatively rare. The type rock is the Table Mountain flow, west of Sonora on the western slope of the Sierra

Nevada, California (1). Phenocrysts of labradorite with some of pale augite and of olivine lie in a hyalopilitic groundmass of plagioclase laths, granules of augite, some iron-ore and apatite, and a residuum of glass. The Dardanelle flow is similar, but the phenocrysts are smaller, there is less olivine; and the proportion of glass in the groundmass is somewhat greater.

Among the late products of the Tuscan magma phase in northern Latium, Italy, latites are to be found: they are the dominant lavas of the early Quaternary Cimino volcano (2), transitional to rhyolite on the one hand and to olivine–augite trachyte on the other. The lavas of the lava domes have some 40 per cent of phenocrysts including labradorite (An_{55} to An_{50}), less frequent but large crystals of sanidine, biotite, hypersthene, augite, and rare olivine. The groundmass is largely cryptocrystalline with occasional very fine granophyric areas. These latites are potentially quartz-bearing and are transitional to rhyolites.

The somewhat later lava flows have large phenocrysts of feldspars, biotite and pyroxene like the lavas of the domes but with highly resorbed margins. There is a second generation of much smaller phenocrysts of clinopyroxene and olivine, and the groundmass is made up of microlites of feldspars and of pyroxene. These lavas are transitional to trachyte.

Augite latite with phenocrysts of pale green augite and of plagioclase sometimes margined by potash feldspar, and *augite–biotite latite* with biotite as an additional phenocrystal mineral occur as flows in the Rossland volcanic group of the Rossland district, British Columbia. Both had originally a hyalopilitic groundmass, now largely altered with the formation of secondary minerals. A *hornblende–augite latite* from the same group has acicular phenocrysts of hornblende with a rough fluidal alignment and subordinate prisms of augite. These are set in a finely crystalline groundmass of the same dark minerals with microlites of potash feldspar and plagioclase.

Latite-andesites are commoner than latites. They are not, however, usually quoted as such in the literature but are termed 'andesite' or 'latite'.

Thus, chemical analysis shows that some of the rocks described as 'pyroxene andesite' from the Matra district of Hungary are really latite-andesites with phenocrysts of zoned plagioclase, orthopyroxene, augite and often a little olivine in a hyalopilitic groundmass of andesine laths, grains of pyroxene and iron-ore, a little apatite, sometimes a little tridymite, and brown glass. Other rocks from the same district falling here have phenocrysts of plagioclase, biotite and oxyhornblende (both showing magmatic resorption).

Similar rocks are to be found elsewhere as, for instance, in the San Franciscan volcanic field, Arizona (San Francisco Mountain; Kendrick Peak) (3), among the 'latites' of the Hinsdale formation in the San Juan region, Colorado (4), in the Aegean Islands of Milos, Oxylithos and Metelin, and at Stromboli.

The San Franciscan rocks are pyroxene latite-andesites with augite, minor orthopyroxene, labradorite laths and abundant iron-ore set in brownish glass. Some have phenocrysts of labradorite, and some a little olivine. The rocks from the Hinsdale formation are also pyroxene latite-andesites with a little olivine, but here the place of glass is taken by a cryptocrystalline paste, probably largely made up of oligoclase and potash feldspar. Those from Stromboli may have biotite as an additional phenocrystal constituent accompanying the pyroxenes and plagioclase, while the groundmass ranges in texture from trachytic to vitrophyric. Chemically, they lie close to latite.

Somewhat altered augite latite-andesites of Silurian age in the Tortworth area of Gloucestershire (5) have zoned plagioclase rimmed with alkali feldspar and altered in a varying degree to sericite and calcite, while much of the augite is replaced by chlorite, calcite and clay minerals.

There are also lavas of more alkaline nature, corresponding with alkali monzonites, monzodiorites or monzogabbros. These fall into two main groups:

(i) A relatively potassic series comprising ***tristanite, trachyandesite*** and ***trachybasalt*** such that tristanites fall in the field of alkali monzonite, trachyandesites lie partly in this field and partly in the field of alkali monzodiorite, while trachybasalts are more or less equivalent to alkali monzogabbros.

(ii) A more sodic series, ***benmoreite, mugearite*** and ***hawaiite***, all of which are equivalent to various types of alkali monzodiorite (fig. 8.1).

Trachybasalt, trachyandesite and tristanite

Trachybasalts have a differentiation index of less than 40, averaging about 34, and the normative plagioclase is a labradorite, averaging An_{55}.

Trachyandesites have a differentiation index lying between 40 and 60, averaging 47, and their normative plagioclase is andesine, averaging An_{44}.

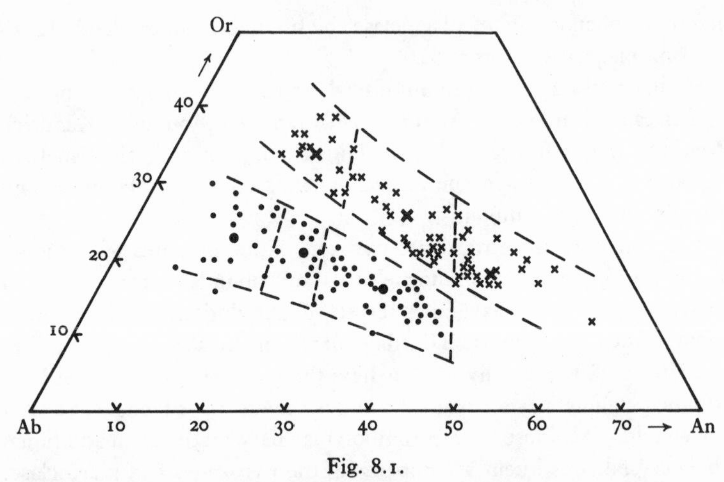

Fig. 8.1.

Plot of normative Or, Ab, and An for rocks of the trachybasalt → tristanite series (crosses) and the hawaiite →benmoreite series (dots). The approximate fields of trachybasalt, trachyandesite and tristanite are indicated, and the average composition of each type is represented by a large cross. The fields of hawaiite, mugearite and benmoreite are also distinguished, and the average composition of each represented by a large dot.

Tristanites have a differentiation index of 60 to 75, averaging 69, and their normative plagioclase is oligoclase or oligoclase-andesine, averaging An_{26}.

The occurrence of these on Gough Island in the Atlantic Ocean (6) will be used to provide typical examples. The trachybasalts here are commonly porphyritic with phenocrysts of olivine and titaniferous augite, and a few of plagioclase strongly zoned from labradorite to andesine, lying in a groundmass of rather more sodic zoned plagioclase laths, little subhedral crystals of pale purple-brown titaniferous augite, rounded grains of olivine (about Fa_{20}) (partly altered to iddingsite), laths of ilmenite and grains of magnetite with interstitial alkali feldspar, crowded with tiny needles of apatite.

The trachyandesites are developed abundantly on the island and vary considerably in texture. Basically, they have the same mineralogy as the trachybasalts but the zoned plagioclase is somewhat more sodic, the interstitial alkali feldspar more abundant, and the olivine now averages Fa_{26}.

The tristanite (7) also has phenocrysts of olivine, clinopyroxene and feldspar but the olivine is more fayalitic (about Fa_{40}), the pyroxene is a pale grey-green and the plagioclase is zoned from labradorite to andesine or oligoclase and often has a thin rim of alkali feldspar. The groundmass has grains of olivine and titaniferous pyroxene together with potash andesine and interstitial alkali feldspar.

Rocks like these are also found on the Tristan da Cunha group of islands, where they may carry a little leucite.

Trachybasalts or trachyandesites and tristanites occur elsewhere, as in Malagasy, the Auvergne district of France (Mt. Dore, Chaine des Puys), and the island of Ischia. The trachyandesites of Ischia have phenocrysts of zoned plagioclase, sometimes soda-sanidine, pale greenish augite, partially resorbed biotite and sporadic olivine. The groundmass has laths of sanidine and plagioclase, little crystals of augite and accessory iron-ore, often relatively abundant, and apatite, with variable amounts of glass. In the tristanites, phenocrysts of soda-sanidine are dominant over those of sanidine-mantled plagioclase, the groundmass feldspar is soda-sanidine, and sphene appears among the accessory constituents.

Hawaiite, mugearite and benmoreite (8)

These rocks are typically aphyric, though porphyritic types do occur, and have a trachytic texture that gives them a marked fissility in hand specimen.

Hawaiites have strongly zoned feldspar occurring as micropheno-crysts and as laths in the groundmass, ranging from potassic calcic andesine to lime anorthoclase, the two being present in roughly equal amounts. The normative plagioclase is andesine averaging An_{40}, and the differentiation index varies from about 35 to 56, averaging 45.

In ***mugearites*** the plagioclase is more sodic and lime anorthoclase is now the dominant feldspar. In some instances there is some interstitial soda-sanidine. The normative plagioclase is oligoclase averaging An_{28} and the differentiation index varies from about 56 to 65, averaging 58.5.

Benmoreites resemble mugearites but have anorthoclase or soda-sanidine as the dominant feldspar. The normative plagioclase averages An_{17} and the differentiation index lies between 65 and 77, averaging 72.

In all three types the usual ferromagnesian constituents are a ferrifer-ous olivine, changing from about Fa_{40} to Fa_{70} in passing from hawaiite to benmoreite, and a lime-rich clinopyroxene. The latter is feebly pleo-chroic from pale yellow to brown and somewhat titaniferous in hawaiites and mugearites: in benmoreite, on the other hand, the clinopyroxene is usually a pale green ferroaugite.

The olivine, which may be partly altered to iddingsite, occurs as microphenocrysts, frequently elongated along the *x*-axis, and as grains in the groundmass, while the pyroxene is commonly present as little grains but may also occur as small subophitic plates. Olivine is usually present in excess of pyroxene in both hawaiite and mugearite. In these two rock types, especially, titaniferous magnetite and apatite are prominent accessory minerals.

In some of these rocks, amphibole and biotite are present in small amount as additional ferromagnesian constituents. Occasionally, they become important.

Hawaiite, mugearite and benmoreite form a well-marked series grading into alkali basalt on the one hand, and into anorthoclase trachyte on the other. Numerous examples are to be found on oceanic islands such as the Hawaiian islands (9), Mauritius, Réunion, St. Helena and Ascension Islands, and the Azores, but they are also to be seen in continental areas such as in the E. Otago volcanic province, New Zealand (Tertiary), the British Tertiary province (particularly in the islands of Skye and Mull) (10) (fig. 8.2A), in Syria (Tertiary to Quaternary), among the Carboniferous igneous rocks of the Midland Valley of Scotland and, in a somewhat altered state, on Skomer Island, Pembrokeshire (Ordovician).

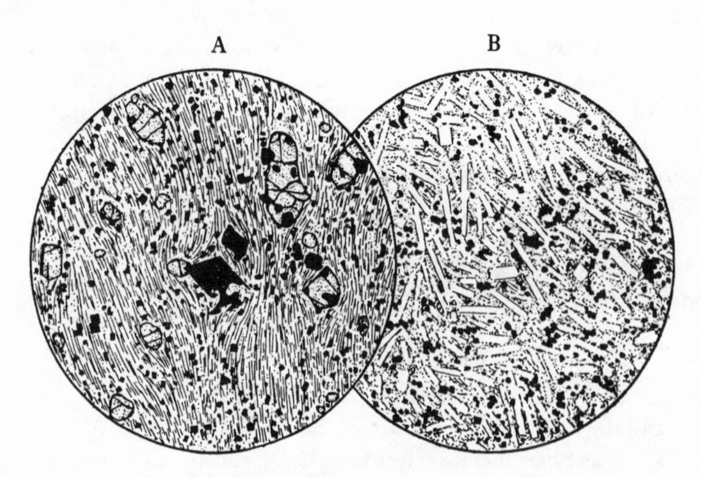

Fig. 8.2. × 20.

A. Mugearite, Druim na Criche, Skye. Densely packed laths of lime-anorthoclase showing flow structure, olivine, a little clinopyroxene and abundant accessory iron-ore.

B. Spilite, Port Isaac, Cornwall. Composed largely of albite with much secondary chlorite and calcite and abundant little crystals of iron-ore.

Chemistry

Average chemical compositions of the lavas mentioned in this chapter are given in table 7.1, columns 3–4 and 7–12. Apart from their higher ratios of Fe_2O_3:FeO, average latite (column 3) and latite-andesite (column 4) are very similar in chemical composition to average monzonite and average monzodiorite (with monzogabbro) respectively (columns 1 and 2).

Note the usual presence of normative *ne* and absence of normative *hy* in the more alkaline lavas (columns 7–12), as in the norms of alkali monzonite and alkali monzodiorite (with monzogabbro) (columns 5 and 6). Note, also, the more potassic nature of the trachybasalt–tristanite series when compared with the hawaiite–benmoreite series. This feature is well brought out by an inspection of the ratio of normative *or* to normative *ab* + *an* in the two series.

Average tristanite (column 7) resembles average alkali monzonite (column 5) fairly closely. The remaining lavas here (columns 8–12) have representatives among the alkali monzodiorites and monzogabbros, the average chemical composition of all these together being: SiO_2, 50.52; TiO_2, 2.41; Al_2O_3, 17.14; Fe_2O_3, 3.40; FeO, 6.78; MnO, 0.18; MgO, 4.25; CaO, 7.38; Na_2O, 4.52; K_2O, 2.26; H_2O+, 0.59; P_2O_5, 0.57 – very similar to that of average alkali monzodiorite with monzogabbro (column 6).

Hypabyssal equivalents

The hypabyssal equivalents of the monzonites and latites are the micromonzonites, and of the monzodiorites and latite-andesites, the micromonzodiorites.

A porphyritic *micromonzonite* has been described under the name 'syenite porphyry' from the Rossland district, British Columbia (11). It has abundant phenocrysts of soda-orthoclase, andesine (often surrounded by a shell of alkali feldspar), biotite and augite, in a fine-grained subhedral granular groundmass composed largely of orthoclase with minor oligoclase and some interstitial quartz.

A more mafic example is found in the Gwanda district, S. Rhodesia, in connection with the Colleen Bawn igneous complex (12). Phenocrysts of plagioclase, augite and olivine lie in a fine-grained groundmass of orthoclase and plagioclase with numerous grains of augite and magnetite, and abundant apatite.

Porphyritic *micromonzodiorites*, termed 'diorite-porphyry', occur as dykes in the southern area of the Crazy Mountains, Montana (13). One, north of Sweet Grass Creek, has phenocrysts of oligoclase and brown hornblende in a groundmass of oligoclase, some potash feldspar and a little quartz, hornblende, biotite, augite, magnetite and apatite. Another, on the northern slope of Big Timber Creek, has small phenocrysts of andesine, orthopyroxene, augite, biotite and a little green hornblende. Others, more or less aphyric, are associated as dykes with the Lausitz and Riesengeberge granodiorites, Germany.

Alkali micromonzonites may be illustrated by the rocks known as 'gauteite' occurring in the Rongstock district of Bohemia, and elsewhere. Those found as dykes and sheets at a few localities in the Highwood Mountains, Montana (14), are rather fine grained with phenocrysts of a sodic amphibole, some large crystals of apatite, considerable iron-ore in scattered grains, and laths of oligoclase and soda-orthoclase. A trace of isotropic material may be analcite or sodalite.

Alkali micromonzodiorites and *micromonzogabbros* occur as dykes at Bear Creek, Spanish Peaks region, Colorado (15). One variety consists largely of tabular potash oligoclase with interstitial soda-orthoclase, some titanaugite, biotite, magnetite and apatite, and has sporadic phenocrysts of augite and amphibole. Another has small phenocrysts of magnesian olivine, abundant augite, tabular labradorite (An_{65}), interstitial alkali feldspar and some analcite, with abundant accessory magnetite and acicular apatite.

References and notes

(1) Ransome, F. L., 1898. Some lava flows of the western slopes of the Sierra Nevada, California. *Am. J. Sci.* (4) **5**, 355–75.

(2) Mittempergher, M. & Tedesco, C., 1963. Some observations on the ignimbrites, lava domes and lava flows of M. Cimino (Central Italy). *Bull. Volc.* **25**, 343–58.

(3) Robinson, H. H., 1913. The San Franciscan volcanic field, Arizona. *Prof. Pap. U.S. geol. Surv.* **76**, 140–5.

(4) Larsen, E. S. & Cross, W., 1956. Geology and petrology of the San Juan region, south-western Colorado. *Ibid.* **258**, table 25, anals, 6–10, and pp. 200–2.

(5) Van de Kamp, P. C., 1969. The Silurian volcanic rocks of the Mendip Hills, Somerset; and the Tortworth area, Gloucestershire, England. *Geol. Mag.* **106**, 548–9.

(6) Le Maitre, R. W., 1962. Petrology of volcanic rocks, Gough Island, South Atlantic. *Bull. geol. Soc. Am.* **73**, 1309–40. (The rocks described in this paper as 'basalts' are trachybasalts, as 'trachybasalts' are trachyandesites, and as 'trachyandesites' are tristanites.)

(7) Tilley, C. E. & Muir, I. D., 1964. Intermediate members of the oceanic basalt–trachyte association. *Geol. Fören. Stockh. Förh.* **85**, 436–44.
(8) For the distinction between hawaiite and mugearite see Muir, I. D. & Tilley, C. E., 1961. Mugearites and their place in alkali igneous rock series. *J. Geol.* **69**, 186–203. For benmoreite see Tilley & Muir, 1964.
(9) Macdonald, G. A., 1949. Hawaiian petrographic province. *Bull. geol. Soc. Am.* **60**, 1541–96. (Hawaiites are referred to in this paper as 'andesine andesites' and mugearites as 'oligoclase andesites'.)
(10) Anderson, F. W. & Dunham, K. C., 1966. The geology of Northern Skye. *Mem. geol. Surv. U.K.* 116–22. Thompson, R. N., Esson, J. & Dunham, A. C., 1972. Major element chemical variation in the Eocene lavas of the Isle of Skye, Scotland. *J. Petrology* **13**, 219–53.
(11) Daly, R. A., 1912. Geology of the North American cordillera at the Forty-Ninth Parallel. *Mem. geol. Surv. Canada* **38**, Part 1, 363–5.
(12) Tyndale-Biscoe, R., 1939. Notes on a monzonitic complex and associated rocks in the Gwanda district, Southern Rhodesia. *Trans. geol. Soc., S. Africa* **41**, 97–8.
(13) Wolff, J. E., 1938. Igneous rocks of the Crazy Mountains, Montana. *Bull. geol. Soc. Am.* **49**, 1587–90.
(14) Pirsson, L. V., 1905. Petrography and geology of the igneous rocks of the Highwood Mountains, Montana. *Bull. U.S. geol. Surv.* **237**, 132–6.
(15) Knopf, A., 1936. Igneous geology of the Spanish Peaks region, Colorado. *Bull. geol. Soc. Am.* **47**, 1743, 1756–8.

9
Diorites

It is customary to divide plutonic rocks having a lime-bearing plagioclase as their only essential light constituent into two groups, diorites and gabbros. Petrographers are agreed on this, but the method of division varies. It was the practice in earlier time to use the nature of the dark minerals, rocks with dominant amphibole being classed as diorites, those with dominant pyroxene as gabbros.

Nowadays the composition of the plagioclase is usually regarded as the most important feature for distinction though, in certain classifications of igneous rocks, diorites are divided from gabbros on the basis of colour index.

The composition of the plagioclase will be used here and diorites will be defined as medium to coarse grained igneous rocks composed essentially of lime-bearing plagioclase ranging in composition from oligoclase to calcic andesine. Most diorites have a considerable percentage of ferromagnesian constituents in addition, the average colour index of forty-seven rocks being 43.

Diorites, in contrast to gabbros, rarely form intrusions of any large size. They occur as small independent bodies or, perhaps more commonly, as a phase, often a border phase, of larger more siliceous intrusions.

General mineralogy

The *lime-bearing plagioclase* tends to build subhedral crystals that are often zoned. Its average composition for diorites as a whole is about An_{40}.

Brown *biotite* is a common constituent of many diorites. It is normally found as flakes or small tablets when occurring alone or in association with hornblende, but in pyroxene diorites it may also occur as poikilitic plates. It is often to be seen forming partly at the expense of earlier hornblende or pyroxene.

The *hornblende* is commonly present as subhedral, prismatic crystals, usually brownish green in thin section, though it may be brown in the more mafic diorites.

Pyroxene, when present, may include both ortho- and clino-varieties. The orthopyroxene is frequently subhedral, prismatic, but is occasionally poikilitic with respect to plagioclase. The clinopyroxene is an augite, practically colourless or very pale green in thin section, and is frequently moulded by hornblende. *Olivine* occurs as a minor constituent in a number of diorites and occasionally becomes important.

The main *accessory constituents* are apatite, zircon, sphene, orthite and iron-ores. The latter include ilmenite, titaniferous magnetite with exsolution lamellae of ilmenite, and magnetite. Sulphides such as pyrrhotite, chalcopyrite and pyrite may also occur, the commonest being pyrrhotite. Many diorites have a little quartz and this may be accompanied by a little potash feldspar.

Texture and structure

The typical texture is the subhedral-granular, but sometimes there is an approach to the poikilitic texture, the ferromagnesian constituents being moulded upon, and partially or wholly enclosing, the plagioclase. Occasionally, diorites develop an orbicular structure (1).

Main varieties with examples

Hornblende–biotite diorite is one of the commonest types and is well represented among the intrusions of the Garabal Hill–Glen Fyne igneous complex, Argyllshire, Scotland (2) (fig. 9.1A). One variety is medium grained with an average colour index of 32 and is composed essentially of zoned plagioclase averaging An_{30}, subhedral brownish green hornblende, flakes of biotite closely associated with the amphibole, with accessory interstitial quartz and less microperthite. Another variety is also medium grained but is richer in dark minerals and lacks accessory quartz and potash feldspar. A third variety tends to be coarser grained and has plagioclase of composition An_{42}, brown margined by green or brownish green hornblende, subordinate biotite, and a colour index of about 60.

Similar rocks are to be found as a minor phase among the other Caledonian major intrusions of Scotland as, for instance, at Arrochar (3) and Portencorkrie (4).

The 'appinites' occurring to the north and west of the summit of

Ardsheal Hill, near Kentallen, Argyllshire (5), are largely ***hornblende diorites***, varying from leucocratic to melanocratic and having a schlieren-like arrangement in which the various schlieren differ in texture as well as in colour index. The more or less euhedral prismatic hornblende is greenish brown, while the strongly zoned plagioclase occurs as tabular crystals. There is a little interstitial potash feldspar and quartz, usually in micrographic intergrowth, and accessory biotite.

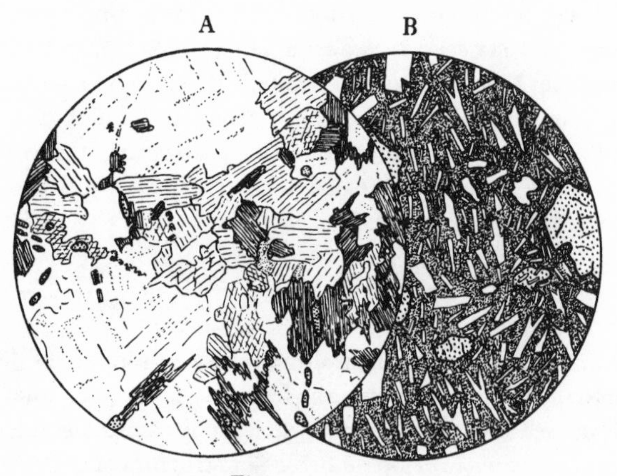

Fig. 9.1. × 20.

A. Hornblende–biotite diorite, Garabal Hill, Argyllshire, Scotland. Plagioclase (andesine), brownish green hornblende, partly chloritised biotite, accessory iron-ore and apatite. Accessory sphene and zircon are present elsewhere in the section.

B. Basaltic andesite, lava of March 1943, Parícutin, Mexico. Phenocrysts of olivine and plagioclase in a hyalopilitic groundmass of plagioclase, ortho- and clinopyroxene, olivine, iron-ore, and glass.

Diorites like these are common as satellite intrusions of, or as minor early phases in, many of the great granitic batholiths. Thus, the diorites in the satellite intrusions of the Prince of Wales–Chichagof belt west of the main Coast Range batholith in S.E. Alaska (6) have colour indices varying from 9 to 73 for rocks whose modes are given, averaging a little over 30. Some are hornblende diorites, with or without accessory biotite; others are hornblende–biotite diorites in which biotite occasionally exceeds hornblende. The plagioclase is andesine and there is commonly a little interstitial quartz and a trace of potash feldspar.

Some hornblende diorites and hornblende–biotite diorites have accessory pyroxene but, in others, it becomes an essential constituent.

The 'bojite' of Pfaffenreuth, Passau, Bavaria, is a *hornblende–augite diorite* with brown hornblende about twice as abundant as augite and a colour index of 60. The plagioclase (An_{40}) is practically unzoned and there is a little accessory biotite.

Pyroxene–biotite diorites are relatively common and are found in a number of the Caledonian igneous complexes of Scotland (Arrochar, Portencorkrie, Garabal Hill, and elsewhere). These have both ortho- and clinopyroxene and normally a little magnesian olivine, with a small amount of interstitial quartz and potash feldspar. They are usually mesotype, but there are rarer melanocratic rocks that may be rich in magnesian olivine and are *olivine–pyroxene–biotite diorites* (Glenn Falloch).

The so-called 'quartz–biotite norites' of the Trondhjem and Stavanger districts, S. Norway, are pyroxene–biotite diorites with strongly zoned plagioclase averaging An_{45}, abundant hypersthene and augite which may be partially replaced by pale green or brownish green hornblende, and accessory interstitial quartz.

Sometimes orthopyroxene falls to the status of an accessory constituent, or fails entirely, as in the *augite–biotite diorite* found at the eastern end of the Newry igneous complex, N. Ireland (7).

A type of diorite entirely different from any of those mentioned so far is found in some of the differentiated basic layered intrusions and is characterised by richness in iron. This type will be distinguished as *ferrodiorite*.

Such ferrodiorites, described as 'ferrogabbro', are found, for instance, in the Skaergaard intrusion, E. Greenland, where they occur in the Main Layered Series and lie above the Middle Gabbro (8). Immediately above this horizon a ferrodiorite is found with plagioclase (An_{40}), olivine (Fa_{61}), and clinopyroxene with sheet-like inclusions of hypersthene, as the main constituents. Above this comes a ferrodiorite in which the clinopyroxene is free from inclusions of hypersthene and is associated with olivine (Fa_{85}) and plagioclase (An_{37}). A little interstitial micropegmatite makes its appearance and apatite becomes relatively abundant. Passing further upwards, the 'purple band' is reached. This is a quartz-bearing ferrodiorite with fayalite (Fa_{96}), plagioclase (An_{30}), clear green clinopyroxene enclosing the other constituents poikilitically, and interstitial quartz in part intergrown with alkali feldspar. Apatite and iron-ore remain as important accessory constituents.

These ferrodiorites show pronounced rhythmic banding. The mineral composition remains the same but the bands vary from leucocratic to melanocratic.

Chemistry

Average chemical compositions of diorites are given in table 9.1. Note the great contrast in chemical composition between diorite and ferrodiorite, especially the lower Al_2O_3, much lower MgO, higher P_2O_5 and much higher iron of the latter. These differences are well brought out in the norm. It will be recalled that ferrodiorites are found in an environment quite different from that of ordinary diorites.

TABLE 9.1 *Average chemical compositions of diorites and andesites*

		1	2	3	4
SiO_2		51.86	47.32	54.90	51.43
TiO_2		1.50	2.06	1.02	2.60
Al_2O_3		16.40	14.42	17.73	13.05
Fe_2O_3		2.73	3.62	2.36	3.36
FeO		6.97	17.49	5.56	9.74
MnO		0.18	0.25	0.15	0.19
MgO		6.12	1.97	4.93	5.28
CaO		8.40	7.54	7.88	8.78
Na_2O		3.36	3.26	3.70	3.18
K_2O		1.33	0.61	1.11	1.04
H_2O^+		0.80	0.61	0.39	0.87
P_2O_5		0.35	0.85	0.26	0.48
qz		0.3	–	4.3	3.2
or		7.8	3.3	6.7	6.1
ab		28.3	27.8	31.4	27.2
an		25.8	22.8	28.4	18.1
di	*wo*	5.6	3.8	3.7	9.2
	en	3.4	0.7	2.3	4.8
	fs	1.9	3.4	1.2	4.1
hy	*en*	11.9	3.9	10.0	8.4
	fs	6.6	20.9	5.5	7.0
ol	*fo*	–	0.2	–	–
	fa	–	1.3	–	–
mt		3.9	5.3	3.5	4.9
il		2.9	4.0	2.0	5.0
ap		0.8	2.0	0.7	1.1
DI		36.4	31.1	42.4	36.5

1. Diorite (average of 50) (*Bull. Geol. Soc. Am.* **65** (1954), 1019, table 6, no. I).
2. Ferrodiorite (average of 10).
3. Andesite (average of 50).
4. Tholeiitic andesite (average of 26) (*Ibid.*, p. 1019, table 6, no. III).

Alteration

Most diorites show at least traces of hydrothermal alteration, and some are strongly affected. Augite, orthopyroxene and hornblende are commonly unchanged but augite may be partly replaced by pale green actinolitic amphibole (uralite) and orthopyroxene by cummingtonite. Biotite may be replaced by chlorite, olivine by serpentine, and plagioclase may be saussuritised and/or sericitised.

Diorite-pegmatites

True diorite-aplites do not seem to be at all common if, indeed, they exist but diorite-pegmatites are known from a number of localities. Some of these have hornblende, as is the case with the diorite-pegmatite cutting diorite at Garabal Hill, and with those cutting peridotites in the Ariège district, Pyrenees, and at Ampangabe, Malagasy Republic. At other times the main ferromagnesian mineral is biotite, as illustrated by a diorite-pegmatite dyke cutting amphibolite at Kragerø, Norway.

The large crystals of feldspar in these pegmatites may be either oligoclase or andesine, and the usual minor constituents are sphene, iron-ore and apatite. There is an absence of the rare minerals that characterise some granite- and syenite-pegmatites, and an absence of complex pegmatites in which hydrothermal replacement has been active.

References and notes

(1) Mourant, A. E., 1932. Orbicular rocks in the Channel Islands. *Geol. Mag.* **69**, 77–83. (The rocks described here as 'gabbro' have andesine as their feldspar, and are therefore diorites in the classification used in this book.)

(2) Nockolds, S. R., 1941. The Garabal Hill–Glen Fyne igneous complex. *Q. Jl. geol. Soc. Lond.* **96**, 465–70.

(3) Anderson, J. G. C., 1935. The Arrochar intrusive complex. *Geol. Mag.* **72**, 276.

(4) Holgate, N., 1943. The Portencorkrie complex of Wigtownshire. *Ibid.* **80**, 179–81.

(5) Walker, F., 1927. The igneous geology of Ardsheal Hill, Argyllshire. *Trans. Roy. Soc. Edinb.* **55**, 147–57.

(6) Buddington, A. F. & Chapin, T., 1929. Geology and mineral deposits of southeastern Alaska, *Bull. U.S. geol. Surv.* **800**, 202–4.

(7) Reynolds, D. L., 1934. The eastern end of the Newry igneous complex. *Q. Jl. geol. Soc. Lond.* **90**, 609–11.

(8) Wager, L. R. & Deer, W. A., 1939. Geological investigations in East Greenland: part III. The petrology of the Skaergaard intrusion, Kangerdlugssuaq, East Greenland. *Med. om Gronland* **105**, no. 4, 98–112, 117–20.

10

Andesites and their hypabyssal equivalents

Andesites are the volcanic equivalents of diorites, commonly porphyritic and relatively dark coloured in hand specimen.

The *feldspar* is often found as phenocrysts as well as tiny laths in the groundmass. The phenocrysts are usually zoned and show a considerable variation in composition even, sometimes, in the same rock. Cores of crystals may be as calcic as An_{80} to An_{60} and outer zones are andesine or even oligoclase. Zoning may be normal or oscillatory, and these phenocrysts may be full of tiny inclusions of glass, pyroxene and iron-ore. The laths of the groundmass are, in general, less calcic and may show some zoning within the andesine range (1).

Pyroxenes are the characteristic ferromagnesian minerals of the andesites. *Clinopyroxene* is very common both as phenocrysts and in the groundmass and is almost colourless or pale green or brownish in thin section. It seems to be a fairly calcium-rich augite when associated with orthopyroxene but in some rocks, particularly those transitional to basalt, the clinopyroxene has a wide range of composition in the individual rock, ranging from augite to pigeonite. In this case much of it is probably quench pyroxene formed metastabily through rapid cooling. *Orthopyroxene* (about Of_{20}–Of_{40}) accompanies clinopyroxene in many andesites and may be present as phenocrysts, or in the groundmass, or in both situations.

Magnesian *olivine* is an essential constituent in some of the more basic andesites, though never abundant. Occurring as phenocrysts, it is also sometimes present in the groundmass. *Hornblende* (commonly oxy-hornblende) and *biotite* are rare, a contrast with the diorites. When found they are usually in small amount, forming phenocrysts that show marked resorption.

Accessory constituents include iron-ore (both ilmenite and magnetite) and apatite, and these may be present both in a phenocryst generation and in the groundmass. Zircon has also been recorded from some andesites.

The groundmass textures and structures of andesites are varied. Some andesites have a trachytic texture; most have a hyalopilitic texture with an ultimate residuum of glass (2); in a few, the groundmass is almost wholly vitreous. Vesicular and amygdaloidal structures are not uncommon.

Some examples

Two of the commonest types of andesite are *basaltic andesite*, a type transitional to basalt with normative plagiocase from An_{48} to An_{52} and commonly olivine-bearing, and *pyroxene andesite* in which the normative plagioclase is less calcic and both ortho- and clinopyroxene are essential constituents.

Both are to be found among the lavas of Parícutin, Mexico (3). This volcano, born on 20 February 1943, continued to erupt until 4 March 1952. Lavas erupted in February and March 1943 were olivine-bearing basaltic andesites (fig. 9.1B) with some phenocrysts of olivine, calcic plagioclase and very rare clinopyroxene in a hyalopilitic groundmass of plagioclase, olivine, ortho- and clinopyroxene, iron-ore and glass. By November 1943 the composition of the magma had changed sufficiently for the normative plagioclase to be An_{46}, though the general mineral assemblage remained the same, thus producing olivine-bearing pyroxene andesite.

After 1944, plagioclase phenocrysts are virtually absent from the lavas erupted, and from 1945 onwards olivine phenocrysts are mantled with orthopyroxene. In later eruptions olivine becomes scarcer while orthopyroxene in the groundmass becomes more abundant. Orthopyroxene also appears as phenocrysts in lavas erupted in 1947 and these are pyroxene andesites with only accessory olivine. No olivine is found in the groundmass of lavas erupted after 1947 and these have now become relatively siliceous with 10 to 11 per cent of normative quartz. They are pyroxene andesites transitional to dacites.

The norms of all chemically analysed specimens of the olivine-bearing lavas at Parícutin show some normative quartz, indicating that the olivine is *reactional* and would not have appeared had cooling been slower and equilibrium established. This is the case with most olivine-bearing andesites from other localities.

Andesites similar to those found at Parícutin form the majority of the late Pleistocene to Recent flows of the whole Parícutin region, and are widespread among the Quaternary lavas of the High Cascades, U.S.A. The latter show considerable variety of texture. Thus among the High

Cascade lavas of the north-central Cascade Mountains, Oregon (4), the basaltic andesites may be aphyric with a more or less trachytic texture (Coffin Mountain 'basalt'), have numerous phenocrysts of plagioclase with small ones of olivine (Black-and-White type), or have a seriate porphyritic texture with phenocrysts of plagioclase, ortho- and clino-pyroxene and olivine (Minto type). The amount of glass in the ground-mass is variable, and in some of the lavas orthopyroxene is absent.

Particularly interesting in connection with the Cascade lavas is some work carried out on a basaltic andesite and an andesite from Medicine Lake (5). The basaltic andesite of the Callahan flow, with normative plagioclase An_{50}, has clinopyroxene covering the complete range in composition from augite to magnesian pigeonite and no orthopyroxene.

The pyroxene andesite of the Schonchin flow, somewhat more siliceous and with normative plagioclase An_{41}, has augite and ortho-pyroxene.

Such basaltic andesites and pyroxene andesites are the main mafic lavas found in orogenic belts and island arcs, as in the Aleutian Islands (Buldir, Semisopochnoi, Unalaga, etc.), the Aegean Islands (e.g. Santorini), the West Indies (St. Kitts), Indonesia, Hungary (e.g. the Cserhat region) and the Andes.

In some of these andesites orthopyroxene is absent, or present only in very small amount, giving rise to *augite andesites*, with or without olivine.

True *hornblende* and *hornblende–biotite* andesites, having these minerals as dominant ferromagnesian constituents, appear to be very rare. Most of the rocks described as such are too siliceous to belong here and should be classed as dacites or rhyodacites. However, some of the more siliceous andesites, transitional to dacite, have pheno-crystal hornblende, commonly red-brown and showing partial to complete magmatic resorption (e.g. Santorini). Occasionally, it may become dominant over pyroxene, as in the andesite flow at 2415m on the south slope of Brokeoff Peak, Lassen Volcanic National Park, U.S.A.

Andesites are also common in older geological periods but these tend to be somewhat altered. Examples of these in Britain are the andesites found among the Ordovician volcanic rocks of Wales and the Lake District, and those among the Old Red Sandstone lavas of Scotland. Thus in the Old Red Sandstone lavas of the western Sidlaw Hills, Perthshire (6), the basaltic andesites and pyroxene andesites are some-times fairly fresh, though the olivine in olivine-bearing types is always altered and the orthopyroxene is partly converted to 'bastite' serpentine.

But in other rocks here all the orthopyroxene is altered, the plagioclase phenocrysts are highly albitised, and the augite granules of the groundmass are pseudomorphed in carbonate and iron-ore.

The alteration of some of these lavas is a reminder that andesite and other lavas associated with them may suffer *propylitisation*; the process whereby original plagioclase is altered to epidote, sericite, and secondary albite and the ferromagnesian minerals to chlorite, calcite, epidote, secondary iron-ore, etc. (7).

The andesites described so far are frequently associated with minor amounts of calc-alkali basalt, with latite-andesites, and with much rhyodacite or dacite. A quite different type of andesitic lava, *tholeiitic andesite*, is found in association with abundant tholeiitic basalt.

These tholeiitic andesites are usually more or less aphyric with laths of andesine, grains of subcalcic augite with or without pigeonite, and grains of iron-ore. Many have varying amounts of interstitial brownish glass charged with granules of iron-ore. Some have olivine as an additional constituent, and sometimes the pyroxene is ophitic or subophitic rather than granular. Unlike the andesites, the ferromagnesian minerals of the tholeiitic andesites are relatively rich in iron.

Typical, though somewhat oxidised, examples of Tertiary age have been described from the Thingmuli volcano, Iceland, as 'tholeiites' and 'basaltic andesites' (8). They have sporadic phenocrysts of feebly zoned labradorite and rare ones of augite in a very fine grained groundmass of zoned andesine laths, granules of augite and pigeonite, and a small amount of interstitial glass enclosing minute grains of iron-ore. The normative plagioclase of the 'tholeiites' ranges from An_{46} to An_{39}, and is about An_{35} in the 'basaltic andesites'. A few of the 'tholeiites' carry a little olivine. A holohyaline variety of tholeiitic andesite consisting of a brownish glass with about one per cent of crystals, mostly plagioclase, occurs as fragments in a volcanic ash erupted from Vatnajøkull, Iceland, in 1922 (9).

Further good examples are the Yakima 'basalts', variants of the Columbia River basalt, and associated with somewhat earlier tholeiitic basalts (Picture Gorge type) (10). These Yakima rocks have plagioclase close to An_{50} but they contain 20 per cent or more of glass, and the normative plagioclase averages An_{45}.

Similar tholeiitic andesites occur at other localities in Iceland and in N.W. Spitzbergen (with phenocrysts of plagioclase), among the Karroo lavas of Southern Rhodesia and the Deccan Traps of India, in south Brazil, among the western Cascade lavas of the north-central Cascade Mountains, Oregon, and elsewhere.

Chemistry

Average chemical compositions of andesite and tholeiitic andesite are given in table 9.1. Average andesite has a fairly high Al_2O_3 content, moderate amounts of iron, MgO, CaO and Na_2O, and a low content of K_2O. Average tholeiitic andesite, on the other hand, has a low content of Al_2O_3, rather high iron, and higher TiO_2 than average andesite.

The differences are well brought out in the norms where average andesite has some 70 per cent of salic normative minerals, but average tholeiitic andesite has only 55 per cent. Normative diopside and hypersthene are both a good deal richer in the $FeSiO_3$ component and normative magnetite and ilmenite are more abundant in average tholeiitic andesite when compared with average andesite.

Average andesite and average diorite have a recognisable relationship, but the latter is rather more mafic in character. This is due to the presence among diorites of melanocratic rocks, cumulative in character, that have no counterpart among the andesites.

There is a noticeable contrast between average ferrodiorite and average tholeiitic andesite, notably the much higher iron and much lower MgO of the former. The ferrodiorites averaged, however, are the highly differentiated products of intrusions where there has been much enrichment in iron, and most tholeiitic andesites are not differentiated to anything like the same degree.

Hypabyssal equivalents

Microdiorites are the hypabyssal representatives of the diorites and andesites and may be aphyric or porphyritic. They occur usually as dykes, frequently associated with others of micromonzodiorite and microgranodiorite, in connection with intrusions of granodiorite, etc., as exemplified by those associated with the Riesengebirge and Lausitz granodiorites in Germany, or by those present in the Ben Nevis (11) and Etive (12) dyke swarms of Scotland.

The aphyric ones are composed typically of rather stumpy little crystals of plagioclase with flakes of biotite, little prisms of green or brownish hornblende with accessory iron-ore and apatite, sometimes sphene, and often a little quartz and alkali feldspar. The relative proportions of biotite and hornblende vary widely, and in those more or less free from biotite some augite may appear. In the porphyritic microdiorites, plagioclase, frequently zoned, hornblende and biotite may be present as phenocrystal constituents.

More rarely, the ferromagnesian constituents are rhombic pyroxene and augite, as is the case with a microdiorite dyke cutting the Chosica pluton, Peruvian Andes, and with some feldspathic microdiorite dykes in the Etive swarm, where chlorite pseudomorphs after rhombic pyroxene occur in association with minor fresh augite.

Hypabyssal equivalents of tholeiitic andesites and ferrodiorites are represented by certain dolerites that have andesine instead of labradorite and are to be found in differentiated quartz-dolerite sills and as dykes (see p. 134).

References and notes

(1) In some basic andesites containing glass in the groundmass the plagioclase laths are of labradorite and the albite molecule is concentrated in the glass.

(2) Very many rocks with a partially glassy groundmass described as andesites in the past are found, when chemically analysed, to have a relatively high content of silica leading to as much as 15 per cent or more of quartz in the norm. Thus these are really dacites or rhyodacites, depending on the potash content of the analyses. See, also, in this connection, Chayes, F., 1971. *Yb. Carnegie Instn. Wash.* **70**, 204.

(3) Wilcox, R. E., 1954. Petrology of Parícutin Volcano, Mexico. *Bull. U.S. geol. Surv.* **965**-C, 281–353.

(4) Thayer, T. P., 1937. Petrology of the Later Tertiary and Quaternary rocks of the North-Central Cascade Mountains in Oregon, with notes on similar rocks in Western Nevada. *Bull. geol. Soc. Am.* **48**, 1625–34.

(5) Smith, A. L. & Carmichael, I. S. E., 1968. Quaternary lavas from the Southern Cascades, western U.S.A. *Contr. Miner. Petrol.* **19**, table 2, nos. 10 & 11 and pp. 215, 221, 223, 226.

(6) Harry, W. T., 1956. The Old Red Sandstone lavas of the western Sidlaw Hills, Perthshire. *Geol. Mag.* **93**, 43–56.

(7) Wilshire, H. G., 1957. Propylitization of Tertiary volcanic rocks near Ebbetts Pass, Alpine County, California. *Univ. Calif. Publ. geol. Sci.* **32**, 243–72.

(8) Carmichael, I. S. E., 1964. The petrology of Thingmuli, a Tertiary volcano in eastern Iceland. *J. Petrology* **5**, 438–41.

(9) Barth, T. F. W., 1937. Volcanic ash from Vatnajøkull. *Norsk. Geol. Tidss.* **17**, 31–8.

(10) Waters, A. C., 1961. Stratigraphic and lithologic variations in the Columbia River Basalt. *Am. J. Sci.* **259**, 583–611.

(11) Anderson, J. G. C., 1935. The marginal intrusions of Ben Nevis; the Coille Lianachain complex; and the Etive dyke swarm. *Trans. geol. Soc. Glasgow* **19**, 260–1.

(12) Bailey, E. B., 1960. The geology of Ben Nevis and Glen Coe. *Mem. geol. Surv. U.K.* 228–9.

11

Gabbros and related rocks

The *gabbros* are coarse or medium grained igneous rocks composed essentially of a calcic plagioclase feldspar and one or more ferromagnesian constituents. The average colour index (average of 150) is 47, but there is much variation and the colour index of individual rocks may have any value between 10, below which the rock becomes an anorthosite, and 90, above which the rock is classed as ultramafic.

Gabbros are found as large and smaller intrusions. Some of the former are very large indeed and, unlike granitic intrusions of comparable size, are more or less concordant. Certain gabbros, the *alkali gabbros*, have a more alkaline character. These are less widely distributed and do not build large intrusions.

Mineralogy

The *lime-bearing plagioclase* in these rocks may, by definition, vary in composition from labradorite to anorthite. In practice, rocks with anorthite are relatively rare, and the average composition of the plagioclase for gabbros as a whole is An_{65}. Gabbros without olivine have an average plagioclase (An_{62}) rather more sodic than gabbros with olivine (An_{70}). The feldspar occurs as more or less equant anhedral grains, or as subhedral crystals tabular on $\{010\}$. Twinning is usually well displayed: in addition to the usual Carlsbad and albite twins, pericline twinning is frequently present and may be better developed than the albite twinning.

Sometimes the plagioclase contains numerous minute rods or plates, opaque or translucent (and then brown in colour), regularly arranged along definite crystallographic directions. These are believed to be ilmenite.

The most important ferromagnesian constituents of gabbros are pyroxenes of one kind or another. *Orthopyroxene*, sometimes faintly pleochroic in section, may be subhedral prismatic or, more rarely, anhedral and poikilitic. Orthopyroxenes in some gabbroic rocks are characterised by fine exsolution lamellae of clinopyroxene parallel to {100} (orthopyroxene of Bushveld type): others have inverted from pigeonite and have relatively coarse plates of clinopyroxene parallel to the {001} plane of the original pigeonite (orthopyroxene of Stillwater type) or blebby clinopyroxene inclusions without any definite crystallographic orientation (1).

Clinopyroxene in gabbros varies from colourless to pale green or brown in section and may show a fairly good cleavage parallel to {010}, and a less perfect one parallel to {100}, in addition to the prismatic cleavage. Twinning on {100} and also on {001} is fairly common and both may be either simple or repeated. The clinopyroxene may be subhedral, but is more commonly anhedral, forming semi-ophitic, ophitic, and poikilitic plates.

It may show exsolution lamellae of two kinds, lamellae of pigeonite parallel to {001}, and lamellae of orthopyroxene parallel to {100}.

Chemically, the clinopyroxenes of normal gabbros are somewhat aluminous augites, with a low content of titanium and ferric iron. They are frequently somewhat deficient in calcium, though the degree of this deficiency is variable, and some would be classed as subcalcic augites.

The clinopyroxene of alkali gabbros, however, is a calcium-rich variety and, typically, a titanaugite. In these alkali gabbros, orthopyroxene is absent.

Olivine is present in many gabbros, usually in the form of subhedral crystals or rather rounded grains, and is commonly one of the first constituents to start crystallising. More rarely, it forms irregular grains or plates, moulded upon feldspar. Small inclusions of spinel, usually chromite, or of iron-ore may be present, also fluid inclusions.

Primary *hornblende* becomes an important constituent in certain gabbros, occurring as irregular poikilitic or semi-poikilitic plates, or as a fringe to crystals of pyroxene. Green or brown in section, it is a variety relatively rich in aluminium. Alkali gabbros may also contain amphibole which is a deep brown variety approaching kaersutite in composition.

Biotite occurs as a minor constituent in many gabbros and some alkali gabbros occasionally becoming more important.

A number of gabbros have a small amount of interstitial *quartz* (quartz-bearing gabbros) and this may be associated with a little *potash feldspar*, the two found frequently in micrographic intergrowth. Some alkali gabbros have a little interstitial *nepheline* or *analcite*.

The commonest of the *accessory constituents* is an opaque ore mineral of some kind. The most frequent is ilmenite, but ilmenite-haematite intergrowths and 'titaniferous magnetite' (magnetite-ulvospinel and magnetite-ilmenite intergrowths) may also be prominent.

Sulphides are present to some extent in almost all gabbroic rocks and include pyrite, pyrrhotite, minor pentlandite and chalcopyrite, with rare bornite. They are sometimes abundant enough to be exploited commercially for nickel, and also for metals of the platinum group (Merensky Reef, Bushveld igneous complex, Transvaal).

Apatite in prisms or needles, is very variable, being quite abundant in some rocks, virtually absent in others. Zircon is found in small amount in some gabbros, normally quartz-bearing gabbros. Spinel is present in a number, frequently green in section (pleonaste) but also brown (picotite or chromite). Sphene is rare and recorded, more especially, from alkali gabbros.

Texture and structure

Some gabbros have a subhedral-granular texture: others have a poikilitic or a subophitic texture in which plates of pyroxene or amphibole partially enclose the crystals of plagioclase. Porphyritic texture is rare.

Many gabbroic intrusions have a well-marked banded or layered structure (e.g. Skye) (2) and a number of the larger intrusions are highly differentiated with products ranging from ultramafic to siliceous (Bushveld, Duluth, etc.). A few gabbros have orbicular structure (3).

Main varieties with examples

Simple *augite gabbros*, with augite as the only essential ferromagnesian constituent, are to be found among the Tertiary gabbros of W. Scotland (fig. 11.1) and N. Ireland. For instance, among the layered Later Gabbros of the Carlingford complex, Co. Louth, Eire (4), there are feldspathic gabbros composed essentially of feebly zoned calcic plagioclase and subordinate subophitic augite with accessory iron-ore and apatite and virtually free from orthopyroxene and olivine.

Gabbros of this kind are sometimes quartz-bearing, as in the peninsula of Ardnamurchan (5) (Older Quartz-Gabbro of Beinn Bhuidhe; gabbros of Faskadale and of Meall am Tarmachain summit), the interstitial quartz being intergrown commonly with alkali feldspar. Sometimes, on the other hand, they carry a little olivine (island of St. Kilda).

Olivine–augite gabbro, with olivine as an essential and sometimes abundant constituent, is a common type, and many of the Tertiary

Fig. 11.1. Augite gabbro, Glen an t-Suidhe, Arran; × 20. Showing subophitic habit of the clinopyroxene.

gabbros of W. Scotland belong here. The 'eucrite' of Beinn nan Ord, in the peninsula of Ardnamurchan (6), is a moderately coarse rock with fairly abundant olivine, plagioclase ranging from labradorite to bytownite in composition, and ophitic augite associated with large crystals of iron-ore. Orthopyroxene is present only as an accessory constituent, fringing the olivine. The Great Eucrite, the Outer Eucrite and the Gabbro of Plocaig, all ring-dykes of Centre 3, Ardnamurchan, are other examples. The 'eucrite' of Ben Buie in the island of Mull (7) is very like these Ardnamurchan gabbros but shows locally a well-developed banding in part textural, in part compositional. Much of the gabbro in the layered basic complex of Carlingford belongs here (fig. 11.2A).

Pyroxene gabbros, with both ortho- and clinopyroxene as essential constituents, are very common. Many of the so-called 'norites' of the great Bushveld igneous complex, Transvaal, S. Africa, are of this type, especially the 'norites' of the Main Zone. These have a colour index of from 25 to 40 and are composed essentially of plagioclase (about An_{65}) and two pyroxenes. The orthopyroxene is normally subhedral, while the clinopyroxene is anhedral. Orthopyroxene may be found completely enveloped by augite, and the two pyroxenes may also be intergrown. Orthopyroxene is usually in excess of augite but in some cases this relation is reversed.

Pyroxene gabbro with labradorite, two pyroxenes and a little hornblende makes up the bulk of the large gabbro–norite massif of the Lulua–Bushimaïe basins, Congo, and melanocratic rocks of this kind with the dark minerals enclosed poikilitically in large plates of bytownite

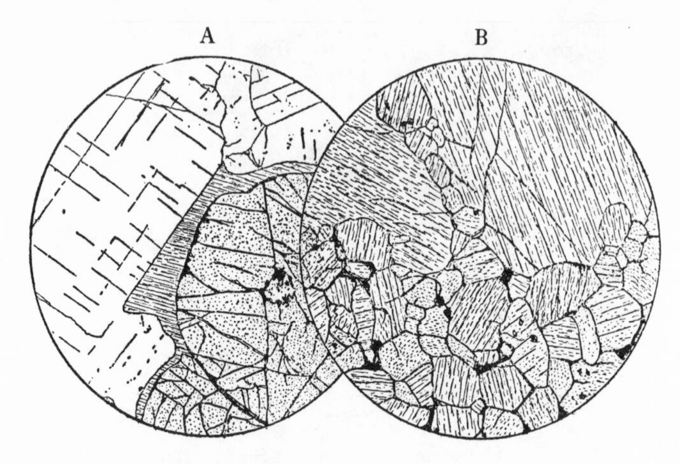

Fig. 11.2. × 16.

A. Olivine–augite gabbro, Slieve Foy, Carlingford, Eire. This is a coarse-grained rock with accessory orthopyroxene fringing olivine, as shown here.

B. Pyroxenite, Webster, N. Carolina, U.S.A. Both clino- and orthopyroxene are present, the former more abundant but the latter in larger crystals.

are recorded from Malagasy. Other well-known pyroxene gabbros include the quartz-bearing 'norite' of Raana, N. Norway, with appreciable biotite; the 'noritic' facies of the 'eucrite' at Rådmansö, Sweden; and the 'augite norite' of Radautal, Harz.

A few of the Tertiary olivine gabbros of W. Scotland are rich enough in orthopyroxene to be classed as *olivine–pyroxene gabbros*, for instance, the main 'hypersthene gabbro' of Ardnamurchan Point (8) and the 'eucrite' of the isle of Rhum. Some of the olivine gabbros of Aberdeenshire, Scotland, also fall here and an interesting example is provided by that of Haddo House (9). This is a mesotype rock with some 15 per cent of olivine and with hypersthene largely predominant over augite. The plagioclase is bytownite (An_{75}) and there is accessory brown hornblende, biotite, and iron-ore. It shows very fine reaction structures, illustrating Bowen's discontinuous reaction series and, in the most complete examples, a core of olivine is followed by more or less regular zones of the successive minerals of the series.

Olivine–pyroxene gabbros from the Skaergaard intrusion, E. Greenland, have been described in great detail (10) and are found in the Transitional Layered Series, the lower part of the Main Layered Series, and in the Western and Northern Border Groups. Pyroxene gabbros rich in olivine make up the Basal Zone of the Insizwa intrusion and related intrusions in E. Griqualand and Pondoland.

The 'olivine gabbros' of the Insch mass, Aberdeenshire (11), are rich in iron and titanium and poor in magnesia. Those richest in iron are actually ferrodiorites, but the remainder are olivine–pyroxene *ferrogabbros* with iron-rich olivine.

Gabbros in which orthopyroxene is the only essential ferromagnesian mineral are given a special name, *norite*. The norites of Norway have long been known owing to their association with deposits of nickeliferous pyrrhotite. They are essentially labradorite–hypersthene rocks, sometimes quartz-bearing (as at Erteli), usually mesotype, but sometimes melanocratic. Hornblende norite, olivine–hornblende norite, pyroxene gabbro and olivine gabbros are found with these simple norites at various localities.

In some other well-known norites, those of the Brocken massif, Harz Mountains, Germany, the plagioclase is bytownite-anorthite or calcic labradorite, the orthopyroxene is enstatite and a small amount of biotite is often present.

Although, as already observed, much of the so-called 'norite' in the Bushveld igneous complex is pyroxene gabbro, true norites are present as well, confined mainly to the Basal and Critical Zones. The subhedral plagioclase varies in composition according to the position of the rock in which it occurs. It is An_{80} to An_{85} in rocks from the lower part of the Critical Zone but An_{70} to An_{75} in rocks from the upper part. Similarly, the anhedral to subhedral orthopyroxene has the composition Of_{20} in the lower part of the zone but its iron content increases gradually upwards. Few of these norites are entirely free from accessory augite but this mineral is very irregularly distributed.

Similar norites are found in other great layered gabbroic complexes, for instance in the Norite Zone of the Stillwater complex, Montana (12).

In Scotland, norites occur in Aberdeenshire, Banffshire and elsewhere. One from Towie Wood, near Ellon, consists essentially of labradorite and orthopyroxene with accessory biotite, brown hornblende, and iron-ore (fig. 11.3). Others have a little augite as well. Few norites are to be found among the Tertiary plutonic rocks of Britain, though beautiful examples, with and without accessory augite, are found near Glenloig, Arran (13).

Some norites have accessory olivine, but *olivine norites*, in which it features as an essential mineral, appear to be rare. A rock of this kind from the Upper Radautal, Harz, has bytownite, bronzite and olivine with accessory biotite and magnetite. Those present among the components of the great gabbroic complex of Sierra Leone have labradorite with olivine in excess of hypersthene, and are transitional to troctolite.

In some gabbros and norites primary hornblende appears as an essential constituent, giving rise to hornblende–pyroxene gabbro, hornblende gabbro and hornblende norite.

A *hornblende–pyroxene gabbro* that has long been known occurs at Pavone, near Ivrea, Italy. It is a melanocratic rock with about equal amounts of brown hornblende and augite both in irregular plates, subordinate hypersthene, and relatively abundant magnetite and green spinel. The feldspar is bytownite. The mesotype gabbro of the Bükkgebirge, Hungary, has zoned plagioclase, averaging labradorite, augite as the chief dark mineral, about equal quantities of brown hornblende and hypersthene, and accessory titaniferous magnetite.

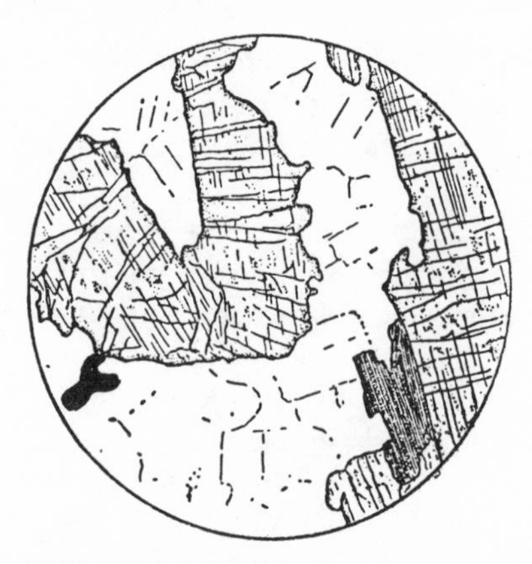

Fig. 11.3. Norite, Towie Wood, Ellon, Aberdeenshire; × 20.
Labradorite and orthopyroxene with accessory iron-ore, biotite or, in other parts of the section, brown hornblende.

Hornblende gabbro, with brown hornblende as the only essential ferromagnesian mineral, occurs as a facies of the Bükkgebirge gabbro. That at Susimaki, Finland, has olive-green hornblende and accessory biotite.

Hornblende norite is the main intrusive type at Bluff Peninsula, South Island, New Zealand (14). The rock is medium grained and somewhat foliated, with zoned plagioclase (averaging An_{64}), euhedral to subhedral hypersthene (Of_{50}) showing exsolved lamellae of clinopyroxene, and subordinate brownish green hornblende.

Troctolites are gabbroic rocks made up essentially of olivine and of plagioclase whose composition may range from labradorite to anorthite. They show great variation in colour index and are found as bands, lenses, or small masses associated with intrusions of gabbroic or ultramafic character. A few examples will illustrate these points.

The troctolite of the Lizard igneous complex, Cornwall, with rounded crystals of partly serpentised olivine enveloped by plagioclase (An_{60}), forms small dyke-like masses cutting serpentinites. The troctolites of the lower zones of the Freetown complex, Sierra Leone, have plagioclase An_{69} to An_{60}, widely varying proportions of olivine and feldspar, and often a little pyroxene.

Troctolite from the Basal Zone of the Insizwa intrusion, E. Griqualand, has plagioclase of composition about An_{80}, as has that found as bands and lenses in the pyroxene gabbro of the Raana norite field, N. Norway. The olivine gabbro at Rådmansö, Sweden, frequently grades into leucocratic troctolite with plagioclase (An_{94}) moulded by olivine:

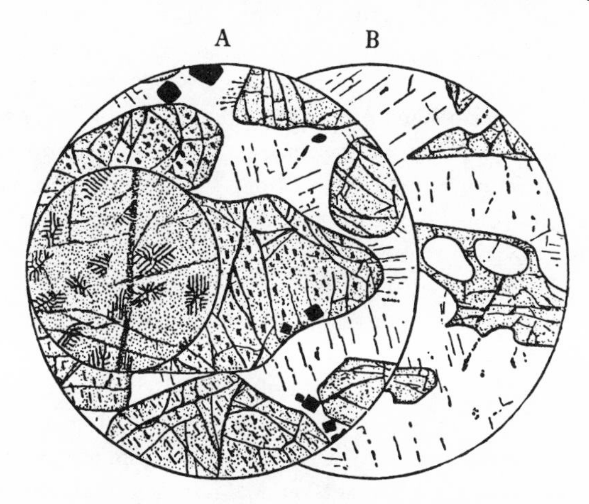

Fig. 11.4. × 20.

A. Troctolite ('allivalite'), Allival, Rhum. The abundant fresh olivine has dendritic inclusions of iron-ore, shown on a larger scale (× 100) in the small inset circle. The opaque octahedra are of chromite.

B. Another variety from the same locality. This is rich in plagioclase, euhedral towards the olivine.

the leucocratic troctolite occurring as a local facies of olivine norite in connection with the San Marcos gabbros, S. California, also has anorthite as its feldspar.

The troctolites ('allivalites') (15) of the isle of Rhum, W. Scotland (16), found as layers interbedded with peridotite, have a very variable colour index. When they are rich in olivine, that mineral is euhedral with respect to plagioclase (fig. 11.4A); when they are rich in plagioclase, the reverse is the case (fig. 11.4B).

Alkali gabbros differ from gabbros in having lime-rich titaniferous augite as their usual pyroxene, often accompanied by a brown amphibole which may be femaghastingsite, magnesiohastingsite or kaersutite. In some cases the amphibole is dominant or even occurs to the exclusion of pyroxene. Most have olivine as a further constituent, and biotite is not infrequent. Some have accessory nepheline and potash feldspar. They occur as small intrusions associated with other alkali rocks.

These alkali gabbros have received various names in the literature. Some have been called simply 'gabbro', some 'essexite', and some 'essexite-gabbro'.

The 'essexite-gabbros' of the Oslo region, Norway, provide good examples. Those of the Gran district (17), for instance, are medium to coarse-grained with either subhedral-granular or ophitic textures. The essential constituents are plagioclase, ranging in composition from about An_{50} to An_{70} and sometimes zoned, and a violet titaniferous augite, usually with appreciable olivine and some biotite in addition. Amphibole is rare but both apatite and iron-ore are often relatively abundant, and a little potash feldspar appears in some. The colour index is variable and there are transitions, through melanocratic types, to an alkali pyroxenite.

Some of the alkali gabbros from the Monteregian Hills, Quebec province, Canada, have pyroxene as their only ferromagnesian mineral, but in others considerable amounts of deep brown magnesiohastingsite or kaersutite are present as well. Some have olivine and a little biotite in addition. The feldspar is usually labradorite, but in the rocks from Rougemont Mountain it is anorthite. Sphene may appear among the accessory constituents and a few have a little interstitial nepheline.

Alkali gabbro is associated with the nepheline-bearing rocks of Nosy Komba, Malagasy. The main dark mineral is femaghastingsite, the others being slightly titaniferous augite, olivine and biotite.

Chemistry

Table 11.1 shows that average gabbro has low SiO_2, moderately high Al_2O_3 and MgO, high CaO; and low alkalies with a strong preponderance of Na_2O over K_2O.

Average alkali gabbro has lower SiO_2 and Al_2O_3, higher total iron, and higher TiO_2 and P_2O_5 than average gabbro.

Note the presence of a fair amount of normative hypersthene in the norm of average gabbro; its absence, and the presence of a little normative nepheline in average alkali gabbro.

TABLE 11.1 *Average chemical compositions of gabbro and related rocks*

		1	2	3	4
SiO_2		48.36	50.28	43.84	43.94
TiO_2		1.32	0.89	0.67	2.86
Al_2O_3		16.84	17.67	13.46	14.87
Fe_2O_3		2.55	1.30	2.20	4.35
FeO		7.92	7.46	9.24	7.80
MnO		0.18	0.14	0.13	0.16
MgO		8.06	9.27	19.71	9.31
CaO		11.07	9.72	8.10	12.37
Na_2O		2.26	1.96	1.33	2.32
K_2O		0.56	0.63	0.58	0.92
H_2O^+		0.64	0.47	0.50	0.66
P_2O_5		0.24	0.21	0.18	0.44
or		3.3	3.3	3.3	5.6
ab		18.9	16.8	11.5	12.1
an		34.2	37.8	28.9	27.5
ne		–	–	–	4.0
di	*wo*	7.8	3.7	4.4	13.0
	en	4.8	2.3	3.1	9.3
	fs	2.5	1.2	0.9	2.5
hy	*en*	9.5	19.1	–	–
	fs	5.0	9.4	–	–
ol	*fo*	4.1	1.3	32.3	9.8
	fa	2.5	0.6	10.2	2.9
mt		3.7	1.9	3.2	6.3
il		2.4	1.7	1.4	5.4
ap		0.6	0.5	0.4	1.0
DI		22.2	20.1	14.8	21.7

1. Gabbro (average of 160) (*Bull. Geol. Soc. Am.* **65**, 1020, no. I).
2. Norite (average of 39) (*Ibid.*, p. 1020, no. II).
3. Troctolite (average of 9) (*Ibid.*, p. 1020, no. VI).
4. Alkali gabbro (average of 42) (*Ibid.*, p. 1020, no. III).

Later alteration

Many gabbros show the effects of later hydrothermal alteration to a varying degree. The changes that may take place can best be illustrated by tracing the alteration of the individual minerals. *Olivine* may be converted to serpentine, talc, or cummingtonite, normally with excretion of iron-ore. *Orthopyroxene* changes to the bastite variety of serpentine, or to cummingtonite which may be replaced in more altered rocks by a pale green fibrous amphibole. *Clinopyroxene* usually alters to pale green, rather fibrous amphibole (often termed 'uralite') but seems to be replaced occasionally by a secondary brown amphibole. These secondary amphiboles differ both in appearance and chemistry from the primary amphiboles of gabbroic rocks. *Primary hornblende* often remains unchanged, but in more altered rocks it is replaced by secondary amphibole. *Biotite* may also remain more or less unaffected but, when alteration is severe, it is partially bleached or converted to chlorite. Secondary sphene, epidote or prehnite may be associated with the altered mica, the latter two sometimes forming lenses along the cleavage planes. *Ilmenite* may be partly or wholly replaced by leucoxene and *magnetite* by haematite. *Plagioclase* may be little affected, even when the ferromagnesian constituents have been replaced completely. At other times it is altered to an aggregate of sericite flakes and granules of zoisite or epidote lying in an albitic base. In some cases the place of an epidote mineral is taken by prehnite or by calcite; in other cases, calcium is removed during the alteration process so that sericite alone remains embedded in the albite.

Apatite and zircon, also quartz and potash feldspar, when present, remain unchanged.

Some gabbros are hardly touched by such hydrothermal alteration but, when alteration has been severe, rocks little resembling the original may be produced. Some of the mafic intrusions of Aberdeenshire afford striking instances of this. Thus, the gabbro of Torre Hill, Morven, is converted to an amphibole–plagioclase rock with both secondary green amphibole and cummingtonite. Chemical analyses show that there is virtually no change in bulk composition other than a small increase in the water content. Similarly, the quartz-bearing pyroxene–hornblende–biotite gabbro of Haddo House is converted to an aggregate of sericitised plagioclase, quartz, secondary amphibole and bleached biotite, with a little chlorite, epidote and iron-ore.

More striking effects are produced when the hydrothermal fluids contain chlorine, as is sometimes the case. One of the best illustrations is provided by the gabbroic intrusions of the N'changa district, Zambia (18). The fresh rocks are augite gabbros and pyroxene gabbros, with or without a certain amount of olivine, and these have been partially or wholly converted into *amphibole–scapolite rocks*. Original pyroxene is changed to pale green amphibole, frequently margined by a blue-green sodic variety when in contact with scapolite, with associated granules of clinozoisite and epidote. Labradorite is partially, sometimes wholly, replaced by a mosaic of scapolite grains with inclusions of clinozoisite and epidote. Quartz is usually present as little blebs and occasionally occurs in micrographic intergrowth with secondary albite. Chemical analyses indicate that water, chlorine and sodium were introduced during the process.

Gabbro-pegmatites

True gabbro-aplites do not appear to exist, but gabbro-pegmatites are associated with a number of gabbroic intrusions, occurring sometimes as dykes, sometimes as segregations grading into the surrounding gabbro.

Those associated with the Tertiary Cuillin gabbro complex, Isle of Skye (19), form segregation veins, chiefly in the heart of the complex. Augite builds more or less euhedral crystals up to twenty centimetres long and there are some pseudomorphs after orthopyroxene. The feldspar is labradorite and the interstices between the pyroxene and feldspar crystals are occupied by titaniferous iron-ore, micropegmatite, and a few small crystals of sphene. Similar rocks are found in some of the great mafic intrusions (Duluth, Sierra Leone, etc.).

Pegmatites found as narrow dykes cutting the Jay anorthosite sheet in the Adirondack Mountains, U.S.A. (20), offer several features of interest. In addition to the usual pyroxene and plagioclase, a number have essential sphene in crystals two to five centimetres long with a little apatite, pyrrhotite and pyrite. The pegmatite minerals are oriented perpendicular to the walls of the dykes and some of the rocks are banded. These have the dark minerals concentrated at the margins, leaving an almost purely feldspathic centre.

The place of pyroxene is taken by primary hornblende in some gabbro-pegmatites. Such is the case with those cutting dunite and pyroxenite at the various platinum centres in the Ural Mountains, U.S.S.R. These are usually made up of about equal amounts of brownish green hornblende and a plagioclase whose crystals have cores of calcic labradorite surrounded by a border of andesine. Apatite in relatively large crystals is particularly abundant.

Anorthosites

Anorthosites are coarse to medium grained igneous rocks in which plagioclase, ranging in composition from andesine to bytownite, makes up more than 90 per cent by volume of the rock. They occur in two different ways (21): as large independent intrusions, apparently of sheet-like form; and as bands or lenses forming part of intrusions that are dominantly of gabbroic or ultramafic character. The type of anorthosite is different in the two modes of occurrence and each will be considered separately (22).

The independent intrusions are, as far as is known, all of Precambrian age. Many are very coarse grained but this is frequently obscured because they have suffered granulation. The plagioclase varies in composition from about An_{40} to An_{65} in different occurrences, and averages about An_{53}. A general feature of the feldspar is the presence of antiperthitic spindles and blebs of potash feldspar, and it sometimes holds numerous regularly arranged minute rods of iron-ore. Late-formed discordant or concordant ilmenite–magnetite bodies are enclosed commonly within these anorthosite intrusions.

The main anorthosite massif of the Adirondack Mountains (23), occupying an area of some 3000km^2, may be taken as a typical example. Two main facies are recognised, the Whiteface and the Marcy. The former is practically confined to the borders of the mass and is predominantly a highly leucocratic gabbro; the Marcy facies forms the core and is a true anorthosite.

This Marcy anorthosite is coarse to very coarse grained, massive or with an indistinct foliation, and with a variable, but subordinate, 'groundmass' of granulated plagioclase. The large plagioclase crystals range from An_{40} to An_{50}, averaging An_{45}: the grains formed by granulation of the large crystals also average An_{45}. Accessory dark minerals are augite, hypersthene and a little iron-ore, and there is a trace of apatite.

Sometimes more than one variety of anorthosite is present, as is the case with the composite St. Urbain anorthosite, Charlevoix district, Quebec, Canada (24). An earlier labradorite (near An_{60}) anorthosite is intruded by a later andesine anorthosite with plagioclase varying from An_{45} to An_{36}.

Anorthosites found as bands and lenses in gabbroic and ultramafic intrusions have, in general, a plagioclase of more calcic composition and antiperthitic spindles or blebs of potash feldspar are lacking. Recorded compositions range from about An_{60} to An_{90} in different occurrences, averaging about An_{75}.

Moreover, the composition of the plagioclase may show a close relationship to that of the other rocks with which the anorthosite is associated. This may be illustrated by considering the anorthosites in the Bushveld igneous complex of the Transvaal (25). These form bands at various levels within the complex and consist of coarse interlocking anhedral grains of plagioclase, usually with accessory pyroxene.

The anorthosite bands of the Lower Critical Zone have plagioclase of composition about An_{85}, those in the Upper Critical Zone carry An_{75}. Near the base of the Main Zone the feldspar is still about An_{75}, but in anorthosite from the top of this zone it has a composition An_{65} to An_{60}.

The accompanying gabbroic rocks at the appropriate horizon have, in every case, plagioclase of a composition almost identical with that of the anorthosites.

References and notes

(1) For a good discussion of exsolution lamellae in orthopyroxene see Hess, H. H., 1960. Stillwater igneous complex, Montana. *Mem. geol. Soc. Am.* **80**, 23–34.

(2) Good illustrations of layering can be found in Rossman, D. L., 1963. Geology and petrology of two stocks of layered gabbro in the Fairweather Range, Alaska. *Bull. U.S. geol. Surv.* **1121**-F.

(3) Miller, F. S., 1938. Hornblendes and primary structures of the San Marcos Gabbro. *Bull. geol. Soc. Am.* **49**, 1228–9. Satterly, J., 1940. An orbicular gabbro from Tremeer Lake, Kenora District, Ontario. *Univ. Toronto geol. Ser.* **44**, 75–82.

(4) Le Bas, M. J., 1960. The petrology of the layered basic rocks of the Carlingford complex, Co. Louth. *Trans. Roy. Soc. Edinb.* **64**, 169–200.

(5) Richey, J. E. & Thomas, H. H., 1930. Geology of Ardnamurchan. *Mem. geol. Surv. U.K.*, 249, 283, 318.

(6) Richey & Thomas, 1930. *Ibid.*, 263.

(7) Bailey, E. B. *et al.*, 1924. Tertiary and post-Tertiary geology of Mull. *Mem. geol. Surv. U.K.*, 250.

(8) Wells, M. K., 1954. The structure and petrology of the Hypersthene-Gabbro intrusion, Ardnamurchan, Argyllshire. *Q. Jl. geol. Soc. Lond.* **109**, 367–95.

(9) Read, H. H., 1935. The gabbros and associated xenolithic complexes of the Haddo House district, Aberdeenshire. *Ibid.* **91**, 598–601.

(10) Wager, L. R. & Deer, W. A., 1939. Geological investigations in East Greenland: part III. The petrology of the Skaergaard intrusion, Kangerdlugssuaq, East Greenland. *Med. om Gronland* **105**, no. 4, 87–94, 151–4.

(11) Read, H. H., Sadashivaiah, M. S. & Haq, B. T., 1961. Differentiation in the olivine-gabbro of the Insch mass, Aberdeenshire. *Proc. Geol. Ass. Lond.* **72**, 394–8.

(12) Hess, 1960, pp. 70–5.

(13) Tyrrell, G. W., 1928. The geology of Arran. *Mem. geol. Surv. U.K.*, 180.

(14) Service, H., 1937. An intrusion of norite and its accompanying contact metamorphism at Bluff, New Zealand. *Trans. Roy. Soc. N.Z.* **67**, 208–12.

(15) The name 'allivalite' was given to what were thought to be anorthite–olivine rocks associated with peridotite, as distinct from 'troctolite', regarded as a labradorite–olivine rock associated with gabbro. As, however, the type rock is actually a bytownite–olivine rock, and as rocks of this kind with even more calcic plagioclase are associated with gabbro, there seems to be no reason for retaining the name.

(16) Brown, G. M., 1956. The layered ultrabasic rocks of Rhum, Inner Hebrides. *Phil. Trans. Roy. Soc.*, Ser. B., **240**, 1–53.

(17) Barth, T. F. W., 1945. Studies on the igneous rock complex of the Oslo region. II. Systematic petrography of the plutonic rocks. *Skrift. Det Norske Vidensk.-Akad. Oslo.* I. *Mat.-Nat., Kl. 1944*, no. **9**, 44–7.

(18) Jackson, G. C. A., 1932. The geology of the N'Changa district, Northern Rhodesia. *Q. Jl. geol. Soc. Lond.* **88**, 494–502.
(19) Harker, A., 1904. The Tertiary igneous rocks of Skye. *Mem. geol. Surv. U.K.*, 116.
(20) Buddington, A. F., 1939. Adirondack igneous rocks and their metamorphism. *Mem. geol. Soc. Am.* **7**, 35–7.
(21) Buddington, 1939. *Ibid.*, 208–9.
(22) It has been suggested in a recent paper (Romney, W. D., 1968. An evaluation of some 'differences' between anorthosites in massifs and in layered complexes. *Lithos* **1**, 230–41) that there are all gradations between the two types.
(23) Buddington, 1939, chapters 3 and 11.
(24) Mawdsley, J. B., 1927. St. Urbain area, Charlevoix district, Quebec. *Mem. geol. Surv. Canada* **152**, 18–28.
(25) Lombard, B. V., 1934. On the differentiation and relationships of the rocks of the Bushveld complex. *Trans geol. Soc. S. Africa* **37**, table 10, p. 36.

12

Basalts and dolerites

Basalts, the volcanic equivalents of the gabbros, are dark coloured rocks of very fine grain whose characteristic minerals are calcic plagioclase (labradorite to anorthite), pyroxene and iron-ore, with or without olivine. They may be vesicular, or the vesicles may be filled with secondary minerals in which case the rock is said to show amygdaloidal structure. Basaltic magma, erupted under water, forms pillow-like masses giving rise to a basaltic pillow lava. The individual pillows have a glassy crust and a more or less crystalline interior which may show radiating contraction cracks. With thick submarine flows only the upper ten metres consists of pillows and below this normal lava is present.

Two main types may be recognised, namely, *tholeiitic basalt* and *alkali basalt* (1). A third type, associated with the andesites and related rocks of orogenic regions and island arcs, will be distinguished as *calc-alkali basalt*, and mention will also be made of a fourth type, *basaltic komatiite*.

1. Tholeiitic basalts

The chief pyroxenes of a tholeiitic basalt (fig. 12.1) are *clinopyroxenes* poor, or relatively poor, in calcium. Commonly both *subcalcic augite* and *pigeonite* are present but, in some, only subcalcic augite occurs. In certain rapidly cooled lavas a single *quench clinopyroxene*, intermediate in composition between subcalcic augite and pigeonite, is found.

Subcalcic augite, or augite zoned to subcalcic augite, sometimes occurs as phenocrysts as well as in the groundmass. Pigeonite is commonly confined to the groundmass.

Hypersthene, with or without accompanying pigeonite, is found with subcalcic augite in the groundmass of some tholeiitic basalts. Occasionally it forms phenocrysts.

In many tholeiitic basalts, *olivine* is either absent or present only in small amount. In some, however, the *tholeiitic olivine basalts*, it becomes an essential phase and may, indeed, become a dominant constituent (*tholeiitic picrite-basalt*). Normally it is present as phenocrysts or microphenocrysts but may also appear in the groundmass.

Plagioclase may be present both as phenocrysts and in the groundmass. The phenocrysts are usually in the bytownite-labradorite range and somewhat zoned, but anorthite phenocrysts have been recorded from some tholeiitic lavas (e.g. Hakone volcano, Japan). The feldspar laths of the groundmass are composed of labradorite.

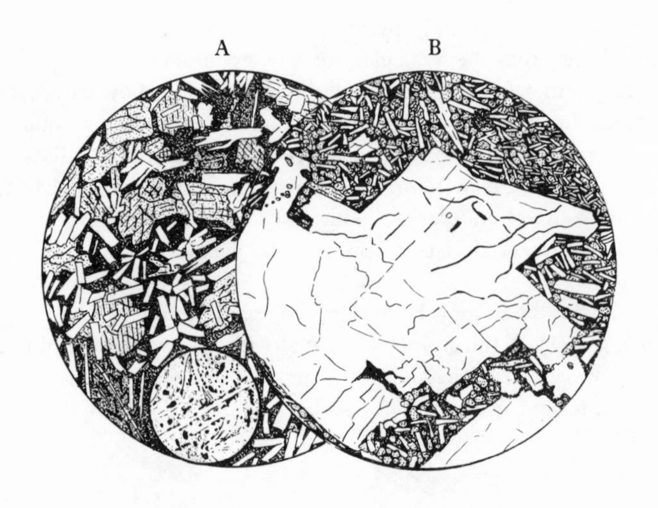

Fig. 12.1. Tholeiitic basalt; × 15.

A. Kangaroo Island, S. Australia. Pyroxene and plagioclase in a glassy groundmass. Under high magnification (inset) the latter shows incipient crystallization of feldspar and iron-ore.

B. Bunbury, W. Australia. Showing a glomero-porphyritic group of large phenocrysts of bytownite in a groundmass of labradorite, granular pyroxene and dark interstitial glass.

Iron-ores include titaniferous magnetite and ilmenite though quite commonly the latter is absent as a separate phase. The former may show partial oxidation and exsolve ilmenite. Under highly oxidising conditions, such as were present in the neighbourhood of vesicles or pores, it may break down to haematite and pseudobrookite. The ores build little grains in the groundmass and, sometimes, microphenocrysts.

The groundmass has commonly a varying amount of interstitial brown *glass*, giving rise to the intersertal texture so often seen in tholeiitic basalts.

Examples

The first examples will be taken from the volcanoes Kilauea and Mauna Loa, in the island of Hawaii, where the tholeiitic lavas have been studied in considerable detail (2).

The lavas of Kilauea are mainly tholeiitic olivine basalts with subordinate tholeiitic basalt, in which olivine is almost lacking, and minor tholeiitic picrite-basalt (3) in which phenocrysts of olivine and sometimes, to a lesser extent, of augite are concentrated. A sample of the 1921 lava, about 3km south of Volcano Observatory, may be taken as representative of the dominant type. It has some phenocrysts of olivine and microphenocrysts of augite set in a groundmass of labradorite laths, grains of olivine, zoned subcalcic augite and iron-ore, with interstitial dark pigmented glass. The prehistoric flow from the National Park quarry, on the other hand, provides an example of a more or less aphyric tholeiitic basalt, with much less olivine and rather more interstitial glass.

Some of the lavas of Mauna Loa are similar to those of Kilauea, but a number differ in having hypersthene as a constituent (4). The 1868 lava, for instance, has large phenocrysts of olivine and occasionally of hypersthene, with microphenocrysts of augite and abundant zoned microphenocrysts of hypersthene, usually jacketed by clinopyroxene. These lie in an extremely fine-grained groundmass of labradorite, iron-ore, and pyroxene whose composition covers the whole range from subcalcic augite to lime-rich pigeonite.

Aphyric tholeiitic basalts form a proportion of the flows among the well-known Deccan 'Traps' of India. Laths of labradorite, subhedral grains of subcalcic augite and iron-ore, occasional altered grains of olivine and varying amounts of interstitial glass are the phases present, and the texture ranges from intergranular to intersertal. Porphyritic types also occur, with phenocrysts of labradorite and rare altered olivine, or with phenocrysts of bytownite, augite and olivine.

A 3000m thick sequence of tholeiitic flood basalts of early Tertiary age in the Faeroe Islands, belonging to the great Brito-Icelandic igneous province, can be divided into three series (5).

The lower series is made up of aphyric and microporphyritic tholeiitic basalts with intergranular texture. They are composed of labradorite laths, grains of subcalcic augite, pigeonite and iron-ore, with a little glass, while the microphenocrysts are of zoned augite, zoned labradorite and iron-ore. Partially altered olivine is present as an accessory constituent in some.

The middle series is composed mainly of porphyritic types in which labradorite phenocrysts, sometimes with a core of bytownite, make up from 10 to 30 per cent of the rock. They are accompanied by much rarer phenocrysts of augite and sparing olivine. The groundmass is similar to that of the lower series rocks.

The upper series is mostly made up of tholeiitic olivine basalts. Most have some phenocrysts of plagioclase (An_{84} zoned to An_{64}) and of rounded and partially resorbed olivine. In a few, olivine is confined to the groundmass. This groundmass of labradorite, augite, olivine and iron-ore is almost devoid of glass, and the texture is commonly ophitic to subophitic rather than intergranular.

An interesting feature of some tholeiitic olivine basalts occurring as pillow lavas is the gravitative settling of olivine in the pillows. This has been noticed in pillow lavas from Iceland and in the tholeiitic picrite-basalt pillow lavas of Cyprus (6).

Tholeiitic basalts like those described are very widely distributed in space and time, often covering very large areas ('flood basalts'). They have been erupted from the Precambrian (e.g. Cuddapah 'Traps', India; Keweenawan basalts, Lake Superior region) onwards and are known in both continental and oceanic environments.

In addition to those already given, the Triassic basalts of New Jersey, the Karroo basalts of South Africa, the Picture Gorge basalts of the Columbia River Basalts (7) and the tholeiitic basalt lavas and dykes of the British Tertiary igneous province (8) with their counterparts in Iceland and Greenland may be mentioned as further examples.

Moreover, it would appear from information so far available that the basalts of the deep ocean floors are largely of this type (ocean-floor tholeiitic basalts) (9) (see p. 127).

Tholeiitic basalts are not, however, the typical basalts of orogenic regions and island arcs, though they may be erupted in the early stages of island arc formation (10).

2. Alkali basalts

Alkali basalts differ essentially from tholeiitic basalts in the nature of their pyroxene which is here a lime-rich, more or less titaniferous augite. Most alkali basalts have olivine as an additional essential constituent. They may be aphyric or, more commonly, porphyritic.

Aphyric alkali basalts may be uniform in grain size or they may have microphenocrysts of olivine, of augite, or of plagioclase. They are composed of olivine, more or less titaniferous augite, labradorite and iron-ore, with accessory apatite which may be relatively abundant.

Olivine, sometimes partly altered to iddingsite or serpentine, occurs as subhedral grains. The *lime-rich augite* varies in colour in thin section from almost colourless to deep lilac in different rocks, depending on its titanium content (11). It may be present as grains or as small ophitic or subophitic plates enclosing the plagioclase. The *plagioclase* laths have a composition within the labradorite range and may show some zoning. The *iron-ore* may be ilmenite or titaniferous magnetite; both may occur together.

The texture of alkali basalts is usually intergranular or ophitic but, in some, there is a varying amount of interstitial yellowish *glass* rich in skeletal crystals of iron-ore; others have been recorded with interstitial colourless glass. When glass becomes a prominent constituent the texture becomes intersertal but it should be pointed out that the presence of glass in alkali basalts is nothing like as common as in tholeiitic basalts, and the typical alkali basalt has none.

Sometimes there is a small amount of interstitial *nepheline* or *analcite*, emphasising the alkaline nature of these rocks. Accessory alkali feldspar appears in a number, marking a transition to the trachybasalts.

The *porphyritic alkali basalts* may have olivine as the only phenocryst constituent. In other rocks augite or plagioclase also occurs and, in some, all three minerals are present in the phenocryst generation. Again, augite may be the most important phenocrystal phase, in others plagioclase becomes dominant. More rarely, phenocrysts of kaersutite are present.

Phenocrystal olivine is less rich in the fayalite molecule than the olivine of the groundmass and may be zoned. Phenocrystal augite may be almost colourless or grey-brown or greenish in thin section: it is frequently zoned and often has a margin of pale lilac titaniferous augite. Alternatively, the phenocrysts, like the groundmass pyroxene, may be of titaniferous augite. The plagioclase phenocrysts are in the calcic bytownite to labradorite range and may show a certain amount of zoning.

Examples

Tholeiitic basalts from certain volcanoes in Hawaii have been described already. Such characterise the early and mature phases in the evolution of a Hawaiian volcano but, during the old phase, alkali basalts associated with hawaiites, mugearites, benmoreites or trachytes are erupted. Mauna Kea, Kohala and Hualalai are Hawaiian volcanoes that have reached this stage and, of these, the last named seems to have erupted the largest amount of alkali basalt.

Typical examples are the prehistoric flow north of Keauhou, rich in microphenocrysts of granular olivine in a base of labradorite laths and granules of titaniferous augite and magnetite, and the 1801 lava which carries scattered phenocrysts of olivine in addition. Other lavas here are richer in phenocrysts of olivine and have some of augite as well.

The island of Rodriguez in the Indian Ocean is made up very largely of rather uniform flows of alkali basalt of Pliocene to Pleistocene age (12). In these olivine is ubiquitous, occurring as small zoned phenocrysts, sometimes skeletal in habit, as well as in the groundmass. Alteration to iddingsite is widespread but is frequently confined to the cores or intermediate zones of the phenocrysts, the outer zones and the groundmass olivine remaining unaffected. Plagioclase, with cores of An_{64}–An_{60} and marginal zones extending to An_{50}, is usually confined to the groundmass, but occurs to some extent as microphenocrysts in a number of the flows. The abundant clinopyroxene, colourless to pale lilac in thin section, is normally present as grains in the groundmass which also contains some magnetite. Apatite and alkali feldspar are generally visible as accessory constituents, and interstitial analcite can be seen in some.

Alkali basalts of recent age are found on many other oceanic islands such as the Polynesian islands of the Pacific; Mauritius and Réunion in the Indian ocean; St. Helena, Ascension, the Azores and the Cape Verde islands in the Atlantic.

Among the basaltic lavas of Tertiary age in Britain, alkali basalts are found, more particularly, in N. Ireland (13) and in the islands of Mull (fig. 12.2) and Skye (14). These are usually more or less aphyric, with olivine as the only microphenocrystal constituent. The plagioclase is labradorite, the titaniferous augite has usually a pale lilac tint and is sensibly pleochroic. The texture is either ophitic or, more rarely, intergranular, and there is normally a small amount of residual base now represented by zeolites and chlorite, or sometimes by analcite.

The Carboniferous alkali basalts of the Midland Valley of Scotland (15) are, by contrast, often porphyritic with phenocrysts of plagioclase and olivine, or of titaniferous augite and olivine, but microphenocrystal types occur as well.

Varieties of alkali basalt very rich in phenocrysts may occur associated with the more normal types. The commonest are ***alkali picrite-basalts*** much enriched in ferromagnesian phenocrysts, with either olivine (oceanite type) or augite (ankaramite type) dominant. Extreme examples of the first are to be found in the Juan Fernandez islands and the Gambier isles where the groundmass may just form an interstitial cement to the large porphyritic olivine grains. The ankaramite type may have large phenocrysts of either slightly greenish or purplish augite, usually accompanied by subordinate olivine.

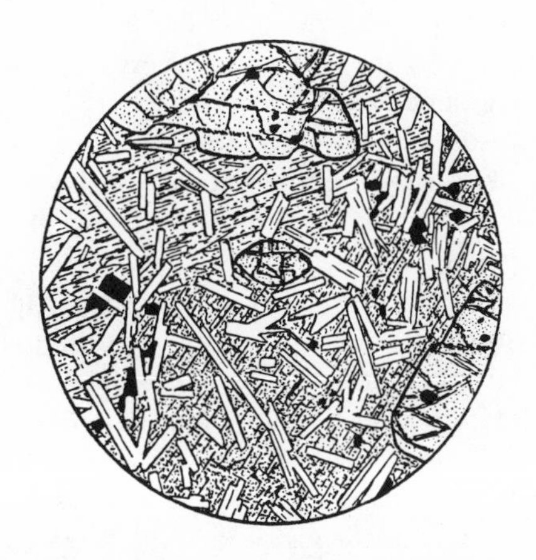

Fig. 12.2. Alkali basalt, Tobermory, Mull; × 20.
With olivine and ophitic clinopyroxene.

Some alkali basalts are rich in phenocrysts of plagioclase. Among the alkali basalts of the older volcanic series in the island of Mauritius (16) there are types rich in phenocrysts of labradorite or bytownite and, in one case, phenocrysts of bytownite make up over 50 per cent of the rock. Here, too, alkali picrite-basalts of both oceanite and ankaramite type may be found.

Much more rarely, kaersutite becomes an important phenocrystal constituent. A good example, with numerous phenocrysts of kaersutite and some of yellowish augite and of plagioclase, has been described from Ehi Mossi, Tibesti.

Alkali basalts have not been erupted in the same volume as tholeiitic basalts but nevertheless large areas have sometimes been covered by them, an example being the great Tertiary to Quaternary volcanic field lying to the south of Damascus.

3. Calc-alkali basalts

Calc-alkali basalts will be defined as those basalts associated with usually more voluminous calc-alkali andesites, dacites and rhyodacites. They resemble tholeiitic basalts to some extent mineralogically, but their ferromagnesian constituents, more especially the pyroxenes, are poorer in iron and plagioclase is, in general, more prominent. Orthopyroxene occurs more frequently and although olivine may feature as a constituent, it never becomes abundant.

According to some petrologists, these basalts may be distinguished from tholeiitic basalts by the occurrence of orthopyroxene rather than pigeonite in the groundmass. Unfortunately, the calc-alkali basalts have not received so far the detailed attention that has been given to tholeiitic and to alkali basalts, but recent work makes it quite clear that typical calc-alkali basalts can have pigeonite, with or without hypersthene, in their groundmass.

These calc-alkali basalts are usually, but not always, porphyritic with phenocrysts of plagioclase ranging from bytownite to labradorite, clinopyroxene, and sometimes orthopyroxene or olivine or both. The groundmass varies from finely holocrystalline to more or less glassy but, on the whole, glass is less noticeable than in tholeiitic basalts.

The mineralogical differences between calc-alkali and tholeiitic basalts are reflected in their chemistry and typical representatives of the two basalt types are quite distinct (p. 129).

Calc-alkali basalts have never been erupted in great volume and, unlike tholeiitic basalts, have never been responsible for great areas of 'flood' basalt.

Some examples

Most of the basalts of Mt. Misery volcano, St. Kitts, West Indies (17), are highly porphyritic with abundant phenocrysts of plagioclase, much less abundant olivine and a few of magnetite, with or without clinopyroxene. Some have only microphenocrysts of these minerals and a few have seriate texture.

The plagioclase phenocrysts (An_{80}–An_{70} with a border zoned to An_{50}) are often clouded with inclusions; olivine phenocrysts may be partly altered to iddingsite and may have a corona of clinopyroxene granules.

The groundmass of these basalts is normally intergranular and composed of plagioclase, pyroxene and iron-ore, with olivine as an additional constituent in some flows. Although the nature of the pyroxene is not specified, norms of the chemical analyses of these basalts show that it must be a somewhat subcalcic augite.

Rather similar basalts occur on the island of Montserrat in the West Indies but here hypersthene is present as well as augite in the phenocrystal or microphenocrystal generation and occurs also in the groundmass.

Among the basalts of Semisopochnoi Island, Western Aleutian Islands (18), one variety has phenocrysts of plagioclase (An_{77}), olivine and augite, with smaller ones of hypersthene and a few of magnetite and apatite, and shows seriate texture. Another has phenocrysts of plagioclase (An_{82}), olivine, 'augite' (which must be subcalcic augite) and magnetite, in a groundmass of plagioclase, 'augite', and magnetite. A third variety has conspicuous phenocrysts of augite with thin pigeonite outer zones lying in a seriate intergranular groundmass of bytownite, pigeonitic augite, olivine and iron-ore.

The basalts of the island of Santorini (19) may be fine grained and microporphyritic, as in the lava domes of the Akrotiri area, or coarsely porphyritic, as in the Main Series flow basalts. The first have microphenocrysts of olivine and, more rarely, of plagioclase (An_{90}–An_{75}) and augite, in an intergranular groundmass of plagioclase (An_{80}–An_{55}), augite, pigeonite and iron-ore. Some have a little interstitial glass.

Typical Main Series basalts have corroded phenocrysts of plagioclase (An_{90}) with narrow rims of An_{75}, slightly resorbed phenocrysts of olivine and, sometimes, abundant augite phenocrysts as well. The groundmass, varying in texture from intergranular to subophitic, is relatively coarse and consists of plagioclase (An_{65}–An_{60}), augite, pigeonite and olivine, commonly rimmed with pyroxene. Interstices between these minerals are occupied by dark brown glass with strings of euhedral magnetite.

Final examples will be taken from the Quaternary calc-alkali basalts of the Southern Cascades, western U.S.A. (20). The basalt of Cinder Cone, Mt. Shasta, has some 6 per cent of olivine phenocrysts (about Fa_{20}–Fa_{30}) and an occasional one of clinopyroxene, and the groundmass pyroxene is a subcalcic augite. At Medicine Lake, the basalt forming the latest flow in the Modoc Lava Beds again has subcalcic augite as its groundmass pyroxene and has phenocrysts of plagioclase with a few of olivine. The Paint Pot Crater basalt flow also has phenocrysts of plagioclase and olivine but the groundmass pyroxene ranges in composition from subcalcic augite to pigeonite.

Pigeonite, when present, has been confined to the groundmass of the basalt in all the occurrences mentioned so far. But among the calc-alkali basalts of the Tromador area, northern Patagonia (21), there are a few that have phenocrysts of this mineral in addition to those of plagioclase and olivine.

It is worthy of note that no picrite-basalts have been recorded associated with calc-alkali basalts, another point of distinction between these and tholeiitic or alkali basalts.

4. Basaltic komatiites

Attention has been directed in recent years to an unusual group of mafic and ultramafic lavas and more or less contemporaneous sills that are widely distributed in early Precambrian terrains in South Africa, Australia, Canada and India. These are the komatiites (22) and include the mafic basaltic komatiite and the ultramafic peridotitic komatiite. They are not, however, confined to the Precambrian as they are now known to occur in the early Palaezoic of Newfoundland and some of Mesozoic age have been found in Greece.

Basaltic komatiites have not yet been found in a completely unaltered condition, having suffered varying degrees of metamorphism, but original textures and structures, such as pillow structure, are often well preserved. In a number, sufficient relicts of original minerals remain to show that these rocks were composed of an aluminous diopsidic pyroxene, with an appreciable content of chromium, accompanied by varying amounts of plagioclase feldspar, magnesium-rich olivine, iron-ore and glass. Some are highly mafic, others have rather more feldspar.

They frequently show unusual textures, due to the presence of the constituent minerals in a variety of skeletal forms (23) and sometimes the pyroxene develops a texture not unlike the spinifex texture (see p. 152) found in the ultramafic komatiites (24). These features apply more especially to thin flows and to the upper and lower portions of thicker flows. In the latter, the central portions may have a more or less granular texture.

These basalts are characterised chemically by their relatively high contents of MgO, Cr, Ni and CaO with low Al_2O_3, TiO_2 and alkalies, and the ratio $CaO:Al_2O_3$ is usually greater than 1. These characters distinguish basaltic komatiites from other basaltic lavas, and they would appear to have been derived from a magma more primitive than that producing the ocean-floor type of tholeiitic basalt which is often associated with them.

Something should perhaps be said about a type of basalt distinguished by some petrologists as *high alumina basalt*. High alumina basalts have been defined as *aphyric* basalts with an alumina content greater than 17 per cent.

The type rock is the Warner basalt, Medicine Lake Highlands, U.S.A., which is an alkali basalt of a mildly alkaline kind carrying a lime-rich clinopyroxene (25). The Hat Creek lava flow of the Lassen region, U.S.A., is almost identical (26), but, as defined, there are also high alumina basalts of tholeiitic type and of calc-alkali type. Thus the term 'high alumina basalt' does not refer to a *specific* basalt type and, if used, the variety of high alumina basalt should be specified or, alternatively, the term should be used only for basalts like the type Warner basalt.

When basalt magma is chilled rapidly it may solidify as a *basalt glass*. Such glasses, black in hand specimen, may be pale green to olive-buff in thin section (*sideromelane*) or they may be slightly oxidised and yellow or brown and more or less opaque (*tachylyte*) (27). As basalt magmas are relatively fluid, even when they have lost their volatiles, glass does not form nearly so readily as with more siliceous lavas, and these basalt glasses are found typically as thin crusts to lava flows, rims to pillows in pillow lavas, or as narrow selvages to very thin dykes.

The commonest basalt glasses seem to be those having the composition of tholeiitic basalts. Completely glassy rocks corresponding in composition with calc-alkali basalts do not appear to have been recorded.

Chemistry

Average chemical compositions of some tholeiitic and alkali basalts, and of calc-alkali basalt, are given in table 12.1. The differentiation index for these ranges from about 24 to 30, except for the cumulate picrite-basalts where it drops to 14 or 15.

Average typical tholeiitic basalt (and dolerite) (table 12.1, column 1) has SiO_2 about 50 per cent, low Al_2O_3, rather high iron coupled with relatively low MgO, low alkalies, and appreciable TiO_2.

Average ocean-floor tholeiitic basalt (table 12.1, column 1a) differs in a number of respects. It is a tholeiitic olivine basalt of somewhat more mafic character, with olivine in the norm instead of a little quartz. In this connection it must be remembered that olivine in tholeiitic basalts bears a reaction relation to the liquid so that normative olivine is less than modal olivine. A tholeiitic basalt may, in fact, have a little modal olivine and yet show a little quartz in its norm.

TABLE 12.1 *Average chemical compositions of basalts*

		1	1a	2	3	4	5	6	7
SiO_2		50.83	49.99	46.56	51.36	46.19	43.76	46.23	51.31
TiO_2		2.03	1.40	1.92	0.82	2.54	1.82	2.20	0.88
Al_2O_3		14.07	15.65	10.44	18.86	15.02	8.98	18.59	18.60
Fe_2O_3		2.88	1.74	1.87	2.92	2.70	2.88	2.71	2.91
FeO		9.06	8.06	10.19	7.28	9.01	9.57	8.17	5.81
MnO		0.18	0.19	0.16	0.18	0.17	0.13	0.13	0.15
MgO		6.34	7.98	18.28	4.89	9.05	22.43	6.23	5.95
CaO		10.42	11.36	8.32	10.67	10.82	7.91	11.21	10.30
Na_2O		2.23	2.70	1.49	2.12	2.78	1.45	2.80	2.93
K_2O		0.82	0.19	0.34	0.39	0.89	0.50	0.84	0.74
H_2O^+		0.91	0.60	0.24	0.41	0.45	0.35	0.52	0.30
P_2O_5		0.23	0.13	0.19	0.10	0.38	0.21	0.37	0.12
qz		3.5	–	–	5.8	–	–	–	0.8
or		5.0	1.1	1.7	2.2	5.6	2.8	5.0	4.5
ab		18.9	23.1	12.6	17.8	19.4	12.1	21.5	24.6
an		25.9	29.8	20.9	40.9	25.6	16.7	35.6	35.3
ne		–	–	–	–	2.3	–	1.1	–
di	*wo*	10.3	10.8	8.1	4.6	10.7	8.8	7.2	6.3
	en	5.8	6.5	5.7	2.5	6.8	6.5	4.3	4.0
	fs	4.1	3.7	1.7	2.0	3.2	1.5	2.5	1.8
hy	*en*	10.0	8.5	12.4	9.7	–	–	–	10.9
	fs	7.1	4.9	3.8	8.1	–	–	–	5.3
ol	*fo*	–	3.4	19.3	–	11.1	34.7	7.9	–
	fa	–	2.1	6.7	–	5.5	8.4	5.2	–
mt		4.2	2.6	2.8	4.2	3.9	4.2	3.9	4.2
il		3.8	2.7	3.7	1.5	4.9	3.5	4.3	1.7
ap		0.5	0.3	0.4	0.3	0.9	0.5	0.9	0.3
DI		27.4	24.2	14.3	25.8	27.3	14.9	27.6	29.9

1. Tholeiitic basalt (and dolerite) (average of 137) (*Bull. Geol. Soc. Am* **65**, 1021, no. VII).
1a. Ocean-floor tholeiitic basalt (average of 27).
2. Tholeiitic picrite-basalt (average of 12).
3. Tholeiitic basalt enriched in plagioclase (average of 47).
4. Alkali basalt (average of 45).
5. Alkali picrite-basalt (average of 13).
6. Alkali basalt enriched in plagioclase (average of 11).
7. Calc-alkali basalt (average of 48).

The most striking features, however, are the very low K_2O and lower TiO_2 and P_2O_5 of average ocean-floor tholeiitic basalt when compared with average tholeiitic basalt.

Average alkali basalt (table 12.1, column 4) has lower SiO_2 and higher

MgO than average tholeiitic basalts, together with somewhat higher alkalies, TiO_2 and P_2O_5.

Both average tholeiitic picrite-basalt and average alkali picrite-basalt (table 12.1, columns 2 and 5) have much higher MgO with lower SiO_2, Al_2O_3, CaO and alkalies, largely reflecting the enrichment in olivine.

Note the presence of considerable hypersthene in the norms of all the average tholeiitic basalt types (table 12.1, columns 1, 1a, 2, 3) and its absence in the norms of the average alkali basalt types (table 12.1, columns 4, 5, 6) where, indeed, a little normative nepheline may appear. This is one of the main distinctions between tholeiitic and alkali basalts (28).

Average calc-alkali basalt (table 12.1, column 7) bears little resemblance to average alkali basalt, and differs markedly from average tholeiitic basalt. The most striking differences are the higher Al_2O_3 and alkalies, and the lower iron and TiO_2. These differences show themselves normatively in the considerably higher normative feldspar, the lower *fs* contents of the normative diopside and hypersthene, and the lower normative ilmenite when compared with the norm of average tholeiitic basalt.

Some tholeiitic and alkali basalts are enriched in plagioclase and it is instructive to compare average compositions of these with the composition of average calc-alkali basalt (compare table 12.1, columns 3 and 6 with column 7). It will be seen that all three averages are quite distinct and that, although the Al_2O_3 contents are all about the same, the tholeiitic and alkali basalt types still retain many of their characteristic features. Observe, in particular, the lower alkalies and MgO, and the higher iron of plagioclase-enriched tholeiitic basalt when compared with calc-alkali basalt. These differences are well brought out in the norm of the former with its lower *ab* and higher *an*, and its higher ratio of *fs* to *en* in both normative diopside and hypersthene.

When the chemical composition of basalts is compared with that of the corresponding gabbros (table 11.1) the agreement is not very close. Certain features are characteristic of both, such as normative hypersthene in both gabbros and tholeiitic basalt, normative nepheline and no normative hypersthene in both alkali gabbro and alkali basalt. But, in general, the corresponding gabbros are more basic in character and have lower differentiation indices.

One reason for this is that all varieties of gabbro and alkali gabbro have been lumped together in making the averages whereas picrite-basalts and plagioclase-enriched basalts have been separated from more typical basalts and averaged separately. But another and perhaps more important reason is that many of the gabbros used in making the averages come from layered or similar differentiated intrusions and are cumulate rocks. This is the case to a still greater degree with norite and troctolite (also anorthosite) for which volcanic equivalents are unknown.

Alteration of basalts

The mineral most susceptible to alteration in basalts is olivine, and this is frequently replaced by red-brown 'iddingsite' or by pale green serpentine or chlorite, even when the other constituents of the rock remain unchanged.

Sometimes alteration goes further, as in the Tertiary basalts of Mull (29) where, in addition to alteration of olivine, there may be alteration of clinopyroxene to chlorite and of plagioclase to albite and epidote.

More interesting is the alteration of basalts, particularly those showing pillow structure, to *spilites*, rocks in which the original labradorite has been replaced by albite. Spilites are well developed in the Upper Devonian of north Cornwall (fig. 8.2B) and south Devon (30), and are also found among the Ordovician volcanic rocks of Scotland and Wales, while Precambrian ones occur in Anglesey and Caernarvonshire. They are known in numerous other countries and among these may be mentioned those of eastern Oregon (31) and of the North Island of New Zealand (32).

In connection with the formation of spilites a dredge haul of spilitic pillow lavas from the Carlsberg Ridge, Indian Ocean (33), is of particular importance. Here, not only spilites but rocks intermediate between spilite and basalt were present. Original calcic plagioclase is replaced *in situ* by albite, olivine and basalt-glass by chlorite, iron-ore by granular sphene, and augite may remain unchanged or may be converted in part to actinolite. Cores of pillows, having little original glass, do not have much chlorite, but the original glassy margins of pillows are converted almost entirely to this mineral. There is thus a great mineralogical and chemical contrast between the core and margin of a pillow (34) involving large-scale local transport of material. Taking the lava pile as a whole, however, the main change has been a gain of much water and the loss of substantial amounts of calcium with some aluminium. The changes that have taken place are thought to be due not to autometamorphism during cooling, as has often been suggested, but to the burial and consequent heating up of the lava pile (see also p. 424 and fig. 39.1).

Some spilites are very similar to those from the Carlsberg Ridge, but in many of the older spilites original ferromagnesian minerals have been replaced completely by chlorite, calcite and epidote.

The spilites so far mentioned are all what may be termed soda spilites but, just as rhyolitic lavas may alter to soda or to potash keratophyres, so may basalts alter to soda or to potash spilites. The latter are less common and the best-known are those of Permian age from Timor, where the original labradorite of the basalt has been partly or wholly converted to adularia.

Basalt glass may be changed to *palagonite*. This is an alteration product of more or less resinous appearance which is yellow to orange and isotropic or feebly birefringent in thin section. Much of the original calcium and sodium and some of the aluminium and silicon in the glass is lost during the alteration, the iron is largely oxidised to the ferric state, and a good deal of water is gained (35).

It has already been mentioned that many basalts are vesicular. Such may have their vesicles filled with secondary minerals such as zeolites, chlorite, calcite, etc. This has been ascribed to the activity of hydrothermal solutions associated with a late stage in the cooling history of a volcanic pile, but recent work indicates that these minerals may be deposited long after the eruption of the lavas by the action of circulating waters, largely of meteoric origin (36).

Although weathering has been excluded when considering the later alteration of igneous rocks, an exception should perhaps be made for the sea-floor weathering of ocean-floor basalts. In a typical case (37), glassy selvages of pillows alter first to palagonite and then to montmorillonite while, within, calcic cores of plagioclase phenocrysts are replaced by potash feldspar, and pyroxene phenocrysts and the matrix are replaced by chlorophaeite and chlorite. This alteration of the flow interiors involves the almost complete oxidation of ferrous to ferric iron and a considerable increase in K_2O and H_2O with corresponding losses in MgO and CaO.

Hypabyssal equivalents

The *dolerites* (38) are the hypabyssal representatives of the gabbros and the basalts, finer grained than the former but coarser grained than the latter. They occur as dykes and sills, more rarely as little bosses. They may also form the central portion of thick basalt lava flows and have then sometimes been mistaken for intrusive sills. Their texture is frequently ophitic or subophitic, though some are subhedral-granular or intergranular. A number are porphyritic but this is not such a common feature as in other hypabyssal rocks.

Dolerites, like basalts, may be divided into two great groups, the *tholeiitic dolerites* (often called quartz dolerites) and the *alkali dolerites*.

Tholeiitic dolerites have pyroxene and calcic plagioclase, typically labradorite, as essential constituents, with or without some olivine, and a variable amount of interstitial residuum consisting of alkali feldspar and quartz, often intergrown. The accessory constituents are iron-ore and apatite with, in some, a little hornblende or biotite. The pyroxenes that may be present include augite or subcalcic augite, orthopyroxene, and pigeonite, and the kind of assemblages that may occur may be illustrated by considering the four main types of Karroo dolerite (39). Two of the types have essential but variable amounts of olivine accompanied in one case by zoned clinopyroxene fringed by ferriferous pigeonite and in the other by clinopyroxene and orthopyroxene or pigeonite. A third type has augite with frequent cores of magnesian pigeonite with or without olivine, while in the fourth type orthopyroxene is accompanied by augite, with or without pigeonite, and olivine is absent.

Tholeiitic dolerites are widely distributed in various parts of the

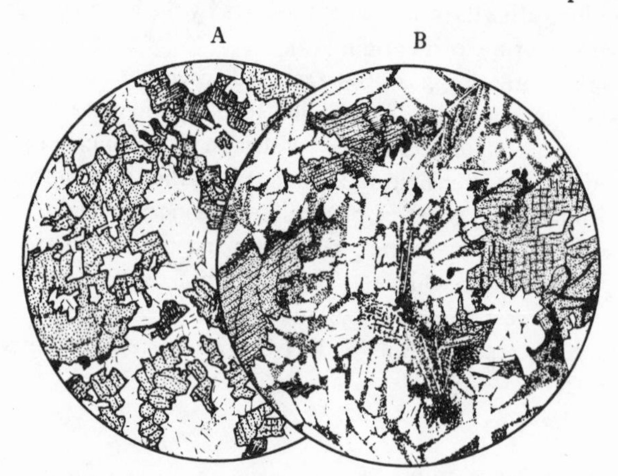

Fig. 12.3. Tholeiitic dolerites ('quartz dolerites') from the differentiated Mount Wellington Sill, Tasmania; × 14.

A. From the magnesia-rich zone. Somewhat altered augite and pigeonite and clear orthopyroxene moulded on labradorite; very small areas of quartzo-feldspathic mesostasis.

B. From near summit of mountain. Large crystals of augite and pigeonite mould labradorite, and large areas of mesostasis contain altered potash feldspar, needles of quartz, chlorite and small grains of iron-ore.

world, and have been recorded from the Precambrian (40) onwards, but one of the greatest periods was the Jurassic when the widespread Karroo dolerites of South Africa and those of Tasmania (fig. 12.3) and Antarctica were intruded. Those of late Carboniferous (41) and Tertiary (42) age in Britain deserve mention.

Alkali dolerites (fig. 12.4) have a single lime-rich more or less titaniferous augite as their pyroxene. They normally have olivine as a further essential constituent and not infrequently have a little interstitial analcite. They may be illustrated by those occurring in the Tertiary igneous province of Scotland.

Those forming most of the early basic cone sheets of Mull (43) are aphyric with zoned labradorite, lilac titanaugite in ophitic relationship with the feldspar, olivine that may be subhedral but is quite often ophitic, and accessory titaniferous magnetite in skeletal crystals.

Many of the dykes of the Mull dyke swarm are of this type and these frequently have some interstitial analcite, as do the alkali dolerites of the Islay–Jura dyke swarm (44) and those forming sills in the southern part of Arran (45) (fig. 12.4B). In these the clinopyroxene takes on a purplish tinge with distinct pleochroism, and apatite may become a noticeable accessory constituent. Some of the Mull rocks have phenocrysts of olivine; more rarely they have large phenocrysts of plagioclase.

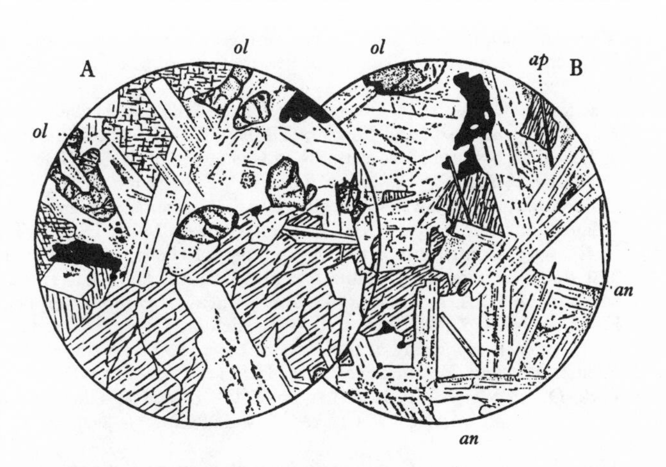

Fig. 12.4. Alkali dolerites; × 20.

A. Alkali dolerite, Fair Head, Antrim. Olivine (*ol*), subophitic clinopyroxene, labradorite and accessory iron-ore.

B. Alkali dolerite, Dippin, Arran. A more alkaline type, transitional in composition towards teschenite, with purplish titaniferous augite, interstitial patches of clear analcite (*an*) and long needles of apatite (*ap*).

Alkali dolerites of Carboniferous–Permian age also occur in Scotland, particularly in Fife where they may form large sills. Some of these are porphyritic. Further examples come from the Midlands of England, as at Rowley Regis, near Birmingham; in the Clee Hills, Shropshire; and in Derbyshire.

Some of the larger bodies of tholeiitic dolerite may show considerable differentiation leading to the formation of late-stage ferrogranophyre (46). There may be marked absolute enrichment in iron in the intermediate stages of the process and, when this is accompanied by some enrichment in sodium, dolerites equivalent to tholeiitic andesite and ferrodiorite are produced (47).

Differentiated bodies of alkali dolerite (48) behave in a different way. There is little or no absolute enrichment in iron in the intermediate stages, and the final product is a rock of syenitic type with abundant alkali feldspar and sometimes a good deal of analcite and nepheline (49).

Another feature of differentiated dolerite intrusions is the presence of irregular patches or veins of *dolerite-pegmatite*, considerably coarser than the rocks in which they occur. They may be found in both tholeiitic and alkali dolerite intrusions.

References and notes

(1) There are also transitional basalts intermediate in composition between tholeiitic and alkali types but these seem to be relatively rare.

(2) Macdonald, G. A., 1949. Petrography of the Island of Hawaii. *Prof. Pap. U.S. geol. Surv.* **214**-D, 56–74. Tilley, C. E., 1960. Differentiation of Hawaiian basalts: some variants in lava suites of dated Kilauean eruptions. *J. Petrology* **5**, 47–55. Tilley, C. E. & Scoon, J. H., 1961. Differentiation of Hawaiian basalts; trends of Mauna Loa and Kilauean historic magma. *Am. J. Sci.* **259**, 60–8. Muir, I. D. & Tilley, C. E., 1963. Contributions to the petrology of Hawaiian basalts: II. The tholeiitic basalts of Mauna Loa and Kilauea. *Ibid.* **261**, 111–28. Macdonald, G. A. & Katsura, T., 1964. Chemical composition of Hawaiian lavas. *J. Petrology* **5**, 82–133. Richter, D. H. & Murata, K. J., 1966. Petrography of the lavas of the 1959–60 eruption of Kilauea Volcano, Hawaii. *Prof. Pap. U.S. geol. Surv.* **537**-D, 1–12. Peck, D. L., Wright, T. L. & Moore, J. G., 1966. Crystallization of tholeiitic basalt in Alae Lava Lake, Hawaii. *Bull. Volc.*, ser. II, **29**, 629–55.

(3) Muir, I. D. & Tilley, C. E., 1957. Contributions to the petrology of Hawaiian basalts: I. The picrite basalts of Kilauea. *Am. J. Sci.* **255**, 241–53.

(4) Tilley, C. E., 1961. The occurrence of hypersthene in Hawaiian basalts. *Geol. Mag.* **98**, 257–60. Muir, I. D. & Long, J. V. P., 1965. Pyroxene relations in two Hawaiian hypersthene-bearing basalts. *Miner. Mag.* **34**, 358–69.

(5) Noe-Nygaard, A. & Rasmussen, J., 1968. Petrology of a 3,000 metre sequence of basaltic lavas in the Faeroe Islands. *Lithos* **1**, 286–304.
(6) Searle, D. L. & Vokes, F. M., 1969. Layered ultrabasic lavas from Cyprus. *Geol. Mag.* **106**, 515–30.
(7) Waters, A. C., 1961. Stratigraphic and lithologic variations in the Columbia River Basalts. *Am. J. Sci.* **259**, 583–611.
(8) Bailey, E. B. *et al.*, 1924. Tertiary and post-Tertiary geology of Mull. *Mem. geol. Surv. U.K.* (Basalts of 'non-porphyritic Central Type'; important in the Central Group of Lavas (p. 149) and providing many to the dykes ('tholeiites') in the Mull dyke swarm (pp. 370–2).) Holmes, A. & Harwood, H. F., 1929. The tholeiite dikes of the north of England. *Miner. Mag.* **22**, 1–52. Patterson, E. M., 1955. The Tertiary lava succession in the northern part of the Antrim plateau. *Proc. Roy. Irish Acad.* **57**, sect. B, 89–95, 104–111.
(9) Engel, A. E. J., Engel, C. E. & Havens, R. G., 1965. Chemical characteristics of oceanic basalts and the Upper Mantle. *Bull. geol. Soc. Am.* **76**, 719–34.
Ocean-floor basalts actually span the range from tholeiitic to mildly alkaline types but are mainly tholeiitic olivine basalts, and the average chemical composition of 94 specimens of ocean-floor basalts in general is very similar to that given here in table 12.1, column 1a for ocean-floor tholeiitic basalt (see Cann, J. R., 1971. Major element variations in ocean-floor basalts. *Phil. Trans. Roy. Soc. Lond.* A **268**, 495–505).
(10) Baker, P.E., 1968. Comparative volcanology and petrology of the Atlantic island-arcs. *Bull. Volc.*, ser. II, **32**, 189–206. Ringwood, A. E., 1974. The petrological evolution of island arc systems. *Jl. Geol. Soc. Lond.* **130**, 183–204.
(11) Some titaniferous augites have a very small 2V like that of pigeonite, a reminder that optical properties of clinopyroxenes may not be reliable indicators of the species.
(12) Upton, B. G., Wadsworth, W. J. & Newman, T. C., 1967. The petrology of Rodriguez Island, Indian Ocean. *Bull. geol. Soc. Am.* **78**, 1495–1506.
(13) Patterson, 1955, pp. 81–5, 98–104.
(14) Bailey *et al.*, 1924, pp. 136–43 ('Plateau Basalts'). Anderson, F. W. & Dunham, K. C., 1966. The geology of northern Skye. *Mem. geol. Surv. U.K.*, 104–16. Tilley, C. E. & Muir, I. D., 1962. The Hebridean Plateau magma type. *Trans. geol. Soc. Edinb.* **19**, 208–15.
(15) MacGregor, M. & MacGregor, A. G., 1948. The Midland Valley of Scotland. *Brit. Reg. Geol.*, 53–6, and references cited on 61–3.
(16) Walker, F. & Nicolaysen, L. O., 1954. The petrology of Mauritius. *Colonial Geol. Miner. Resources* **4**, 3–43.
(17) Baker, P. E., 1968. Petrology of the Mt. Misery Volcano, St. Kitts, West Indies. *Lithos* **1**, 127–9.
(18) Coats, R. R., 1959. Geologic reconnaissance of Semisopochnoi Island, Western Aleutian Islands, Alaska. *Bull. U.S. geol. Surv.* **1028**-O, 503–6.
(19) Nicholls, I. A., 1971. Petrology of Santorini Volcano, Cyclades, Greece. *J. Petrology* **12**, 74–5.
(20) Smith, A. L. & Carmichael, I. S. E., 1968. Quaternary lavas from the Southern Cascades, Western U.S.A. *Contr. Miner. Petrol.* **19**, 212–38.
(21) Larsson, W., 1941. Petrology of interglacial volcanics from the Andes of northern Patagonia. *Bull. Geol. Inst. Upsala* **28**, 250–78, especially 267–72.

(22) Viljoen, M. J. & Viljoen, R. P., 1969. The geology and geochemistry of the Lower Ultramafic Unit of the Onverwacht Group and a proposed new class of igneous rocks. *Geol. Soc. S. Afr. Spec. Publ.* **2**, 55–85.
(23) Williams, D. A. C., 1972. Archaean ultramafic, mafic and associated rocks, Mt. Monger, Western Australia. *J. geol. Soc. Austral.* **19**, 163–88.
(24) Williams, D. A. C., 1972. *Ibid.*, plate 7D, opp. p. 188.
(25) Tilley, C. E., 1950. Some aspects of magmatic evolution. *Q. Jl. geol. Soc. Lond.* **106**, 55–6. Yoder, H. S. & Tilley, C. E., 1962. Origin of basalt magmas: an experimental study of natural and synthetic rock systems. *J. Petrology* **3**, 360 and table 3, column 16.
(26) Smith & Carmichael, 1968, 221.
(27) Fuller, R. E., 1932. Concerning basaltic glass. *Am. Miner.* **17**, 104–7.
(28) Some mildly alkaline basalts have a very little hypersthene in the norm. When alkali basalts have been altered, considerable amounts of hypersthene may appear in the norm. Tilley & Muir, 1962, pp. 210–11.
(29) Bailey *et al.*, 1924, pp. 141–3. Fawcett, J. J., 1965. Alteration products of olivine and pyroxene in basalt lavas from the Isle of Mull. *Miner. Mag.* **35**, 55–68.
(30) Dewey, H. & Flett, J. S., 1911. British pillow-lavas and the rocks associated with them. *Geol. Mag.* dec. V, **8**, 202–9 and 241–8. Middleton, G. V., 1960. Spilitic rocks in south-east Devonshire. *Ibid.* **97**, 192–207.
(31) Gilluly, J., 1935. Keratophyres of eastern Oregon and the spilite problem. *Am. J. Sci.* (5) **29**, 225–52 and 336–52.
(32) Battey, M. H., 1956. The petrogenesis of a spilitic rock series from New Zealand. *Geol. Mag.* **93**, 89–110.
(33) Cann, J. R., 1969. Spilites from the Carlsberg Ridge, Indian Ocean. *J. Petrology* **10**, 1–19. See also Vallance, T. G., 1974. Spilitic degradation of a tholeiitic basalt. *Ibid.* **15**, 79–96 for the transition from basalt to spilite within a single Deccan flow.
(34) See also Vallance, T. G., 1965. On the chemistry of pillow lavas and the origin of spilites. *Miner. Mag.* **34**, 471–81.
(35) Hay, R. L. & Iijima, A., 1968. Nature and origin of palagonite tuffs of the Honolulu Group on Oahu, Hawaii. *Mem. geol. Soc. Am.* **116**, 331–76.
(36) Walker, G. P. L., 1951. The amygdale minerals in the Tertiary lavas of Ireland: I. The distribution of chabazite habits and zeolites in the Garron plateau area, County Antrim. *Miner. Mag.* **29**, 773–91. Walker, G. P. L., 1960. Zeolite zones and dike distribution in relation to the structure of the basalts of eastern Iceland. *J. Geol.* **68**, 515–28. Nashar, B. & Davies, M., 1960. Secondary minerals of the Tertiary basalts, Barrington, New South Wales. *Miner. Mag.* **32**, 480–91.
(37) Mathews, D. H., 1971. Altered basalts from Swallow Bank, an abyssal hill in the N.E. Atlantic and from a nearby seamount. *Phil. Trans. Roy. Soc. Lond.* A**268**, 551–71.
(38) These rocks are usually termed *diabase* by American and many Continental petrologists. British petrologists use the term 'dolerite' for fresh, and 'diabase' for altered, rocks of this kind.
(39) Walker, F. & Poldervaart, A., 1949. Karroo dolerites of the Union of South Africa. *Bull. geol. Soc. Am.* **60**, 591–706.
(40) *E.g.*, Schwartz, G. M. & Sandberg, A. E., 1940. Rock series in diabase sills at Duluth, Minnesota. *Ibid.* **51**, 1135–72.
(41) Holmes, A. & Harwood, H. F., 1928. The age and composition of the Whin Sill and related dikes of the north of England. *Miner. Mag.* **21**,

493–542. Walker, F., 1935. The late Palaeozoic quartz-dolerites and tholeiites of Scotland. *Ibid.* **24**, 131–59.

(42) Richey, J. E. *et al.*, 1961. Scotland: the Tertiary volcanic districts. *Brit. Reg. Geol.*, and references cited there. For a description of typical Tertiary tholeiitic dolerites see Bailey *et al.*, 1924, pp. 300–4.

(43) Bailey *et al.*, 1924, pp. 239–40.

(44) Walker, F., 1960. The Islay–Jura dyke swarm. *Trans. geol. Soc. Glasgow* **24**, 121–37.

(45) Tyrrell, G. W., 1928. The geology of Arran. *Mem. geol. Surv. U.K.*, 112–22.

(46) For differentiation of olivine-bearing tholeiitic dolerite, with formation of olivine-rich cumulates, see Walker, K. R., 1969. The Palisades Sill, New Jersey: a reinvestigation. *Spec. Pap. geol. Soc. Am.* **111**. For differentiation of olivine-free tholeiitic dolerite, with formation of cumulates rich in orthopyroxene, see Gunn, B. M., 1962. Differentiation in Ferrar dolerites, Antarctica. *N.Z. Jl. Geol. Geophys.* **5**, 820–63.

(47) *E.g.*, the ferrohypersthene dolerite and the ferrodolerite of the Palisades Sill (Walker, K. R., 1969, *op. cit.*, table 8, pp. 76–7).

(48) See, for instance, Wilshire, H. G., 1967. The Prospect alkaline diabase–picrite intrusion, New South Wales, Australia. *J. Petrology* **8**, 97–162.

(49) Walker, F. & Patterson, E. M., 1959. A differentiated boss of alkali dolerite from Cnoc Rhaonastil, Islay. *Miner. Mag.* **32**, 140–52.

13
Ultramafic igneous rocks

Ultramafic igneous rocks are hypermelanic, composed essentially of ferromagnesian minerals. *Dunites* have olivine as the only essential mineral; *peridotites* have dominant olivine associated with other ferromagnesian minerals. These rocks may occur as layers or lenses in layered mafic intrusions and may also form small independent intrusions (with or without associated mafic rocks), dykes and sills. Large, sometimes very large, ultramafic bodies are to be found among what are known as 'Alpine type peridotites'. These are ultramafic masses occurring in the folded rocks of orogenic belts as large inclined sheet-like bodies or as swarms of smallish lenses in arcuate belts and they are frequently serpentinised. Some of these Alpine type ultramafics are associated with a sequence of gabbros, dolerite dykes and mafic pillow lavas, often more or less altered, to form the so-called *ophiolite complexes*. These complexes closely resemble the known igneous rock succession from the ocean floor and are thought to represent fragments of old ocean crust and upper mantle (1).

Perknites are composed essentially of ferromagnesian minerals other than olivine, the most important being *pyroxenites* and *hornblendites*. These usually occur as small intrusions of one kind or another, but the former may also be present as layers in layered mafic intrusions.

Mineralogy

The *olivine* found here is commonly, but not exclusively, a magnesian variety rich in the forsterite molecule. It may build more or less euhedral crystals (fig. 13.2) unless it is so abundant that the crystals interfere, when it is granular (fig. 13.1). It characteristically shows a network of cracks rather than cleavage, but sometimes the cleavage is quite well developed. It shows a preferred orientation in some ultramafic rocks (2) and may also exhibit a rather obscure lamellar twinning due to strain.

Orthopyroxene, relatively rich in magnesia and normally falling in the enstatite–bronzite range, is a common constituent in a number of peridotites and perknites. In certain peridotites, more especially, it is an aluminous orthopyroxene with up to 6 per cent or so of alumina (3).

The *clinopyroxene* of ultramafic rocks is typically a somewhat aluminous diopsidic variety either colourless in thin section or, when comparatively rich in chromium, pale green (chrome-diopside). The clinopyroxene of alkali peridotites and perknites, however, is titaniferous augite.

Primary *hornblende* is present in some peridotites and is the dominant mineral of the hornblendites. It may be brown or green in thin section. Alkali peridotites may carry *kaersutite*. *Biotite* occurs in small amount in a number of rocks, occasionally becoming important. It is usually a phlogopitic variety, rich in magnesium.

Pyrope garnet is an essential constituent of certain peridotites; *melanite garnet* may be present in some of the more alkaline types of ultramafic rock.

The characteristic accessory minerals are members of the spinel group, commonly *chromite* (deep brown to opaque in section), *picotite* (coffee brown) or *pleonaste* (green), though *magnetite* occurs also. These are generally euhedral, except when segregated to form virtually monomineralic layers or lenses, as is sometimes the case. *Sulphides*, similar to those occurring in gabbros, may be present in small quantity, also native *platinum*.

Apatite is not found in peridotite, but is present as an accessory constituent in alkali peridotites. *Perofskite* is confined to the more alkaline ultramafic rocks.

A little *calcic plagioclase* is present in some rocks, marking a transition to melanocratic gabbros, while in some alkaline types there is a little interstitial *nepheline* or *analcite*.

Textures and structures

Peridotites have a granular texture if olivine is very abundant. When abundant olivine occurs with minor pyroxene, the latter may build relatively large crystals, giving rise to a pseudoporphyritic texture (fig. 13.1). When olivine is rather less abundant its crystals may be enclosed in plates of pyroxene or hornblende, resulting in a poikilitic texture (fig. 13.2). Some peridotites, especially Alpine type peridotites, may show a development of cataclastic textures and becomes schistose, or even mylonitised (4). Perknites usually have a more or less granular texture.

A large-scale banded structure, or layering, is noticeable in some peridotite intrusions (e.g. isle of Skye).

Dunite and peridotite

The original locality for *dunite* is Dun Mountain, Nelson, South Island, New Zealand, but equally good examples occur at many other localities as, for instance, among the Tertiary plutonic rocks of the isle of Skye (5), the Ural Mountains, U.S.S.R., and St. John's (Zeberged) Island in the Red Sea. All have accessory spinel in addition to fresh olivine and this may be chromite (Dun Mountain, Ural Mountains) or chromite, picotite, or pleonaste in different varieties (Skye). The St. John's Island dunites are notably coarse grained and have long been famous as a source of peridot, the gem variety of olivine.

These dunites, and others associated with gabbroic or other ultramafic rocks, have a magnesian olivine ranging from about Fa_7 to Fa_{20} in different cases. Certain dunites, however, have olivine relatively rich in the fayalite molecule. The best known are probably the hortonolite dunites forming pipe-like masses in the great Bushveld igneous complex of the Transvaal.

Dunites rich in iron-ore may be referred to as *ore-dunites*. In these the iron-ore usually forms a cement for the olivine grains as in the example from Susimaki, Vampula, Finland, where rounded grains of olivine (Fa_{45}), a little brown hornblende and accessory iron-rich spinel are embedded in a mass of ore made up of ilmenite and titaniferous magnetite with accessory pyrite and pyrrhotite.

The dunites and ore-dunites considered so far have been associated with normal gabbroic or ultramafic rocks. Those associated with alkali rocks are apparently less common but they do occur. Thus the dunites and ore-dunites making up the central part of the Lesnaya Varaka intrusion in the south-eastern part of the Kola Peninsula, U.S.S.R., are associated with alkali pyroxenites and a nepheline-bearing pyroxene–melanite garnet rock. A feature emphasising the alkaline associations of these dunites is the presence in them of accessory perofskite which becomes abundant locally.

Peridotite with orthopyroxene ('harzburgite') is quite a widely distributed variety. *Enstatite peridotite* (fig. 13.1) is a common component of the Permian ultramafic intrusions in the South Island of New Zealand (6). Olivine (Fa_7 to Fa_9) is associated with enstatite (Of_7 to Of_9) showing exsolution lamellae of Bushveld type (p. 103), and accessory chromite.

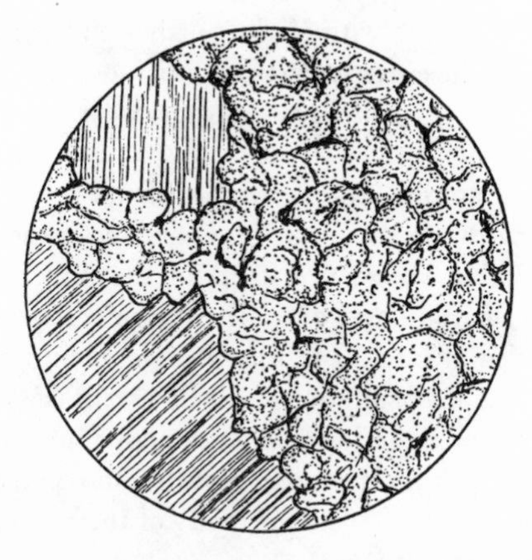

Fig. 13.1. Enstatite peridotite, Skulvik, near Tromsö, Norway; × 20. A granular aggregate of olivine with irregular pseudoporphyritic crystals of partly altered enstatite.

Most of the rocks in the Great Serpentine Belt of New South Wales were originally enstatite peridotites, now largely serpentinised. Indeed, enstatite peridotite grading into dunite seems to be the commonest type in bodies of Alpine type peridotite. *Bronzite peridotites*, with orthopyroxene of composition Of_{10} to Of_{25}, are to be found in the Bushveld complex and in the ultramafic zone of the Stillwater complex, Montana (7), while *hypersthene peridotite* is present in the igneous complex of Freetown, Sierra Leone.

Augite peridotite is associated with dunite in the isle of Skye (8) usually carrying a little calcic plagioclase, and both this and the augite are interstitial to the olivine, the augite being sometimes poikilitic (fig. 13.2B). A similar association is found in the Tulameen district of British Columbia (9) where two bodies of dunite are surrounded by pyroxenite and augite peridotite forms a transitional type between the dunite and the pyroxenite. Here the augite occurs frequently as large crystals up to 10 cm long, embedded in a matrix of olivine grains and abundant magnetite and chromite.

Pyroxene peridotite ('lherzolite'), with both ortho- and clinopyroxene, is one of the commonest varieties. The type rock from Lake Lherz in the Pyrenees is one of a number of similar occurrences in that region. Dominant olivine is associated with orthopyroxene, green chrome-diopside, accessory picotite and, in most specimens, a little brown hornblende.

The primary mineral assemblage of the Lizard peridotite, Cornwall

(10), is that of a lherzolite with olivine, ortho- and clinopyroxene, and olive-green spinel. The enstatite has exsolved lamellae of clinopyroxene, and both it and the chromiferous diopside are relatively rich in aluminium.

In both these examples, the lherzolite occurs in the form of small intrusions. It may, however, occur on a large scale, as in the Zlatibor district of W. Serbia where it makes up the bulk of an ultramafic intrusion having a surface area of some 1300km^2.

Lherzolite is also of interest in that it is the main igneous rock type found as inclusions in a number of alkali basalts and basanites. It is believed that most of these inclusions, though apparently not all (11), have been carried up from the upper mantle and represent more or less depleted mantle material. A detailed study of their textures and fabrics has been published recently (12).

Many of the rocks described in the literature as 'hornblende peridotite' are olivine hornblendites rather than hornblende peridotites. Those of the Ariège district in the Pyrenees, however, are true *hornblende peridotites* with olivine as the main dark mineral. One variety has large poikilitic crystals of orange-yellow hornblende enclosing the olivine; another has the same hornblende, but the texture is granular; a third has very large crystals of pale greyish green hornblende enveloping olivine and carries a little hypersthene and spinel. Most of these rocks also have small amounts of biotite.

Similar peridotites with varying proportions of olivine and transitional to olivine hornblendites are found in Anglesey and in Caernarvonshire (fig. 13.2A).

Some pyroxene and hornblende peridotites are rich in iron-ore. Good examples come from the Bükkgebirge, Hungary, where hornblende peridotites, augite–hornblende peridotites, and rocks transitional to dunite, all have a good deal of titaniferous magnetite, frequently making up 20 per cent or more of the rock.

Garnetiferous peridotites (13) occur in orogenic zones (e.g. in Norway and Switzerland), commonly as a component of small, more or less lensoid masses in which they may be associated with dunite and normal peridotites. They are also found as xenoliths in kimberlite (*see* p. 144) pipes in South Africa. The pyrope-rich garnet may vary from a trace to as much as 20 per cent or so and, in addition to olivine and garnet, these peridotites may have orthopyroxene, or more or less chromiferous diopside, or both pyroxenes together.

Typical *alkali-peridotites* differ from other peridotites in having lime-rich titaniferous augite as their usual pyroxene, in the frequent presence

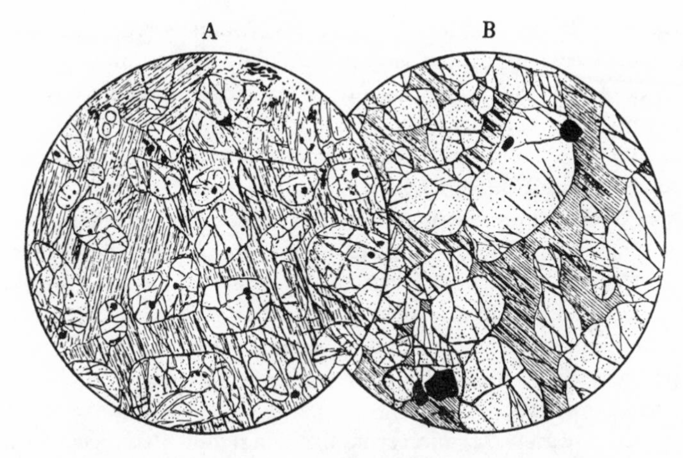

Fig. 13.2. Peridotites; × 16.

A. Hornblende peridotite, Mynydd Penarfynydd, Caernarvonshire, Wales. Large crystals of brown hornblende enclosing numerous olivine grains.

B. Augite peridotite, Cuillin Hills, Skye. Here the olivine crystals are enclosed in a large plate of augite.

of red-brown alkali amphibole (kaersutite), in the almost constant presence of apatite, and in the absence of orthopyroxene.

The 'hornblende peridotite' making up more than half of the Lugar sill, Ayrshire, Scotland (14), is a good example. Medium to coarse grained, it is composed essentially of rounded crystals of olivine; titanaugite in clusters of crystals wedged between the olivine grains, and red-brown kaersutite as irregular poikilitic plates enclosing some of the olivine and groups of pyroxene crystals.

This rock is associated with mafic alkali rocks and many other occurrences of sills are known with a similar association. Some of the peridotites in these sills are similar to the Lugar rock; others have only olivine and titanaugite as essential constituents. Many have a little calcic plagioclase and analcite.

The 'hornblende picrite' intrusions of Devonian age in Cornwall (15) have a comparable mineralogy. These have partially serpentinised olivine crystals enclosed in plates of titaniferous augite that is in process of replacement by kaersutite. The accessory constituents are biotite, apatite and magnetite.

Mica peridotite provides a special variety of alkali peridotite, composed essentially of olivine, with or without monticellite, and phlogopitic mica.

The mica peridotite that makes up the main part of the small chonolith at Haystack Butte, Highwood Mountains, Montana (16), has monticellite in excess of olivine and the poikilitic phlogopite shows the reversed pleochroism often shown by at least some of the mica in other mica peridotites. The accessory constituents are apatite, perofskite and magnetite. In the medium to very fine grained mica peridotites found as narrow dykes and sheets intrusive into the Lower Gondwana rocks of Bengal, olivine is present as phenocrysts and as small crystals (usually serpentinised) associated with plates of mica that have numerous inclusions of olivine and apatite. There is commonly a little reddish yellow amphibole and sometimes some colourless to pale green augite. The amount of apatite is variable but, in extreme cases, may make up some 10 per cent of the rock. The other accessories are iron-ore, chromite, and rare perofskite.

A special variety of micaceous alkali peridotite has been termed ***kimberlite***, a name given by H. C. Lewis in 1887 to the porphyritic peridotite which is the source rock of diamond at Kimberley, South Africa. It has been redefined recently (17) as containing more or less rounded phenocrysts of olivine, phlogopite, magnesian ilmenite, pyrope and chromium-rich pyrope, set in a fine-grained groundmass of second generation olivine and phlogopite together with carbonate, serpentine and/or chlorite with accessory magnetite, apatite and perofskite. Clinopyroxene (especially chrome-diopside) and occasional nodular aggregates of enstatite may also be present but are probably xenocrystal. Some have diamond as an additional accessory constituent; some carry xenoliths of peridotite, including garnet peridotite, and rare glimmerite (18).

Olivine ranges in composition from Fa_6 to Fa_{18} and, whereas the phenocrysts may be fresh, the groundmass olivine is usually more or less completely altered to serpentine. The phlogopite may show either normal or reversed pleochroism, both types being sometimes present in a single zoned crystal. It may be replaced to a varying extent by chlorite. Some of the carbonate is secondary but some, occurring as anhedral grains, is almost certainly primary (19). Both ortho- and clinopyroxenes are present in some kimberlites but, at least in many cases, are probably xenocrystal.

Kimberlite occurs as small pipes, narrow dykes or, more rarely, sills (19) mainly in Cape Province and other parts of Africa (20), the Yakutia region of the U.S.S.R. and at various localities in North (21) and South America.

Chemistry

Average chemical compositions of dunite, peridotites and alkali peridotite are given in table 13.1, columns 1, 2, and 3. Most alkali peridotites are somewhat altered and the nearest it is possible to get to their original composition is by deducting CO_2 and excess H_2O and recalculating to 100. This has been done in column 3a.

All peridotitic rocks have a very low differentiation index, very low SiO_2 and high MgO, the latter being extremely high in dunite. Al_2O_3 is negligible in average dunite and low in average peridotite, but becomes appreciable in average alkali peridotite where aluminous titanaugite and kaersutite are prominent constituents. Alkalies are very low throughout but are more prominent in alkali peridotite. The higher CaO content of this alkali type is due largely to the presence of lime-rich clinopyroxene and absence of orthopyroxene. P_2O_5 is very low in average dunite and peridotite but is present in appreciable amounts in average alkali peridotite. TiO_2, also, is more prominent in the latter.

Note the presence of normative hypersthene in average peridotite, its absence in average alkali peridotite and the presence of normative nepheline in the latter.

Later alteration

The main process of alteration suffered by dunites and peridotites is that of *serpentinisation*, the olivine in these rocks being particularly susceptible. Many, perhaps most, rocks show some degree of this but the process has not infrequently gone to completion, resulting in a serpentine rock or *serpentinite*.

The Burro Mountain, California, ultramafic complex (22) will be taken as an example. Here, dunites and peridotites have been partially or completely serpentinised and all stages of alteration can be seen. The primary minerals are dominant olivine, with some enstatite, less diopsidic augite and accessory picotite. *Olivine* alters to green irregular platy serpentine combined with areas of cross- and slip-fibre serpentine, giving rise to a mesh structure (cf. fig. 13.3A). The serpentine here proves to be a mixture of lizardite and chrysotile, either of which may be dominant in any one sample. Ferriferous brucite or magnetite, or both, may be associated with the serpentine. *Enstatite* is pseudomorphed by colourless to pale green 'bastite' serpentine (cf. fig. 13.3B), composed only of lizardite, and the *clinopyroxene* apparently alters to serpentine as

TABLE 13.1. *Average chemical compositions of some ultramafic igneous rocks*

	1	2	3	3a	4	5	6
SiO_2	40.16	43.54	40.27	41.96	50.50	41.55	42.00
TiO_2	0.20	0.81	1.30	1.35	0.53	3.31	2.86
Al_2O_3	0.84	3.99	7.29	7.59	4.10	7.25	11.39
Fe_2O_3	1.88	2.51	4.28	4.46	2.44	6.80	5.27
FeO	11.87	9.84	9.08	9.46	7.37	7.77	10.30
MnO	0.21	0.21	0.25	0.26	0.13	0.20	0.24
MgO	43.16	34.02	24.31	25.32	21.17	13.02	12.35
CaO	0.75	3.46	7.15	7.45	12.00	16.93	11.31
Na_2O	0.31	0.56	1.06	1.10	0.45	1.38	1.80
K_2O	0.14	0.25	0.62	0.65	0.21	0.70	0.84
H_2O^+	0.44	0.76	3.67	–	0.47	0.50	1.31
P_2O_5	0.04	0.05	0.38	0.40	0.09	0.59	0.33
CO_2	–	–	0.34	–	–	–	–
or	0.6	1.7		3.6	1.1	–	5.0
ab	2.1	4.7		6.8	3.7	–	8.9
an	0.6	7.5		14.0	8.6	11.4	20.6
ne	0.3	–		1.4	–	6.5	3.4
lc	–	–		–	–	3.0	–
di *wo*	1.3	3.9		8.4	21.0	27.4	13.9
di *en*	1.0	3.0		6.3	15.7	21.8	9.6
di *fs*	0.1	0.5		1.2	3.2	2.4	3.2
hy *en*	–	11.8		–	24.7	–	–
hy *fs*	–	2.1		–	4.9	–	–
ol *fo*	74.8	49.1		39.9	9.7	7.6	14.8
ol *fa*	15.5	9.6		8.3	2.1	0.8	5.5
cs	–	–		–	–	1.1	–
mt	2.8	3.7		6.5	3.5	10.0	7.7
il	0.5	1.5		2.6	1.1	6.2	5.5
ap	0.1	0.1		1.0	0.2	1.3	0.8
DI	3.0	6.4		11.8	4.8	9.5	17.3

1. Dunite (average of 9 fresh rocks) (*Bull. Geol. Soc. Am.* **65**, 1023, no. 1).
2. Peridotite (average of 23 fresh rocks) (*Ibid.*, p. 1023, no. 1).
3. Alkali peridotite (average of 12) (*Ibid.*, p. 1023, no. II)*.

3a. Average alkali peridotite recalculated to 100, minus H_2O + and CO_2.

4. Pyroxenite (average of 46) (*Ibid.*, p. 1022, no. 1).
5. Alkali pyroxenite (average of 21) (*Ibid.*, p. 1022, no. II).
6. Hornblendite (average of 15) (*Ibid.*, p. 1022, no. 8).

*Excluding the special variety, mica peridotite.

well but, near the centre of the body, tremolite, mixed with talc and locally with serpentine, may occupy positions previously held by enstatite and diopside. *Picotite* is altered to magnetite marginally and along fractures. The final serpentinite is cut by late veins of cross-fibre chrysotile and of fine-grained talc and carbonate.

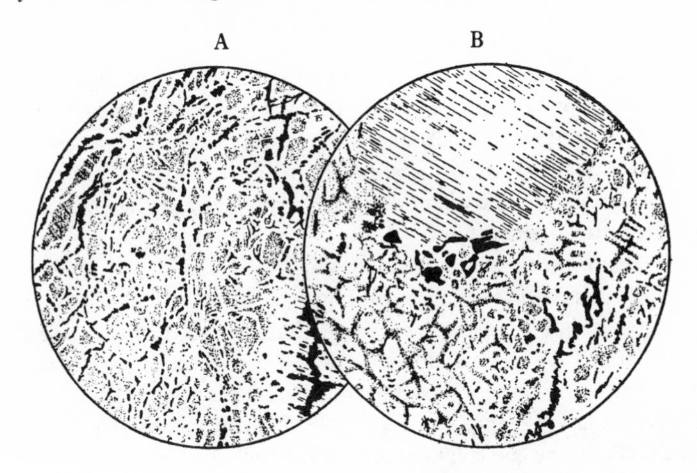

Fig. 13.3. Serpentinite, Lizard, Cornwall; × 20.

A. Cadgwith. Shows mesh structure due to derivation from olivine, and much secondary magnetite.

B. Balk Quarry. This was originally an enstatite peridotite with pseudoporphyritic texture (cf. fig. 13.1) and shows enstatite pseudomorphed by 'bastite'.

The Burro Mountain ultramafic mass is one of Alpine type and, as mentioned at the beginning of this chapter, Alpine type ultramafic bodies are often partially or completely serpentinised. It is clear that many of them have been emplaced tectonically and do not show intrusive contacts, and it has been shown that in the western United States the degree of serpentinisation is related to the tectonic history of each mass (23). Tectonic forces have caused many of these masses to move after their original emplacement, and those having a complex tectonic history are serpentinised completely. It seems likely that serpentinisation takes place gradually over a period of time in many ultramafic masses and may even be occurring at the present day through the agency of meteoric waters acting at low temperatures (24).

The presence of ferriferous brucite, and relative poverty in secondary magnetite, seems to characterise serpentinites derived from Alpine type peridotites and is thought to indicate a low temperature of serpentinisation.

On the other hand, the highly serpentinised peridotites dredged from the Mid Atlantic Ridge and those found sometimes in layered intrusions have much secondary magnetite and no brucite, and are thought to have been serpentinised at a higher temperature (25).

Perknites

Pyroxenites may be composed essentially of orthopyroxene, of clinopyroxene, or of both pyroxenes together. They are found as layers or lenses in layered mafic or ultramafic intrusions or as small masses, dykes or veins associated with mafic or ultramafic rocks. Their texture is commonly that of a mosaic of anhedral grains but the grain size is very variable, often even within the limits of a small outcrop.

Pyroxenites with orthopyroxene alone are found in a number of layered intrusions occurring, for instance, in the ultramafic zone of the Stillwater complex and in the Critical Zone of the Bushveld complex. In the latter (26), those at the base of the zone have orthopyroxene of composition Of_{14} to Of_{17} but the composition ranges from Of_{23} to Of_{33} in different layers higher up in the zone. They normally have chromite, a little clinopyroxene, olivine and interstitial plagioclase as accessory constituents.

Pyroxenites with clinopyroxene alone are rarer, but occur quite extensively in the Tulameen district, British Columbia, associated with dunite and with the augite peridotite already mentioned. While augite is the only essential constituent, magnetite with inclusions of green spinel is an abundant additional constituent in many of the rocks, forming a cement to the crystals of augite.

Pyroxenite with both ortho- and clinopyroxene ('websterite') is the commonest variety, found both in connection with some great layered igneous bodies (e.g. the Great Dyke, S. Rhodesia) and in small intrusions. The type rock, essentially an aggregate of chromiferous diopside, orthopyroxene and sparse disseminated chromite, comes from near Webster, N. Carolina (fig. 11.2B). Pyroxenite of this kind is an important component of the Cortlandt series of igneous rocks at Peekskill, Westchester Co., New York. Here, brown hornblende forming at the expense of augite may be an additional component becoming abundant in some rocks and building large poikilitic plates; these are *hornblende pyroxenites*.

Spinel pyroxenite ('ariègite') forms bands and veins in peridotite not only in the Suc Valley, Ariège, Pyrenees, but at numerous other localities in the region. Clinopyroxene with subordinate orthopyroxene is accompanied by considerable amounts of green spinel that may enclose the pyroxenes poikilitically, and there is often some pyrope garnet. Some occurrences are rich in garnet, while others have some brown hornblende. A number have a little interstitial plagioclase.

Olivine is present in a number of pyroxenites as an essential but subordinate constituent and these *olivine pyroxenites* form a transitional group between pyroxenite and peridotite. They are well represented at the various platinum centres in the Ural Mountains, U.S.S.R., where some of them contain small amounts of native platinum as irregular little patches between the pyroxene crystals. These form one component of zoned ultramafic complexes similar to those occurring in S.E. Alaska. In the latter, however, *hornblende pyroxenite* is also an important component (27).

Alkali pyroxenites are found associated with alkali rocks and have titanaugite, diopside-hedenbergite, or aegirine-augite as their pyroxene with or without an alkali amphibole.

Those having titanaugite are the commonest and the type rock ('jacupirangite') comes from the Jacupiranga district, São Paulo, Brazil. It consists of grains of titanaugite with abundant accessory iron-ore, some apatite, rare perofskite, a little nepheline, and variable amounts of dark mica. Almost identical rocks are found in the Ice River complex, British Columbia, and at Magnet Cove, Arkansas (28). Sphene and melanite garnet appear among the accessory constituents at both these localities, and the sphene may become abundant locally.

Part of the pyroxenite body at Iron Hill, Gunnison Co., Colorado (29), is made up of an alkali pyroxenite which has diopsidic pyroxene with relatively large amounts of perofskite and iron-ore. Sphene and apatite are prominent in places and there is sometimes a little interstitial nepheline.

The pyroxenite from the Salitre Mountains, Minas Geraes, Brazil, has zoned aegirine-augite and some 30 per cent of sphene with abundant accessory apatite and a little microcline.

Alkali pyroxenites of Yamaska type ('yamaskite') have both titaniferous augite and an alkali amphibole, usually a brown kaersutite, as essential constituents. The original locality is Mount Yamaska, Quebec, and similar rocks are found in connection with the other intrusions of the Monteregian Hills, notably at Mount Royal (30). Some have olivine as a further essential constituent (Rougemont Mt., Quebec; Oslo district, Norway; Serra de Monchique, Portugal).

Mica pyroxenites are of two kinds. One is associated with ordinary peridotites, pyroxenites, gabbros, etc., the other is associated with alkali peridotites, alkali pyroxenites, or other alkali rocks. Mica pyroxenites of the first kind are uncommon, but that found at Seeconnell, east end of the Newry complex, Co. Down, N. Ireland (31), in which a good deal of deep greenish brown hornblende is replacing the augite, may be cited as an example.

Mica pyroxenites of the second kind are represented by the rock forming part of the intrusion of alkali rocks at Libby, Montana (32), and that forming the most abundant and average type in the pyroxenite body at Iron Hill, Colorado. The mica in these is a phlogopitic variety, while the pyroxene is a diopsidic variety, perofskite, apatite and iron-ore are the typical accessories.

Most of the recorded mica pyroxenites of this group, however, occur as ejected blocks thrown out from volcanoes. Those from Monte Somma, Vesuvius, and from Villa Senni, Alban Hills, Italy, have usually a little interstitial leucite; that from Soccavo in the Phlegrean Fields carries some olivine; those from Katwe crater and from Lutale, Uganda, may be rich in sphene or in melanite garnet; while those from E. Java have practically nothing other than augite and mica.

The augite in all these is a greenish variety, and the mica a strongly pleochroic phlogopitic variety. All are more or less friable and porous, and some have a little interstitial glass. They are normally associated with leucite-bearing lavas and are to be regarded as subvolcanic rocks having some evident connection with the lavas.

Hornblendites occur as small masses, dykes or veins, in connection with larger intrusions of ultramafic, gabbroic, or more rarely, dioritic character. Their texture is commonly more or less granular but, like the pyroxenites, the grain size is variable. The hornblende may be brown, brownish green or green in section in different occurrences. Some consist almost entirely of hornblende; in others, the incoming of essential amounts of pyroxene, olivine, etc., marks a transition to hornblende pyroxenites, hornblende peridotites and related rocks.

The hornblendites of the Ariège district in the Pyrenees are connected with ultramafic rocks in which they form dykes and veins. The brown hornblende is accompanied, in some instances, by essential amounts of augite, of garnet, or of biotite. The hornblende is also brown in section in the main hornblendite of the Garabal Hill – Glen Fyne complex, Argyllshire, Scotland (33). Here it is clear that much of the amphibole has been formed at the expense of pre-existing pyroxene, and there are all transitions from hornblendite to pyroxenite.

The hornblende of the hornblendite grading into hornblende hypersthenite near Maracas, Bahia, Brazil, is olive green in section and the rock has accessory olivine, while the hornblendite forming veins in gabbro at East Sooke, Vancouver Island, British Columbia, has green hornblende (34). The latter occurrence is a reminder that not all hornblendite dykes and veins are intrusive. The majority of the veins here are only a few centimetres wide but range up to 80 metres, and they are replacement veins grading into unaltered gabbro with no definite walls.

Hornblendites composed of alkali amphibole seem to be rare and to occur only as a minor facies of alkali pyroxenite or related alkali rocks (e.g. Brandberget, Gran, Oslo district, Norway).

The name *glimmerite* is applied to those rocks whose dominant, or only, essential constituent is a dark mica. Such rocks are rare and found only as a local facies of other types, such as mica pyroxenite, or as ejected blocks from volcanoes.

The glimmerite of Iron Hill, Colorado (35), forms a minor facies of an intrusion of alkali pyroxenite and mica pyroxenite. Mica may make up almost the whole of the rock, but a more typical example has 65 per cent phlogopite, 15 per cent apatite, 11 per cent iron-ore and 9 per cent perofskite. In an almost identical occurrence at Libby, Montana (36), the mica has been altered to vermiculite.

Glimmerite is found as rare ejected blocks in a volcanic agglomerate at Villa Senni, Alban Hills, Italy, and as rare xenoliths in the kimberlites of Cape Province. Chromiferous diopside accompanies the mica in some of the latter.

Chemistry

Average chemical compositions of pyroxenite, alkali pyroxenite and hornblendite will be found in table 13.1, columns 4, 5, 6. Average pyroxenite has roughly 50 per cent of SiO_2, low Al_2O_3, high MgO and fairly high CaO, moderate iron, low TiO_2, very low alkalies and negligible P_2O_5. Average alkali pyroxenite, on the other hand, has lower SiO_2 and MgO coupled with higher iron, CaO and alkalies. Both TiO_2 and P_2O_5 are more prominent, a general feature of alkali mafic and ultramafic igneous rocks when compared with their subalkali analogues.

Note the presence of abundant normative hypersthene in average pyroxenite, its absence in average alkali pyroxenite, and the presence of normative nepheline and leucite in the latter.

The chemical composition of average hornblendite shows well the aluminous character of the hornblendes in these rocks. The hornblendites averaged were composed of common hornblende and are associated with other subalkali rocks. It is of interest, therefore, to see the under-saturated nature of average hornblendite as revealed by the presence of olivine, rather than hypersthene, and, more especially, of nepheline in the norm.

Pyroxenites, like other ultramafic rocks, have a very low differentiation index, but that of average hornblendite is appreciably higher.

Volcanic equivalents

There are, in general, no true *volcanic equivalents* of most peridotites and perknites but Russian petrologists have found lavas (*meimechites*) which they regard as the extrusive representatives of kimberlite, and the chemical and original mineral compositions of the peridotitic komatiites already mentioned (p. 126) are comparable with those of augite peridotites or, more commonly perhaps, those of olivine–augite pyroxenite (37).

These *peridotitic komatiites* deserve further notice. Although some may represent more or less contemporaneous sills, there is good evidence that others are true lavas, a number of which show pillow structure. Like their basaltic counterparts, they have been subjected to a varying degree of metamorphism but were composed originally of forsteritic olivine, aluminous diopsidic pyroxene, iron-ore with or without accompanying chromite, and varying amounts of glass.

One of their most remarkable features is the occurrence of a quench texture known as *spinifex texture* (38). This consists, in its commonest form, of bundles of long (sometimes up to several centimetres) sub-parallel lamellae of skeletal olivine, or its alteration products, cut at varying angles by other bundles. The areas between the lamellae consist of skeletal acicular clinopyroxene, and dendritic magnetic sometimes with skeletal chromite as well, embedded in what was originally glass now usually represented by chlorite.

Chilled contacts may show a microspinifex texture with blades of olivine embedded in glass and, in the thicker flows, this is followed by a spinifex zone, fine at the top and coarse at the bottom, and then a zone, which may form the bulk of the flow, of medium to fine-grained rock with some 70 per cent or more of elongated or equant olivine, clinopyroxene, and interstitial glass.

References and notes

(1) Moores, E. M. & Vine, F. J., 1971. The Troodos Massif, Cyprus and other ophiolites as oceanic crust: evaluation and implications. *Phil. Trans. Roy. Soc. London.* A **268**, 443–66.

(2) Brothers, R. N., 1959. Flow orientation of olivine. *Am. J. Sci.* **257**, 574–84.

(3) Such high contents of Al_2O_3 in orthopyroxene have been taken as indicating crystallisation under considerable pressure, but recent work has shown that solubility of Al_2O_3 in enstatite may be mainly a function of temperature. Anastasiou, P. & Seifert, F., 1972. *Contr. Min. Petr.* **34**, 272–87.
(4) Tilley, C. E., 1947. The dunite-mylonites of St. Paul's Rocks (Atlantic). *Am. J. Sci.* **245**, 488–91.
(5) Harker, A., 1904. The Tertiary igneous rocks of Skye. *Mem. geol. Surv. U.K.*, 71–2.
(6) Challis, G. A., 1965. The origin of New Zealand ultramafic intrusions. *J. Petrology* **6**, 322–64.
(7) Hess, H. H., 1960. Stillwater igneous complex, Montana. *Mem. geol. Soc. Am.* **80**, 56–60.
(8) Harker, 1904, pp. 63–73.
(9) Camsell, C., 1913. Geology and mineral deposits of the Tulameen district, B.C. *Mem. geol. Surv. Canada* **26**, 49–58.
(10) Green, D. H., 1964. The petrogenesis of the high-temperature peridotite intrusion in the Lizard area, Cornwall. *J. Petrology* **5**, 134–88.
(11) Littlejohn, A. L. & Greenwood, H. J., 1974. Lherzolite nodules in basalts from British Columbia, Canada. *Can. J. Earth Sci.* **11**, 1288–1308.
(12) Mercier, J.-C. C. & Nicolas, A., 1975. Textures and fabrics of upper mantle peridotites as illustrated by xenoliths from basalts. *J. Petrology* **16**, 454–87.
(13) O'Hara, M. J. & Mercy, E. L. P., 1963. Petrology and petrogenesis of some garnetiferous peridotites. *Trans. Roy. Soc. Edinb.* **65**, 251–314.
(14) Tyrrell, G. W., 1917. The Picrite–Teschenite sill of Lugar (Ayrshire). *Q. Jl. geol. Soc. Lond.* **72**, 111.
(15) Flett, J. S., 1907, *In* The geology of the country around Plymouth and Liskeard, *Mem. geol. Surv. U.K.* 101. Dewey H., 1911. *In* The geology of the country around Tavistock and Launceston. *Mem. geol. Surv. U.K.*, 63–4.
(16) Buie, B. F., 1941. Igneous rocks of the Highwood Mountains, Montana: part III. Dikes and related intrusions. *Bull. geol. Soc. Am.* **52**, 1800–1.
(17) Mitchell, R. H., 1970. Kimberlite and related rocks – a critical reappraisal. *J. Geol.* **78**, 686–704. It should be noted that the phenocrysts of garnet mentioned here, and some of those of olivine, are regarded by others as xenocrysts derived from the disruption of xenoliths of garnet peridotite (see Dawson, J. B., 1972. Kimberlites and their relation to the mantle. *Phil. Trans. Roy. Soc. Lond.* A, **271**, 297–311).
(18) An excellent study of the ultramafic nodules from a single kimberlite pipe will be found in Boyd, F. R. & Nixon, P. H., 1972. Ultramafic nodules from the Thaba Putsoa kimberlite pipe. *Yb. Carnegie Inst. Wash.* **71**, 362–73.
(19) Dawson, J. B. & Hawthorne, J. B., 1973. Magmatic sedimentation and carbonatitic differentiation in kimberlite sills at Benfontein, South Africa. *Q. Jl. geol. Soc. Lond.* **129**, 61–85.
(20) *E.g.* Dawson, J. B., 1962. Basutoland kimberlites. *Bull. geol. Soc. Am.* **73**, 545–60. Grantham, D. R. & Allen, J. B., 1960. Kimberlite in Sierra Leone. *Overseas Geol. Miner. Res.* **8**, 5–25.
(21) *E.g.* Miser, H. D. & Ross, C. S., 1922. Diamond bearing peridotite in Pike County, Arkansas. *Ec. Geol.* **17**, 662–74.

(22) Page, N. J., 1967. Serpentinisation at Burro Mountain, California. *Contr. Miner. Petrol.* **14**, 321–42.
(23) Coleman, R. G., 1967. Low-temperature reaction zones and Alpine ultramafic rocks of California, Oregon and Washington. *Bull. U.S. geol. Surv.* **1247**, 4–8.
(24) Barnes, I. & O'Neil, J. R., 1969. The relationship between fluids in some fresh Alpine-type ultramafics and possible modern serpentinisation, western United States. *Bull. geol. Soc. Am.* **80**, 1947–60.
(25) Aumento, F., 1970. Serpentine mineralogy of ultrabasic intrusions in Canada and on the Mid-Atlantic Ridge. *Pap. geol. Surv. Canada* **69–53**.
(26) McDonald, J. A., 1967. Evolution of part of the lower critical zone, Farm Ruighoek, western Bushveld. *J. Petrology* **8**, 169–72.
(27) Ruckmick, J. C. & Noble, J. A., 1959. Origin of the ultramafic complex at Union Bay, south eastern Alaska. *Bull. geol. Soc. Am.* **70**, 981–1018. Taylor, H. P. & Noble, J. A., 1960. Origin of the ultramafic complexes in south eastern Alaska. *Internat. geol. Congr.* XXI *Session, Norden*, Part **13**, 175–87.
(28) Erickson, R. L. & Blade, L. V., 1963. Geochemistry and petrology of the alkalic igneous complex at Magnet Cove, Arkansas. *Prof. Pap. U.S. geol. Surv.* **425**, 17–20.
(29) Larsen, E. S., 1942. Alkalic rocks of Iron Hill, Gunnison County, Colorado. *Ibid.* **197**-A, 13–22.
(30) Bancroft, J. A. & Howard, W. V., 1923. The essexites of Mount Royal, Montreal, P.Q. *Trans. Roy. Soc. Canada* Ser. 3, **17**, 13–43 especially 29–30.
(31) Reynolds, D. L., 1934. The eastern end of the Newry igneous complex. *Q. Jl. geol. Soc. Lond.* **90**, 606–9.
(32) Larsen, E. S. & Pardee, J. T., 1929. The stock of alkaline rocks near Libby, Montana. *J. Geol.* **27**, 99–104.
(33) Nockolds, S. R., 1941. The Garabal Hill–Glen Fyne igneous complex. *Q. Jl. geol. Soc. Lond.* **96**, 454–5.
(34) Cooke, H. C., 1919. Gabbros of East Sooke and Rocky Point. *Geol. Surv. Canada Mus. Bull.* **30**, 5 and 12–13.
(35) Larsen, 1942, p. 14.
(36) Larsen, E. S. & Pardee, J. T., 1929, p. 99.
(37) See, for example, Viljoen, M. J. & Viljoen, R. P., 1969. Evidence for the existence of a mobile extrusive peridotitic magma from the Komati Formation of the Onverwacht Group. *Geol. Soc. S. Afr. Spec. Publ.* **2**, 87–112. Pyke, D. R., Naldrett, A. J. & Eckstrand, O. R., 1973. Archean ultramafic flows in Munro Township, Ontario. *Bull. geol. Soc. Am.* **84**, 955–77.
(38) Good illustrations of spinifex textures can be found in Nesbit, R. W., 1971. Skeletal crystal forms in the ultramafic rocks of the Yilgarn Block, western Australia: evidence for an Archaean ultramafic liquid. *Geol. Soc. Austral. Spec. Publ.* **3**, 331–47.

14

Nepheline syenites and allied rocks

Nepheline syenites (1) are medium to coarse-grained igneous rocks consisting essentially of alkali feldspar and nepheline, normally with one or more ferromagnesian minerals in addition. They are typically leucocratic with an average colour index of 16, and with the greasy-looking nepheline easily distinguishable in hand specimen. A number, however, are mesotype or even melanocratic.

They occur mostly as stocks or ring-dykes either alone or associated with ultra-alkaline rocks or with peralkaline and alkali syenites. Intrusions in which they are found are usually small, though occasionally comparatively large, as in the case of the Khibina and Lovozero complexes in the Kola Peninsula, U.S.S.R.

Mineralogy

Alkali feldspars that may be present include orthoclase, microcline, orthoclase-microperthite and microcline-microperthite, cryptoperthite and albite. The typical 'cross-hatching' is often absent in the microcline here, and there is either a fine twinning on the albite law, or the mineral is untwinned.

Hypersolvus nepheline syenites have a microperthite or cryptoperthite as their only feldspar; subsolvus nepheline syenites have orthoclase or microcline together with independent albite, and there are also transitional types in which microperthite is accompanied by some independent albite which, in the early stages, may form a fringe to the microperthite crystals.

Nepheline sometimes shows good euhedral or subhedral form (fig. 14.2A) or is wedged between the feldspar crystals, or may even be found intergrown with the feldspar. It has been shown that the composition of the nepheline is different in hypersolvus and subsolvus nepheline syenites (2). In hypersolvus nepheline syenites the composition of the nepheline depends on the bulk composition of the rock, whereas in subsolvus nepheline syenites the nepheline is restricted in composition to that of the Morozewicz-Buerger convergence field.

Nepheline is accompanied in many rocks by *sodalite*, either appearing as euhedral dodecahedra, or anhedral and interstitial. More rarely, *nosean* or *hauyne* may be present, both usually euhedral. Some rocks have some *analcite*, and others have *cancrinite* in small hexagonal prisms or irregular grains.

The pyroxene is normally a green *aegirine-augite* or *aegirine*; more rarely, yellowish *acmite*. Some rocks have *sodic hedenbergite* or, in more mafic types, *titanaugite*.

The common alkali amphiboles that may occur are *ferrohastingsite*, *arfvedsonite* and *barkevikite*. Note the usual absence here of riebeckite, common in peralkaline granites and syenites.

In some nepheline syenites both pyroxene and amphibole are subhedral, prismatic, but in others these minerals are interstitial to the feldspar and nepheline.

Mica, when present, is typically a deep brown or green, strongly pleochroic iron-rich variety. Muscovite is found in some nepheline syenites but is usually due to alteration of the rock and is not a primary constituent. However, muscovite, in part intergrown with dark mica and therefore believed to be primary, occurs in the mica–nepheline syenite of the Ditro stock, Hungary.

Fayalite is a constituent of certain nepheline syenites; *melanite garnet*, which may be zoned and may be euhedral, or anhedral and interstitial, is present in some rocks and may become important.

Among the *accessory minerals*, sphene is very common; apatite and zircon are variable in amount; iron-ores (ilmenite, titaniferous magnetite, and magnetite) are usually only sparingly present. Sulphides are sometimes to be seen in small quantity, the commonest being pyrite, though pyrrhotite, molybdenite, zinc blende and galena have been recorded. Fluorite is widely distributed; calcite, apparently primary, is present in many rocks. There is also a whole host of rarer accessory minerals occurring at individual localities, many of which are titano- and zirconosilicates. Of these, eudialyte is one of the commonest and this mineral becomes a noticeable constituent of some nepheline syenites.

Textures and structures

Many nepheline syenites have a subhedral-granular ('granitic') texture, but in others the feldspar has a marked tabular habit, and the more or less parallel arrangement of these tabular crystals gives rise to a trachytoidal texture. In certain mesotype varieties that have an abundance of acicular crystals of aegirine ('lujavrites'), these and the thin tabular crystals of feldspar both have a parallel arrangement, resulting in a marked fissility or schistosity. A number of nepheline syenites are porphyritic with phenocrysts of feldspar.

Some intrusions of nepheline syenite have a layered structure (3).

Main varieties with examples

Nepheline syenites are very variable rocks and, as a consequence, many different types have been distinguished and named. Some of the names (e.g. 'foyaite', 'ditroite') have been used in more than one sense and many of them are unnecessary. Here, nepheline syenites will be divided into those with microperthite (or cryptoperthite) as the only feldspar (hypersolvus type), those with independent potash feldspar and albite (subsolvus type), those with potash feldspar only (Juvet type) and those with albite only (Mariupol type).

The so-called 'lardalite' of the Oslo district, Norway, is a ***hypersolvus nepheline syenite*** with cryptoperthitic to microperthitic feldspar and nepheline that is partly euhedral against the feldspar, partly interstitial (fig. 14.1). The main pyroxene is aegirine-augite sometimes rimmed with aegirine, but pale violet titaniferous augite is not uncommon. Brown,

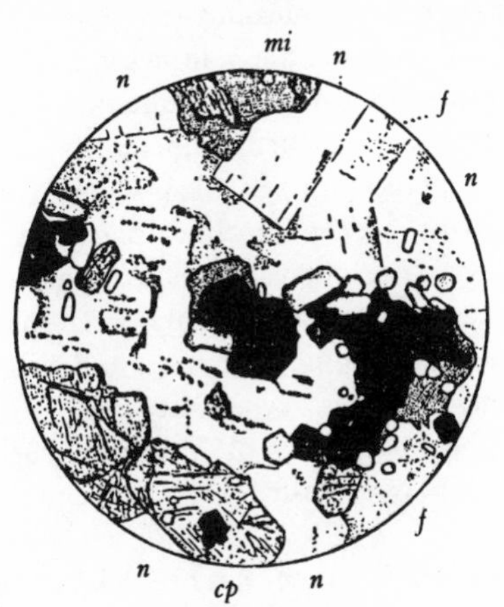

Fig. 14.1. Nepheline syenite ('lardalite'), Lunde, near Larvik, S. Norway; × 20.

Clinopyroxene (*cp*), dark mica (*mi*), cryptoperthitic feldspar (*f*), nepheline (*n*) and abundant accessory apatite and iron-ore.

iron-rich mica, rich in inclusions of apatite and iron-ore, is often poikilitic, and occurs also as rims round grains of iron-ore. Barkevikite is present in some specimens, and there is usually a little sodalite, sometimes some iron-rich olivine.

In the nepheline syenites from Monmouth township, Haliburton–Bancroft province, Ontario (4), microperthite and nepheline are associated with green sodic hedenbergite, subordinate ferrohastingsite, and accessory fayalite, magnetite and apatite.

The 'miaskites' of the Ilmen Mountains, S. Urals, U.S.S.R., are banded and foliated but the normal rock has microperthite and nepheline with some 10 per cent of iron-rich mica and accessory constituents that include zircon, ilmenite, apatite, pyrochlore and calcite. Mica gives place to ferrohastingsite in certain types; these have a higher colour index and carry accessory sphene.

The well-known nepheline syenites of the Serra de Monchique, Portugal, may have microperthite alone or associated with subordinate independent albite, marking a transition to the subsolvus type. The texture of these coarse-grained rocks varies from trachytoidal to granitic and the nepheline is anhedral, moulding the feldspar. Green aegirine-augite is in some cases partnered by a little soda amphibole and iron-rich mica, and accessory euhedral sphene is relatively abundant.

The dominant type of nepheline syenite in the equally well-known Ditro stock, Hungary, has microcline-microperthite and minor albite, and dark brown or dark green mica in varying amount. Normally some cancrinite and muscovite, sometimes a little sodalite, and calcite (associated with the cancrinite) are present in addition.

The *subsolvus nepheline syenite* of Litchfield, Maine, U.S.A. (5), forms schlieren within, and grades into, an alkali syenite. It has albite, partly antiperthitic, dominant over microcline, partly perthitic; a variable amount of nepheline with a composition close to that of the Morozewicz-Buerger convergence field; and a little dark green mica. Cancrinite and sodalite are disseminated through the rock, and magnetite and zircon are prominent among the accessory constituents.

Some of the nepheline syenites of Malagasy, such as those in the vicinity of Anosikely, have cross-hatched microcline and albite as their feldspars.

The *Juvet type of nepheline syenite* has orthoclase or microcline instead of microperthite as the only feldspar. Rocks of this kind are found in connection with some of the alkali igneous rock complexes of eastern and southern Africa, as at Nkalonje Hill, south of Lake Chilwa, Nyasaland (now Malawi) (6). Here, orthoclase slightly exceeds nepheline in amount and was always the last mineral to cease crystallising. The ferromagnesian minerals are euhedral aegirine-augite (with, in one example, cores of titaniferous augite), and flakes of dark mica. There is abundant accessory sphene with some apatite and magnetite; cancrinite, replacing nepheline, occurs in some, and calcite, in poikilitic plates, may be present also.

The only British nepheline syenites, those of the Assynt district, Scotland (7), are largely orthoclase-, or microcline-, bearing types in which the nepheline, nearly always altered, may be euhedral (fig. 14.2A)

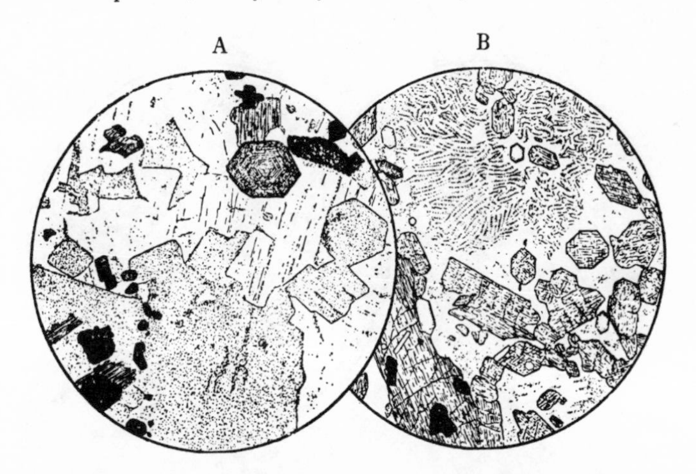

Fig. 14.2. Nepheline syenites, Assynt, Scotland; × 20.

A. Shows altered euhedral nepheline, orthoclase, biotite, deep brown melanite garnet and iron-ore.

B. Elphin Bridge. Shows dactylitic intergrowth of nepheline (altered) and orthoclase, aegirine-augite, and accessory apatite and iron-ore.

or, more commonly, anhedral and interstitial, or in dactylytic intergrowth with the feldspar (fig. 14.2B). Melanite garnet, in crystals or irregular grains, is almost constantly present and is often accompanied by green mica, but aegirine-augite, sometimes zoned, may occur also. One variety has about equal amounts of orthoclase and nepheline with only a little melanite and garnet. Others have dark minerals in greater abundance, pyroxene and garnet varying inversely.

The mesotype to melanocratic nepheline syenites occurring as laccoliths, sills, and dykes in the northern area of the Crazy Mountains, Montana (8), have barium-orthoclase as their feldspar, usually anhedral and tending to enclose the other constituents. Prisms of clinopyroxene, zoned from diopsidic augite to aegirine-augite and commonly bordered by aegirine (fig. 14.3B), are prominent among the ferromagnesian minerals and, in most rocks, there is a little olivine.

Nepheline syenites of Mariupol type have albite as their only feldspar and the best-known examples are from the original locality, Mariupol, Sea of Azov, U.S.S.R. Aegirine is the typical ferromagnesian mineral, but lepidomelane is present in some rocks. Zircon is a characteristic accessory constituent and may be abundant, pyrochlore and fluorite are usually present, sometimes accessory cancrinite and potash feldspar as well.

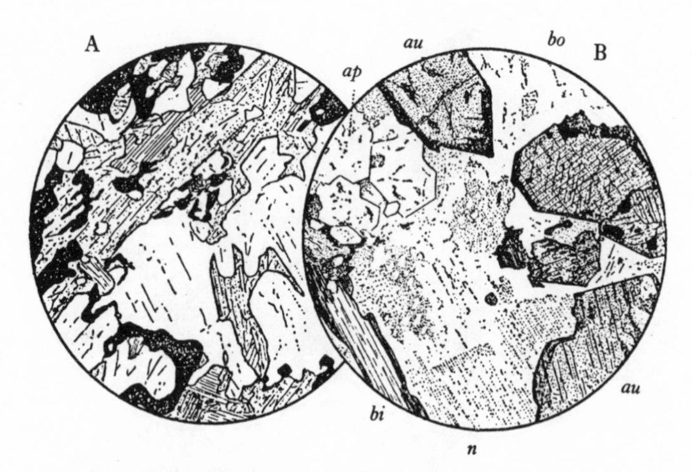

Fig. 14.3. × 20.

A. Ijolite, Magnet Cove, Arkansas, U.S.A. Composed of nepheline, clinopyroxene, some pale biotite and irregular deep brown melanite garnet.

B. Nepheline syenite, Gordon's Butte, Crazy Mts., Montana, U.S.A. More or less anhedral barium-orthoclase (*bo*), nepheline (*n*), zoned augite (*au*) bordered by aegirine, pale biotite (*bi*) with a deep brown border, and noticeable accessory apatite (*ap*).

An 'albite foyaite' occurs to a small extent at Sarna, Sweden, with dominant subhedral nepheline, albite as divergent or subparallel laths, and aegirine needles of varying size, mostly lying in the nepheline.

'Nepheline syenites' with albite (or sometimes oligoclase) as the only essential feldspar, and carrying ferrohastingsite or dark mica, sometimes clinopyroxene and garnet, are widely distributed in parts of the province of Ontario, Canada. Though described originally as igneous rocks, evidence has accumulated that these are metasomatic in origin and a product of nephelinisation (9).

Many nepheline syenites have a little cancrinite, usually formed at a late stage and at the expense of nepheline. Sometimes, however, cancrinite occurs in some quantity as a primary mineral, giving rise to a *cancrinite–nepheline syenite* as at Sarna, Sweden, where cancrinite as elongated prisms or rounded grains is about as abundant as nepheline.

Again, a member of the sodalite group of minerals may become an important constituent or, more rarely, occur to the exclusion of nepheline. *Sodalite–nepheline syenite* is found, for instance, as small veins and patches in the main nepheline syenite of the Ditro stock, Hungary, and *sodalite syenite*, with sodalite greatly predominant over nepheline, has been described from Tupersuatsiak, S. Greenland. Among the alkali igneous rocks of the Los Archipelago, there are sodalite–nepheline syenites, *hauyne–nepheline syenites* and *hauyne syenites*; the 'assyntite' of the Assynt district, Scotland, is a *nosean–nepheline syenite* with nosean about twice as abundant as nepheline (10), and a *nosean syenite* with no nepheline but some sodalite forms part of a small stock-like mass on the north slope of Hsi Kuang T'a Men, Shansi, N. China.

Analcite syenites have a different paragenesis and are found mostly as a minor component in sills and dykes associated with analcite-bearing dolerites, teschenites and similar rocks as, for instance, at South Park, Colorado (11). The syenites here are vuggy rocks with crystals of potash feldspar, sometimes a little albite, abundant interstitial analcite, and with the feldspar replaced to a varying degree by analcite, thomsonite and natrolite. The main dark minerals are amphibole and biotite with minor green augite margined by aegirine-augite. Accessory apatite and iron-ore are relatively abundant and there may be a little olivine.

True *leucite syenites* are known only as subvolcanic rocks in the form of ejected blocks from volcanoes that have erupted leucitic lavas as, for instance, Vesuvius (Monte Somma). The rocks here are usually coarse grained with sanidine and leucite, often phenocrystal in habit, and varying amounts of melanite garnet and amphibole. Apatite is an accessory constituent, and biotite and sphene may be present as additional phases.

Leucite is not found in igneous rocks that have crystallised at depth. There are, however, rocks that have been described as *pseudoleucite syenites*, for example the well-known rock that forms a nearly complete ring in the alkali complex at Magnet Cove, Arkansas (12). The pseudoleucite crystals, varying in abundance, range from one to five centimetres in diameter, and are composed of a fine-grained aggregate of potash feldspar and nepheline that has kept the icositetrahedral form of the original mineral. The remainder of the rock consists essentially of tabular crystals of sodic orthoclase or microperthite, subhedral to anhedral nepheline, sodic pyroxene and melanite garnet. Mention may also be made here of the pseudoleucite 'borolanites' of the Borralan complex, Assynt, Scotland, which have melanite garnet and green biotite as ferromagnesian constituents (13) (fig. 14.4).

A number of nepheline (and analcite) syenites have subordinate lime-bearing plagioclase as an additional constituent and are transitional to nepheline (or analcite) monzonites. The nepheline syenite of Pouzac in

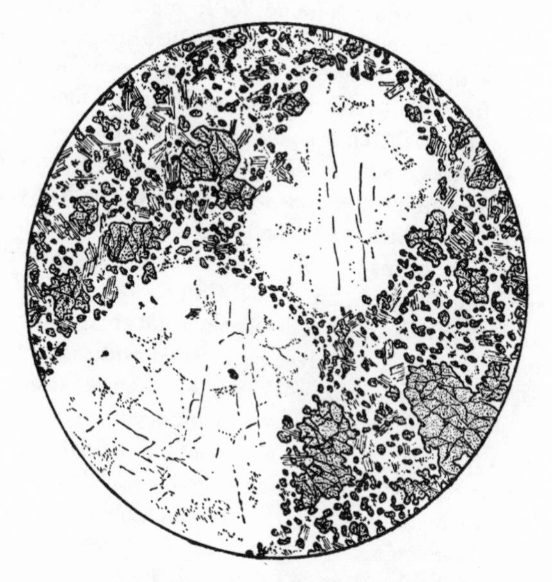

Fig. 14.4. Pseudoleucite syenite (Borolan type), Borralan complex, Assynt, Scotland; × 5.
With rather rounded pseudoleucites, consisting of orthoclase. The ferromagnesian constituents are green biotite and melanite garnet.

the Pyrenees has some andesine, while in the analcite syenites forming the central parts of sills in the San Rafael Swell district, east-central Utah (14), labradorite associated with soda-orthoclase may make up almost a third of the total feldspar content.

Chemistry

Average hypersolvus and subsolvus nepheline syenite (table 14.1, column 1) is characterised by moderate SiO_2, high Al_2O_3, high alkalies (with $Na_2O > K_2O$), low MgO and relatively low CaO. Average nepheline syenite of Juvet type (column 2) is similar except that the position of the alkalies is reversed, with $K_2O > Na_2O$. Note the very high Na_2O and

TABLE 14.1. *Average chemical compositions of nepheline syenites and phonolites*

	1	2	3	4	5	6	7
SiO_2	55.23	53.11	56.02	56.90	53.56	53.70	55.75
TiO_2	0.65	0.83	0.40	0.59	0.75	0.63	0.74
Al_2O_3	21.20	21.39	21.60	20.17	21.29	19.09	20.36
Fe_2O_3	2.46	2.54	3.15	2.26	2.68	2.62	2.22
FeO	2.20	1.59	0.99	1.85	1.04	2.98	2.04
MnO	0.22	0.12	0.17	0.19	0.17	0.17	0.11
MgO	0.52	0.72	0.19	0.58	0.73	0.47	1.07
CaO	1.92	2.89	1.04	1.88	2.88	2.57	3.66
Na_2O	9.24	5.98	12.59	8.72	9.57	8.81	3.71
K_2O	5.03	9.23	2.15	5.42	4.96	5.47	9.55
H_2O^+	0.89	0.79	0.82	0.96	1.18	3.31	0.60
P_2O_5	0.16	0.31	nil	0.17	0.28	0.17	0.19
CO_2	0.13	0.26	0.24	–	–	–	–
Cl	0.12	0.07	0.10	0.23	0.09	–	–
SO_3	0.06	0.18	–	0.13	0.87	–	–
F	–	–	0.18	–	–	–	–
ZrO_2	–	–	0.46	–	–	–	–
or	29.5	54.5	12.2	31.7	29.5	32.8	56.2
ab	33.0	4.7	46.1	36.2	27.8	18.3	10.0
an	2.2	5.0	–	1.7	3.6	–	10.8
ne	23.9	24.1	29.0	18.7	25.6	26.4	11.6
di { *wo*	2.3	2.4	0.9	2.9	2.1	5.0	2.6
di { *en*	1.1	1.8	0.5	1.4	1.8	1.1	1.8
di { *fs*	1.2	–	0.4	0.9	–	4.2	0.5
ol { *fo*	0.1	–	–	–	–	0.1	0.6
ol { *fa*	0.1	–	–	–	–	0.2	0.2
wo	–	–	0.3	–	1.6	–	–
ac	–	–	5.5	–	–	6.5	–
mt	3.7	3.2	1.9	3.3	1.6	0.5	3.3
il	1.2	1.5	0.8	1.2	1.4	1.2	1.4
ap	0.3	0.7	–	0.4	0.7	0.4	0.4
cc	0.3	0.6	0.6	–	–	–	–
hl	0.1	0.1	0.1	0.4	0.1	–	–
th	0.1	0.3	–	0.3	1.6	–	–
fl	–	–	0.3	–	–	–	–
hm	–	0.3	0.7*	–	1.6	–	–
DI	86.4	83.3	87.3	86.6	84.5	77.5	77.8

* *zr*

1. Hypersolvus and subsolvus nepheline syenite (average of 46).
2. Nepheline syenite of Juvet type (average of 10) (*Bull. Geol. Soc. Am.* **65**, 1024, no. I).
3. Nepheline syenite of Mariupol type (average of 4 from Mariupol).
4. Phonolite (average of 47) (*Ibid.*, p. 1024, no. III).
5. Hauyne phonolite (average of 3).
6. Analcite phonolite (average of 5).
7. Leucite trachyte (average of 4).

low K_2O in nepheline syenite of Mariupol type (column 3), and the high differentiation index shown by all nepheline syenites.

Later alteration

It is mainly the light constituents of nepheline syenites that are susceptible to hydrothermal alteration, and many rocks show some of the effects of this. Nepheline, in particular, may be partly or wholly replaced by cancrinite or cancrinite and muscovite; by sodalite; by analcite; by natrolite (15) or natrolite and muscovite; by muscovite, muscovite and calcite, or muscovite and kaolinite.

Sodalite may be altered to analcite, to natrolite with or without calcite, or, less commonly, to cancrinite. Potash feldspar may be sericitised or it may be replaced, at least to some extent, by sodalite, analcite, or cancrinite.

Pyroxenes and amphiboles do not normally appear to be affected, but the dark mica may be converted to chlorite, melanite garnet to sphene, and eudialyte to catapleiite and zircon or to catapleiite and fluorite.

Nepheline syenite-aplites and nepheline syenite-pegmatites

Nepheline syenite-aplites are not common when compared with nepheline syenite-pegmatites, but those occurring as thin dykes cutting nepheline syenite and essexite at Nosy Komba, Malagasy, appear to be good examples. They have a uniform microgranular texture and are composed of potash feldspar and nepheline with a very little aegirine.

Nepheline syenite-pegmatites may be present as segregations within nepheline syenite, passing gradually into the latter, or as veins and dykes cutting the parent rock and the surrounding country rock. Like granite- and syenite-pegmatites, they may be simple or complex.

Simple nepheline syenite-pegmatites have alkali feldspar and nepheline with or without sodalite and analcite, together with some aegirine, sodic amphibole or lepidomelane in different occurrences. Some of those carrying aegirine are very rich in eudialyte (e.g. Khibina, Kola Peninsula, U.S.S.R.), while others have large plates of astrophyllite. Accessory minerals that may be present in noticeable amounts include melanite garnet, fluorite, zircon, sphene and titaniferous magnetite.

Complex nepheline syenite-pegmatites that have undergone various stages of hydrothermal replacement are often the repository of a large number of rare minerals. Here belong the famous nepheline syenite-pegmatites of the Langesundfjord area, in the Oslo district of Norway, and many of those occurring in connection with the Khibina and Lovozero complexes, Kola Peninsula, U.S.S.R.

The kind of changes that occur in complex nepheline syenite-pegmatites are well shown by the complex pegmatites of the Rocky Boy stock, Bearpaw Mountains, Montana (16), where both simple and complex pegmatites are present.

References and notes

(1) A general survey of nepheline syenites (and phonolites) will be found in Allen, J. B. & Charsley, T. J., 1968. *Nepheline-syenite and phonolite.* Inst. Geol. Sci. Miner. Resources Division. 169 pp.

(2) Tilley, C. E., 1958*a*. Problems of alkali rock genesis. *Q. Jl. geol. Soc. Lond.*, **113**, 325–7.

(3) Sörenson, H., 1969. Rhythmic igneous layering in peralkaline intrusions. *Lithos* **2**, 261–83.

(4) Tilley, C. E. & Gittins, J., 1961. Igneous nepheline-bearing rocks of the Haliburton–Bancroft province of Ontario. *J. Petrology* **2**, 38–48.

(5) Barker, D. S., 1965. Alkalic rocks at Litchfield, Maine. *Ibid.* **6**, 1–27.

(6) Dixey, F., Smith, W. Campbell & Bisset, C. B., 1955. The Chilwa series of southern Nyasaland. *Bull. geol. Surv. Nyasaland* **5** (revised), 33–4.

(7) Shand, S. J., 1939. Loch Borolan laccolith, north-west Scotland. *J. Geol.* **47**, 408–20, and references given there. Tilley, C. E., 1958*b*. Some new chemical data on assemblages of the Assynt alkali suite. *Trans. geol. Soc. Edinb.* **17**, 156–64.

(8) Wolff, J. E., 1938. Igneous rocks of the Crazy Mountains, Montana. *Bull. geol. Soc. Am.* **49**, 1604–8.

(9) Tilley, 1958*a*, pp. 341–57, and references cited there. Gittins, J., 1961. Nephelinisation in the Haliburton–Bancroft district, Ontario, Canada. *J. Geol.* **69**, 291–308.

(10) Tilley, 1958*b*, p. 159.

(11) Jahns, R. H., 1938. Analcite-bearing intrusives from South Park, Colorado. *Am. J. Sci.* (5) **36**, 17–25. For some British examples see Tyrrell, G. W., 1928. Dolerite-sills containing analcite-syenite in central Ayrshire. *Q. Jl. geol. Soc. Lond.* **84**, 540–67.

(12) Erickson, R. L. & Blade, L. V., 1963. Geochemistry and petrology of the alkalic igneous complex at Magnet Cove, Arkansas. *Prof. Pap. U.S. geol. Surv.* **425**, 10–15.

(13) Woolley, A. R., 1973. The pseudoleucite borolanites and associated rocks of the south-eastern tract of the Borralan Complex Scotland. *Bull. Brit. Mus. (Nat. Hist.) Mineral.* **2**, 287–333.

(14) Gilluly, J., 1927. Analcite-diabase and related alkaline syenites from Utah. *Am. J. Sci.* (5) **14**, 199–211.

(15) The alteration product of nepheline, sodalite, etc. called 'hydronephelite' has been shown to be largely natrolite (Edgar, A. D., 1965. The mineralogical composition of some nepheline alteration products. *Am. Miner.* **50**, 978–89).

(16) Pecora, W. T., 1942. Nepheline syenite pegmatites, Rocky Boy stock, Bearpaw Mountains, Montana, *Am. Miner.* **27**, 397–424.

15

Phonolites, related volcanics and their hypabyssal representatives

The phonolites are the volcanic representatives of the nepheline syenites and may be light coloured or dark in hand specimen, usually with a greasy lustre due to the presence of nepheline. Some have a fissile character and ring under the hammer when struck, hence the name given to them. Most are porphyritic, the phenocrysts lying in a very fine-grained aphanitic matrix.

Mineralogy

The alkali feldspar of these rocks may be *sanidine*, normally fairly rich in soda, or *anorthoclase*, occurring as little laths in the groundmass and frequently also as phenocrysts. The phenocrysts may show zoning.

Nepheline is commonly confined to the groundmass where it builds little euhedral crystals yielding hexagonal or rectangular sections. More rarely, it is present as phenocrysts and, in some phonolites, these show marked zoning (e.g. Brüx, Bohemia, Czechoslovakia). Nepheline is frequently accompanied by related minerals. The commonest is perhaps *nosean*, with its characteristic dark border, but *hauyne* and *sodalite* are also found, as is *analcite*. All have a strong tendency to be euhedral and phenocrystal in habit though they may occur in the groundmass as well.

The common occurrence of nosean and hauyne in phonolites marks a contrast with the nepheline syenites where these minerals are rare. On the other hand, cancrinite, common in nepheline syenites, is not found in phonolites.

The characteristic ferromagnesian constituent is *aegirine* or *aegirine-augite* forming little prisms or needles in the groundmass or occurring interstitially, but present also as phenocrysts in some rocks. In some instances these phenocrysts may have a core of titaniferous augite. *Soda-amphiboles* are less frequent but, when present, may be found in the groundmass as well as among the phenocrysts. A brown amphibole, probably kaersutite, is found as phenocrysts in some phonolites.

Less strongly peralkaline rocks may carry a slightly sodic *augite* and *fayalitic olivine*.

Melanite garnet is not uncommon in some of the more peralkaline phonolites. It is euhedral, phenocrystal, and often zoned. A rarer constituent is *aenigmatite* ('cossyrite'), present either in the groundmass or as phenocrysts. *Biotite* is also rarely found, confined to the phenocryst generation and usually in process of resorption by the magma.

Among the *accessory constituents*, sphene is characteristic and often belongs, in large part, to the phenocryst generation; apatite and zircon are usually unimportant; iron-ore is variable in amount and often negligible. It is noticeable that many of the accessory constituents of nepheline syenites, such as eudialyte and astrophyllite, are not found in the phonolites.

Textures and structures

Phonolites are normally holocrystalline and glassy types are rare. As already mentioned, they are practically always porphyritic but the texture of the groundmass varies. It may be trachytic when there is a preponderance of little alkali feldspar laths (fig. 15.4B) but this is no longer the case when the groundmass is rich in nepheline and the texture is dominated by the little crystals of that mineral (fig. 15.1B).

Some phonolites are vesicular, others are porous or have a small-scale drusy structure. The druses may be lined with late-stage minerals as is the case with some of the phonolites of the Bohemian Mittelgebirge, Czechoslovakia, where pale brown biotite, soda-orthoclase, aegirine and deep purple fluorite were deposited at a relatively high temperature, followed by analcite and other zeolites, calcite and, finally, by wad.

Varieties with examples

Phonolite-obsidians and even phonolite-pumice, though rare, do occur. Some of the best examples are to be found in Tenerife, one of the Canary Isles. The obsidians here may be black, yellow or green in hand specimen and are relatively rich in phenocrysts of anorthoclase, sometimes with augite and biotite in addition, set in the groundmass of glass. Others have been described from Tibesti and from Kenya.

The *phonolites* of the Black Hills, S. Dakota, have a simple mineralogical composition (fig. 15.3A) and that from the Devil's Tower (1) may be taken as an example. It has abundant phenocrysts of zoned soda-sanidine with occasional small prisms of zoned aegirine-augite rimmed with aegirine and some euhedral crystals of sphene, in a trachytic groundmass of alkali feldspar laths, small nepheline crystals, little prisms and shreds of aegirine, and some small crystals of sodalite.

Such phonolites, with aegirine or aegirine-augite as the only essential ferromagnesian mineral are common elsewhere as in Bohemia, Malagasy, and on some of the Polynesian islands where they sometimes have nepheline as an additional phenocrystal constituent (Ua-Pou) or are aphyric (Rarotonga).

A wide variety of phonolites has been described from Kenya (2) and among these are some characterised by the occurrence of soda-amphiboles (kataphorite or sky-blue amphibole) in the groundmass in addition to aegirine and aenigmatite. A special type, *kenyte*, is more mafic and has abundant large phenocrysts of anorthoclase, smaller ones of nepheline and sometimes some of olivine and pale green pyroxene, set in a matrix of brown glass (devitrified in more altered specimens) that may contain some grains of olivine, or its pseudomorphs, a few minute laths of alkali feldspar, and small crystals of magnetite and apatite.

A number of phonolites have phenocrysts of brown amphibole that may be replaced marginally or completely (fig. 15.1B) by an aggregate

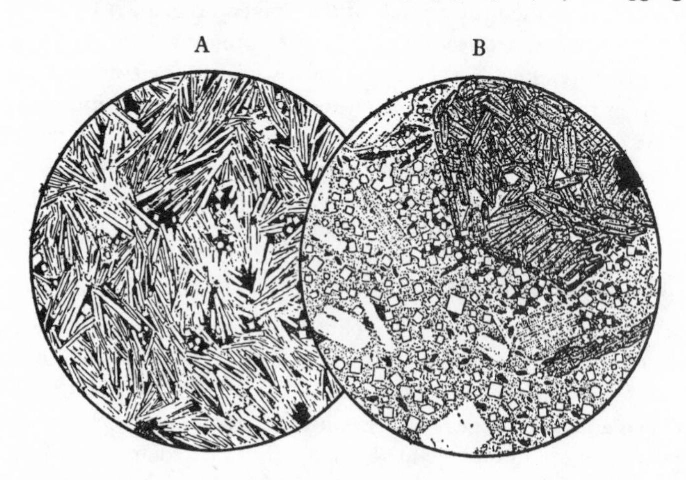

Fig. 15.1. × 20.

A. Peralkaline phonolitic trachyte, Dunedin, New Zealand. Laths of sanidine with interstitial aegirine and a few small euhedra of nepheline.

B. Phonolite, Brüx, Bohemia, Czechoslovakia. This has phenocrystal as well as groundmass nepheline, and shows an original amphibole phenocryst transformed by magmatic reaction to an aggregate of green aegirine.

of aegirine crystals (Bohemia, Cape Verde Islands) or by an aggregate of iron-ore and pyroxene (Comores, Malagasy, Tibesti).

Most phonolites are leucocratic but some are mesotype. The type example of these comes from Muri Point, Rarotonga, Cook Archipelago, and carries phenocrysts of fayalite rimmed by augite and magnetite, of titaniferous augite rimmed by aegirine-augite, of nepheline and of sphene in a matrix of nepheline, aegirine and poikilitic plates of potash feldspar.

Many phonolites have a member of the sodalite group of minerals as an accessory constituent, but such frequently become an important and essential constituent.

The only true phonolite in the British Isles is the *nosean phonolite* of the Wolf Rock, Cornwall (3). This has phenocrysts of nosean, sanidine and nepheline in a groundmass of sanidine, little euhedral crystals of nepheline and little prisms of aegirine (fig. 15.2).

The *hauyne phonolite* forming a flow at Asekrem, Ahaggar, is rich in phenocrysts of hauyne accompanied by some of anorthoclase and aegirine, but there is no hauyne in the groundmass. That from Tachaberg in the Polzen district of Bohemia, on the other hand, has phenocrysts of hauyne and aegirine-augite in a trachytic groundmass of sanidine and some nepheline with smaller crystals of hauyne and pyroxene. Other good examples come from the Cape Verde Islands (4).

More rarely, nepheline is absent, and a *hauyne trachyte* results. This type is found in the Corrize–Cantal region, Auvergne, France, where it has a large number of large crystals of colourless hauyne, rich in

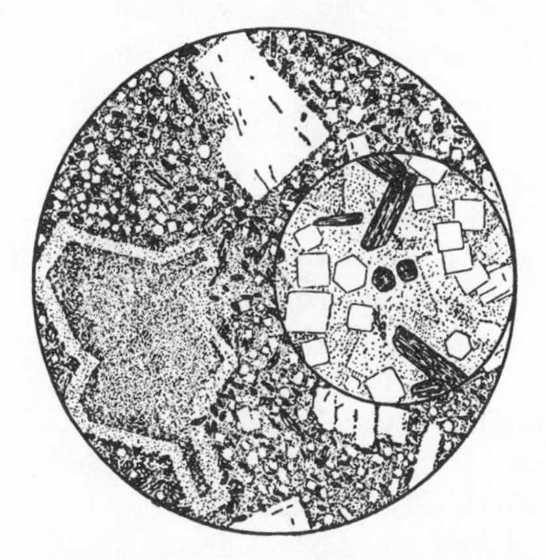

Fig. 15.2. Nosean phonolite, Wolf Rock, Cornwall; × 20.

Phenocrysts of sanidine and turbid nosean (with nepheline also in other parts of the section), in a groundmass of sanidine, euhedral nepheline and little prisms of aegirine (shown more clearly in the small inset circle, × 100).

ferruginous inclusions and rare phenocrysts of anorthoclase, augite and sphene. All these minerals occur, also, as microlites in the groundmass.

Sodalite phonolite appears to be a rarer type, but is known from the island of Ischia, where the sodalite occurs interstitially in the groundmass, and from the Bohemian Mittelgebirge where sodalite is present as phenocrysts, accompanied by nosean, as well as in the groundmass.

With the disappearance of nepheline, sodalite phonolites pass into *sodalite trachyte*, a good example of which has been described from Kosekria Hill, Turkana, Kenya (5). This is an aphanitic rock made up of small laths of alkali feldspar, grains of aegirine, small patches of deep blue soda-amphibole and abundant very small euhedral crystals of sodalite.

The late Cretaceous *analcite phonolites* of Uvalde Co., Texas (6), have phenocrysts of alkali feldspar and nepheline with fewer of aegirine-augite, partly resorbed brown amphibole titanaugite and occasional olivine. These lie in a groundmass of analcite, nepheline, somewhat analcitised feldspar, and small acicular crystals of aegirine.

The analcite phonolite that forms the small Carboniferous intrusion of Traprain Law, Haddingtonshire, Scotland (7), has analcite dominant over nepheline and minor sodalite. The dark minerals are aegirine and fayalite and there are occasional small phenocrysts of perthitic orthoclase and albite-oligoclase.

The best-known *analcite trachytes* are those of the Highwood Mountains, Montana (8). These are mafic rocks with phenocrysts of augite, olivine, analcite and, sometimes, a few of barium sanidine lying in a groundmass of analcite and barium sanidine with more or less pyroxene and iron-ore (9).

Lavas are found, also, in which leucite accompanies nepheline as an essential constituent. These are the *leucite phonolites*, sometimes called 'leucitophyres' (10), known only from a few localities, of which the Eifel district of Germany is the most important. These Eifel rocks have phenocrysts of nosean, and often of sanidine, in a groundmass of leucite, nepheline, sanidine and aegirine. In some, the leucite forms microphenocrysts (fig. 15.3B). Sphene is a typical accessory mineral and there may be a varying amount of interstitial glass.

In the leucite phonolite lava of Mt. Mikeno, Belgian Congo (11), the phenocrysts of leucite are surrounded by a rim of pseudoleucite, made up of fibrous orthoclase intergrown with nepheline. The other phenocrysts are of augite and magnetite and these minerals appear also in the very fine-grained groundmass together with nepheline and soda-orthoclase.

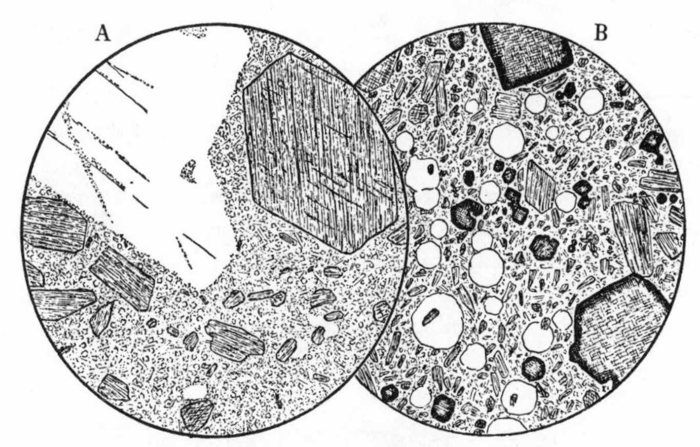

Fig. 15.3. × 20.

A. Phonolite, Black Hills, S. Dakota, U.S.A. Phenocrysts of soda-sanidine and aegirine-augite in a groundmass of sanidine, nepheline and aegirine.
B. Leucite phonolite, Burgberg, near Rieden, Eifel, Germany. Phenocrysts of dark-bordered nosean; microphenocrysts of clear leucite, nosean and green aegirine; in a groundmass of aegirine, sanidine and nepheline.

Leucite phonolites and, more especially, *leucite trachytes* occur at some of the volcanic centres in the Sabatinian, Ciminian, and Vulsinian districts of Italy. Most of these are atypical in that they contain appreciable amounts of labradorite as well as sanidine.

Mafic leucite trachytes are found in the Highwood Mountains, Montana, where they may form fairly thick flows, as on the southwest slope of North Peak (12). These have phenocrysts of augite, olivine and leucite and the groundmass has the same minerals embedded in irregular grains of barium sanidine. Much of the groundmass leucite, and part of the phenocrystal leucite, has been replaced by analcite.

Chemistry

Average phonolite has a chemical composition very similar to that of average hypersolvus and subsolvus nepheline syenite (compare columns 1 and 4, table 14.1). Hauyne phonolite is characterised by the presence of appreciable SO_3, and analcite phonolite by its relatively high content of water.

Note the resemblance of leucite trachyte to nepheline syenite of Juvet type (compare columns 2 and 7, table 14.1). There does not appear to be a volcanic representative of nepheline syenite of Mariupol type.

Hypabyssal equivalents

Nepheline microsyenites are the hypabyssal equivalents of the phonolites and nepheline syenites and intermediate in grain size between these two. A common variety, with a greenish colour in hand specimen due to the presence of aegirine, has been named *tinguaite* (fig. 15.4 A, C). Tinguaites are composed essentially of alkali feldspar, nepheline and aegirine, the latter commonly as little acicular crystals, but sometimes as irregular interstitial patches. A number contain some sodic amphibole and some

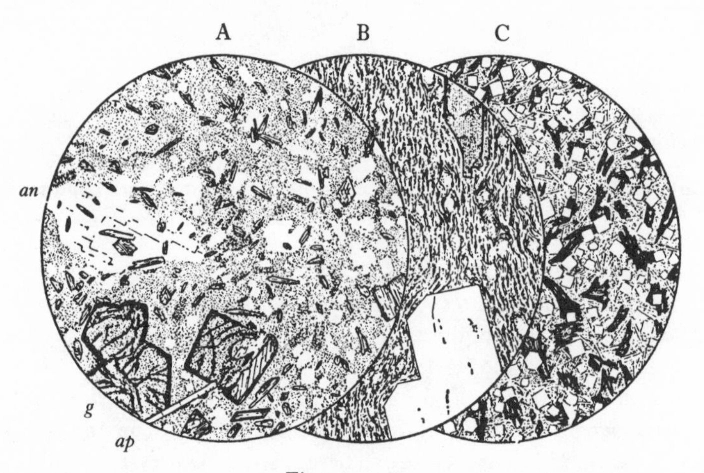

Fig. 15.4. × 20.

A. Tinguaite, Hot Springs, Arkansas, U.S.A. Essentially made up of green aegirine, turbid alkali feldspar and nepheline, with interstitial patches of analcite (*an*) and a little brown melanite garnet (*g*) and apatite (*ap*).
B. Phonolite, Port Cygnet, Tasmania. Phenocrysts of sanidine and aegirine–augite in a groundmass of the same and nepheline.
C. Tinguaite, Mt. Kosciusko, New South Wales. Composed essentially of aegirine, euhedral nepheline and slender laths of potash feldspar.

have a little biotite. A single alkali feldspar may be present, described as anorthoclase, cryptoperthite, microperthite or orthoclase in various occurrences, or both potash feldspar and albite may be present as individual crystals.

The texture may be more or less microgranular or it may be trachytic on a scale coarser than that found in the phonolites. Phenocrysts, when present, are normally sparsely developed and comprise alkali feldspar, more rarely nepheline and aegirine.

These rocks are widely distributed as dykes, or sometimes as sills, and are found commonly in connection with intrusions of nepheline syenite, as in the Oslo district, Norway; Serra de Monchique, Portugal; Khibina and Lovozero complexes of the Kola Peninsula, U.S.S.R. etc. Here too, apparently, belong the 'microfoyaites' of the Lake Chilwa district and elsewhere in Malawi (13).

Some tinguaites have small amounts of a sodalite mineral, of analcite (fig. 15.4A), or of cancrinite. More rarely, one or other of these becomes an important constituent. Such is the case with the ***analcite tinguaite*** of Pickard's Point, near Manchester, Massachusetts (14), which has analcite in excess of nepheline, while in an occurrence on the west shore of Manchester Harbour (15) nepheline is absent, and the rock is an ***analcite microsyenite***.

A dyke of ***cancrinite tinguaite***, with cancrinite in both the phenocryst and groundmass generation, is associated with cancrinite–nepheline syenite at Särna, Sweden. The most striking variants, however, are the ***pseudoleucite tinguaites*** with phenocrysts of pseudoleucite which may be up to 15cm across (Magnet Cove, Arkansas; Beemerville, New Jersey; various localities in Montana; N. China and India). Examples that have been described in some detail include one from the Bearpaw Mountains, Montana (16) and from Ghori, India (17).

Nepheline microsyenites may be markedly porphyritic and are then often described as 'nepheline porphyry' or 'nepheline syenite porphyry'. Some have nepheline and aegirine as phenocrysts (Khibina), and some have crypto- or microperthite in addition (Leeuwkraal, Pretoria), while others may have melanite garnet and rare sphene as well (Predazzo).

References and notes

(1) Pirsson, L. V., 1894. On some phonolitic rocks from the Black Hills. *Am. J. Sci.* **47**, 341–6.

(2) Smith, W. Campbell, 1931. A classification of some rhyolites, trachytes and phonolites from part of Kenya Colony, with a note on some associated basaltic rocks. *Q. Jl. geol. Soc. Lond.* **87**, 229–50.

(3) Tilley, C. E., 1959. A note on the nosean phonolite of the Wolf Rock, Cornwall. *Geol. Mag.* **96**, 503–4.

(4) Part, G. M., 1950. Volcanic rocks from the Cape Verde Islands. *Bull. Brit. Mus. (Nat. Hist.)* **1**, 41–4.

(5) Smith, W. Campbell, 1938. Petrographic description of volcanic rocks from Turkana, Kenya Colony, with notes on their field occurrence from the manuscript of Mr A. M. Champion. *Q. Jl. geol. Soc. Lond.* **94**, 522.

(6) Spencer, A. B., 1969. Alkalic igneous rocks of the Balcones Province, Texas. *J. Petrology* **10**, 276–7.

(7) MacGregor, A. G. & Ennos, F. R., 1922. The Traprain Law phonolite. *Geol. Mag.* **59**, 514–23.

(8) Larsen, E. S., 1941. Igneous rocks of the Highwood Mountains, Montana. Part II. The extrusive rocks. *Bull. geol. Soc. Am.* **52**, 1743–4.

(9) On the composition of the analcite in these relatively potassic rocks see Wilkinson, J. F. G., 1968. Analcimes from some potassic igneous rocks and aspects of analcime-rich igneous assemblages. *Contr. Miner. Petrol.* **18**, 252–69.

(10) The name 'leucitophyre' has been applied to volcanic rocks with essential leucite, nepheline and alkali feldspar, but also to those with leucite and alkali feldspar and no nepheline. The position is further complicated by the fact that the name 'leucite phonolite' has also been applied to the latter assemblage. It seems best to drop the name 'leucitophyre' and to use the name 'leucite phonolite' for a volcanic rock with essential leucite, nepheline and alkali feldspar, and the name 'leucite trachyte' for a volcanic rock with essential leucite and alkali feldspar, following the example of the great German petrographer F. Zirkel.

(11) Bowen, N. L. & Ellestad, R. B., 1937. Leucite and pseudoleucite. *Am. Miner.* **22**, 409–13.

(12) Larsen, 1941, pp. 1744–6.

(13) Dixey, F., Smith, W. Campbell & Bisset, C. B., 1955. The Chilwa Series of Southern Nyasaland. *Bull. geol. Surv. Nyasaland* **5** (revised), 23–4.

(14) Washington, H. S., 1898. Sölvsbergite and tinguaite from Essex County, Mass. *Am. J. Sci.* (4) **6**, 182–7.

(15) Clapp, C. H., 1921. Geology of the igneous rocks of Essex County, Massachusetts. *Bull. U.S. geol. Surv.* **704**, 105–6.

(16) Zies, E. G. & Chayes, F., 1960. Pseudoleucite in a tinguaite from the Bearpaw Mountains, Montana. *J. Petrology* **1**, 86–98.

(17) Sukheswala, R. N. & Sethna, S. F., 1967. Giant pseudoleucites of Ghori, Chhota Udaipur, India. *Am. Miner.* **52**, 1904–10.

16

Nephéline monzonites, essexites and related rocks

The rocks falling here are medium to coarse grained and mesotype or melanocratic, characterised by the presence of essential feldspathoid and lime-bearing plagioclase together with various ferromagnesian constituents.

The main types are *nepheline monzonite* in which lime-bearing plagioclase is accompanied by a roughly equal amount of potash feldspar (60–40 volume per cent of the total feldspar); *essexite*, with only subordinate potash feldspar (40–10 volume per cent of the total feldspar); and *theralite*, in which potash feldspar is absent, or occurs only as an accessory constituent. These all have essential nepheline, but there is another relatively common type, *teschenite*, in which the place of nepheline is taken by analcite.

The nepheline-bearing representatives typically form small intrusions, stocks or plugs, often associated with other, more abundant, alkali igneous rocks. Teschenites occur, more especially, as sills.

Mineralogy

Plagioclase feldspar, normally the most abundant individual constituent, is either andesine or labradorite. It is subhedral, frequently columnar or somewhat tabular in habit. *Potash feldspar*, when present, occurs as rims to the plagioclase or is interstitial. *Nepheline* (or *analcite*) is usually interstitial, but euhedral nepheline is present in some nepheline monzonites. *Sodalite* or *nosean* accompany nepheline in some instances. In certain subvolcanic rocks, found largely as ejected blocks, the feldspathoid present is *leucite*.

Pyroxenes include titaniferous augite (the commonest), an almost colourless augite and aegirine-augite, the latter occurring frequently as margins to titanaugite or augite. The pyroxene crystals are generally well formed and may be truly euhedral. *Amphibole*, when present, is commonly a brown alkali type, often described as barkevikitic but, in those cases where it has been separated and analysed, it has proved to be kaersutite. A certain amount of *biotite* is present in many rocks, and *olivine* becomes an essential constituent of some. The relative proportions of these ferromagnesian minerals vary considerably, but pyroxene is usually dominant.

Accessory constituents include sphene, titaniferous magnetite, ilmenite and apatite, the latter, in particular, being sometimes relatively abundant.

Textures

Taking these rocks as a whole, the commonest texture is probably the subhedral-granular, but some have a trachytoidal texture and others, especially many teschenites, have a coarse ophitic or subophitic texture.

Varieties, with examples

Nepheline monzonites are not common. The type rock from the Bezavona massif, Malagasy, carries titanaugite and brown alkali amphibole as do those occurring as blocks, not *in situ*, in the Papenoo valley, Tahiti. The latter have euhedral nepheline and some analcite, in part primary, and show all transitions to nepheline syenite on the one hand, and essexite on the other.

Similar nepheline monzonites are to be found at Sal in the Cape Verde Islands, while a coarse-grained *nosean monzonite* with dark green amphibole and with nosean taking the place of nepheline has been recorded from Cabri islet, Los Archipelago. Subvolcanic *leucite monzonites* ('sommaites') with leucite in place of nepheline occur as ejected blocks at Mte. Somma, Vesuvius.

Essexites are more widely distributed (1), and good examples are to be found at various localities in Malagasy (2). In some, the plagioclase is tabular and rimmed with potash feldspar; the crystals of brown amphibole and of titanaugite margined by aegirine-augite are euhedral, while nepheline and sodalite, usually with analcite in addition, are interstitial. In others, the same minerals are present but the texture is more or less granular. A little olivine is present in a number as an additional constituent.

Essexites occur in Scotland at Carclout, Patna, Ayrshire, and near Crawfordjohn, Lanarkshire (3). The first has irregular crystals of olivine and laths of labradorite enclosed in large ophitic plates of titanaugite frequently edged with aegirine-augite, patches of titaniferous magnetite associated with biotite, and interstitial potash feldspar, nepheline and analcite. Apatite is a noticeable accessory constituent. The mineralogy of the Crawfordjohn rock is similar but olivine is abundant and the zoned titanaugite occurs as large, well-formed crystals (fig. 16.1).

The well-known essexite (described originally as theralite) filling a volcanic neck at Flurbühls, near Duppau, Germany, is a medium-grained, subhedral-granular rock with a rather iron-rich zoned clino-pyroxene, subordinate brown amphibole, less biotite, and a little olivine.

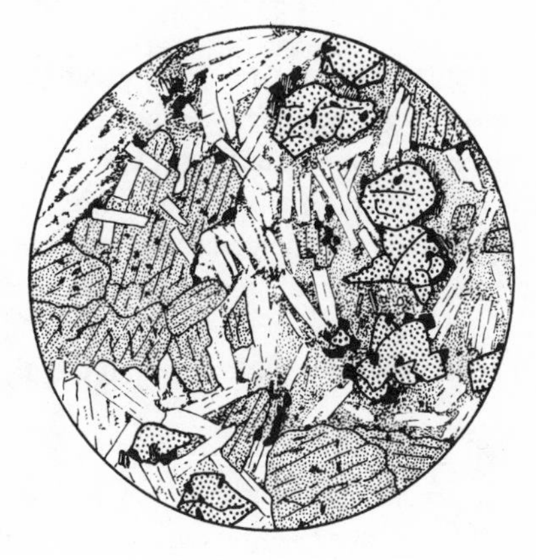

Fig. 16.1. Essexite, Craighead quarry, Crawfordjohn, Lanarkshire, Scotland; × 20.

Large crystals of titaniferous augite with smaller ones of olivine and labradorite; interstitial turbid potash feldspar and nepheline; accessory iron-ore fringed with biotite and needles of apatite. Elsewhere in the section there are small interstitial patches of analcite.

The zoned plagioclase is, in part, mantled by potash feldspar, and nepheline is interstitial. The rock grades into a marginal facies of olivine pyroxenite, and ranges from mesotype to melanocratic. Melanocratic essexite, rich in titanaugite and brown amphibole, is also found associated with nepheline syenite in the Serra de Monchique, Portugal.

Theralites appear to be rather rare but occur, for instance, as dykes cutting nepheline syenite in the Khibina complex, Kola Peninsula, U.S.S.R. These vary from coarse to rather fine grained and are melanocratic rocks rich in well-formed crystals of titanaugite often rimmed with kaersutite, independent crystals of kaersutite, some olivine, and dark red-brown biotite.

Theralites are found also in the Lugar sill, Ayrshire (4). The main type is a rather fine-grained melanocratic rock with pseudoporphyritic crystals of olivine, titanaugite and ilmenite in a base comprising numerous little prismatic grains of titanaugite embedded poikilitically in highly zoned plagioclase, nepheline and a little analcite. Minor biotite is present, partly as independent flakes, partly associated with the iron-ore, and apatite is a noticeable accessory constituent. Another type is similar but contains pseudoporphyritic kaersutite in irregular plates.

A special variety, distinguished as theralite of Lugar type (5), appears as segregations and veins, mainly within the normal theralite. This is coarse grained with large euhedral crystals of purple titanaugite and kaersutite.

Theralites of a quite different kind with green clinopyroxene and hastingsite, and whose plagioclase is an oligoclase-andesine, occur with related rocks in the Haliburton–Bancroft province, Ontario (6).

Subvolcanic *leucite-bearing rocks* corresponding with essexite and theralite, but with leucite taking the place of nepheline, are also known. Thus, among the ejected blocks of 'sommaite' found at Mte. Somma, Vesuvius, there are some with bytownite dominant over sanidine, while others have anorthite associated with leucite and augite and lack potash feldspar.

Teschenites are well displayed among the late Palaeozoic intrusions of Scotland, and the upper and lower teschenites of the Lugar sill, Ayrshire (7), will be used to provide the first examples (fig. 16.2A). These are medium to coarse-grained rocks whose main constituents, in order of abundance, are laths of labradorite, titanaugite with a subophitic relationship to the plagioclase, analcite occurring interstitially and also partially replacing feldspar, minor serpentinised olivine, and iron-ore. Kaersutite occasionally becomes an important constituent and the rocks also contain a variable amount of potash feldspar, usually much altered and partly replaced by analcite. Accessory constituents include biotite in association with iron-ore or forming at the expense of pyroxene, or as independent flakes, and relatively abundant apatite. In some there is also a little altered nepheline.

The normal rocks here are mesotype but there is a melanocratic variety in which olivine becomes more abundant, and a leucocratic variety in which potash feldspar and plagioclase are about equal in amount.

A large proportion of the late Palaeozoic Scottish teschenites have an ophitic or subophitic texture like those at Lugar, but some are subhedral-granular. Some are without kaersutite, but in others it may predominate over titanaugite. Most have essential potash feldspar as well as plagioclase and, in this respect, they differ from the Tertiary teschenites of Scotland in which the former mineral is absent (8).

While some teschenites have no potash feldspar and are the analcitic equivalents of theralite, and while a few have much potash feldspar and are the analcitic equivalents of nepheline monzonite, the majority would appear to be like most of the late Palaeozoic Scottish teschenites and to have essential potash feldspar present but subordinate to plagioclase, thus being the analcitic equivalents of essexite. This is the case with the teschenites in the neighbourhood of Teschen (now Cieszyn), Poland, where the type teschenite occurs, and with the teschenites ('analcite syenogabbros') of the Terlingua–Solitario region, Texas (9).

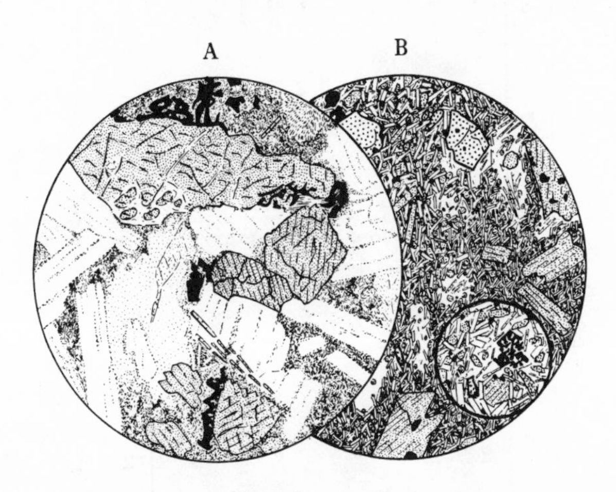

Fig. 16.2. × 20.

A. Teschenite, Lugar Sill, Lugar Water, Ayrshire, Scotland. Showing laths of labradorite, subophitic titaniferous augite, iron-ore, and analcite, both interstitial and also partly replacing feldspar. There is some chlorite and calcite associated in places with the analcite, and a little turbid potash feldspar partially margining some of the plagioclase.

B. Analcite basanite, Pico del Tiede, Tenerife. Phenocrysts of more or less euhedral titaniferous augite and subordinate olivine set in a groundmass in which these two minerals are accompanied by laths of plagioclase, abundant iron-ore, needles of apatite, and interstitial clear analcite (*see* inset).

A number of teschenite sills are differentiated as, for example, the Tertiary Black Jack sill, near Gunnedah, New South Wales (10), and, in some, accumulation of olivine leads to the production of a *picroteschenite* much enriched in that mineral (Stankards sill, W. Lothian; Braefoot sill, Fife, etc.) or even, with complete loss of felsic minerals, to an alkali peridotite (Lugar sill).

TABLE 16.1. *Average chemical compositions of nepheline monzonite, essexite, etc., and of corresponding volcanic rocks*

	1	2	3	4	5	6	7	8	9	10	11
SiO_2	50.78	46.88	45.97	46.03	44.44	47.05	44.12	47.70	51.84	47.76	54.00
TiO_2	2.28	2.81	2.34	2.40	2.95	1.54	2.73	2.39	1.57	3.04	1.90
Al_2O_3	20.53	17.07	16.64	16.16	14.44	16.05	14.56	17.69	19.27	17.90	18.95
Fe_2O_3	2.59	3.62	2.94	3.43	3.88	3.49	4.20	3.38	3.21	2.99	2.26
FeO	3.82	5.94	8.27	7.39	8.11	5.78	7.89	5.73	2.79	7.41	3.59
MnO	0.12	0.16	0.26	0.22	0.17	0.17	0.21	0.17	0.16	0.20	0.19
MgO	2.05	4.85	5.73	5.21	8.84	6.20	8.12	3.85	2.13	3.85	2.41
CaO	5.33	9.49	10.45	8.36	10.66	10.80	10.39	7.76	5.17	7.98	4.58
Na_2O	6.58	5.09	5.06	4.50	4.36	2.35	3.98	5.87	7.02	5.42	7.43
K_2O	4.01	2.64	1.37	2.19	1.08	5.38	2.01	3.23	4.50	2.16	3.82
H_2O^+	1.51	0.97	0.66	3.45	0.57	0.60	1.01	1.23	1.32	0.56	0.39
P_2O_5	0.40	0.48	0.31	0.66	0.50	0.59	0.74	0.81	0.35	0.73	0.48
SO_3	–	–	–	–	–	–	–	0.19	0.67	–	–
or	23.9	15.6	8.3	12.8	6.1	22.2	11.7	18.9	26.7	12.8	22.2
ab	25.2	14.7	13.1	20.7	13.1	–	8.9	18.3	25.7	24.6	34.6
an	14.5	16.1	18.4	17.5	16.7	17.5	15.8	13.3	10.3	18.1	7.0
ne	16.5	15.3	16.2	9.5	12.8	10.8	13.6	16.2	15.6	11.4	15.3
lc	–	–	–	–	–	7.4	–	–	–	–	-
di { *wo*	3.8	11.6	13.2	8.2	13.8	13.6	13.0	8.4	5.5	7.1	5.3
di { *en*	2.8	8.2	7.6	5.0	9.6	9.3	8.9	5.4	4.6	4.0	3.7
di { *fs*	0.7	2.4	5.0	2.8	3.0	3.2	3.0	2.4	0.1	2.8	1.2

		1	2	3	4	5	6	7	8	9	10	11
ol	*fo*	1.6	2.8	4.7	5.6	8.7	4.3	8.0	2.9	0.5	3.9	1.6
	fa	0.4	0.8	3.4	3.5	3.2	1.7	3.1	1.3	tr.	2.9	0.6
mt		3.7	5.3	4.2	4.9	5.6	5.1	6.0	4.9	4.6	4.4	3.3
il		4.4	5.3	4.4	4.6	5.6	2.9	5.2	4.6	3.0	5.8	3.7
ap		1.0	1.2	0.7	1.5	1.2	1.3	1.7	1.9	0.8	1.7	1.1
th		–	–	–	–	–	–	–	0.4	1.3	–	–
DI		65.6	45.6	37.6	43.0	32.0	40.4	34.2	53.4	68.0	48.8	72.1

1. Nepheline monzonite (average of 6).
2. Essexite (average of 15) (*Bull. Geol. Soc. Am.* **65**, 1027, no. I).
3. Theralite (average of 6) (*Ibid.*, p. 1027, no. IV).
4. Teschenite (average of 20).
5. Nepheline tephrite and basanite (average of 14).
6. Leucite tephrite and basanite (average of 31) (*Ibid.*, p. 1030, no. VI).
7. Feldspathoidal* trachybasalt (average of 15).
8. Feldspathoidal* trachyandesite (average of 10).
9. Feldspathoidal* tristanite (average of 10).
10. Nepheline hawaiite (average of 5).
11. Nepheline benmoreite (average of 3).

*Nepheline, analcite, or a member of the sodalite group

Chemistry

Average chemical compositions of nepheline monzonite, essexite, theralite and teschenite are given in table 16.1. All are characterised by rather low contents of SiO_2 coupled with moderate to high Al_2O_3 and relatively high alkalies. On passing from nepheline monzonite to theralite there is a decrease in SiO_2 and Al_2O_3 with an increase in total iron and a relatively marked increase in MgO and CaO. The alkali content, especially that of K_2O, also decreases but remains high throughout compared with corresponding non-feldspathoidal rocks. The decrease in K_2O reflects, of course, the decreasing content of potash feldspar. The remaining changes can be correlated largely with increasing content of ferromagnesian minerals and together these lead to a progressive decrease in the differentiation index.

Note the relatively high TiO_2 in all these alkali rocks and also the high H_2O content of teschenite where analcite replaces nepheline and where, in addition, olivine is usually serpentinised and chlorite and zeolites such as thomsonite and natrolite may be associated with the analcite.

References and notes

(1) Not as widely, however, as a perusal of the literature would imply. The name 'essexite' has been used very loosely by some petrologists and has been applied even to rocks without any nepheline that should be classified as alkali monzodiorites, alkali monzogabbros or alkali gabbros.

(2) Lacroix, A., 1922. *Minéralogie de Madagascar*. Paris: A. Challamet, vol. 2, 628–30.

(3) Scott, A., 1915. The Crawfordjohn essexite and associated rocks. *Geol. Mag.* **2**, 455–61.

(4) Tyrrell, G. W., 1917. The picrite–teschenite sill of Lugar (Ayrshire). *Q. Jl. geol. Soc. Lond.* **72**, 105–7.

(5) Tyrrell, G. W., 1948. A boring through the Lugar Sill. *Trans. geol. Soc. Glasgow* **21**, 167–8.

(6) Tilley, C. E. & Gittins, J., 1961. Igneous nepheline-bearing rocks of the Haliburton–Bancroft province of Ontario. *J. Petrology* **2**, 44–7.

(7) Tyrrell, 1917, pp. 98–105. Phillips, W. J., 1968. The crystallisation of the teschenite from the Lugar Sill, Ayrshire. *Geol. Mag.* **105**, 23–34.

(8) Tyrrell, G. W., 1928. The geology of Arran. *Mem. geol. Surv. U.K.*, 122.

(9) Lonsdale, J. T., 1940. Igneous rocks of the Terlingua–Solitario region, Texas. *Bull. geol. Soc. Am.* **51**, 1609–11.

(10) Wilkinson, J. F. G., 1958. The petrology of a differentiated teschenite sill near Gunnedah, New South Wales. *Am. J. Sci.* **256**, 1–39.

17

Tephrites, basanites, allied volcanics and their hypabyssal representatives

The *tephrites* are volcanic rocks characterised by having essential calcic plagioclase and a feldspathoid; in the *basanites* olivine enters as a further essential constituent. Both normally have abundant clinopyroxene in addition. They vary in colour from grey to black in hand specimen and have frequently a porous character that gives them a rough feel. When the names 'tephrite' and 'basanite' are used without further qualification it is usually implied that the feldspathoid present is nepheline though strictly speaking such rocks are *nepheline tephrites* and *nepheline basanites* and are the volcanic equivalents of the theralites. Sometimes the place of nepheline is taken by analcite, and these *analcite tephrites* and *analcite basanites* are the volcanic equivalents of those teschenites that are devoid of potash feldspar. There are also rarer and more extreme types, *leucite tephrite* and *leucite basanite*, having leucite as their feldspathoidal constituent.

In many of these rocks, more especially in those which would have nepheline, this mineral cannot be detected under the microscope but is occult in the very fine-grained or partly glassy groundmass. They are then often referred to as 'tephritoids' and 'basanitoids', and it is necessary to have a chemical analysis and to calculate the norm before they can be classified. A problem that arises is where to draw the line between tephrites and basanites on the one hand, and alkali basalts on the other. The latter have, not infrequently, a little accessory analcite or nepheline and, even when these are lacking, the presence of titaniferous augite leads to the appearance of a small amount of normative nepheline. A number of petrologists take 5 per cent of normative nepheline as the dividing line between the two, while others take 10 per cent. The latter seems to be a more realistic figure, having regard to the fact that nepheline is an essential, and not an accessory, constituent in tephrites and basanites, and is the one that will be adopted here.

Mineralogy

The *calcic plagioclase feldspar* varies in composition from bytownite to labradorite when occurring as phenocrysts, and is often zoned. The plagioclase laths of the groundmass are normally somewhat richer in the albite molecule. The *nepheline* is usually confined to the groundmass and may be euhedral or anhedral. In the latter case it may occur interstitially and be very hard to see. In some rocks *analcite* partly or wholly takes the place of nepheline. In others, the feldspathoid present is *leucite* occurring typically in more or less euhedral crystals both as phenocrysts and in the groundmass. *Sodalite*, *nosean* and *hauyne* may appear as minor constituents in the form of phenocrysts.

The usual ferromagnesian constituent is a *titaniferous augite*, commonly present both as phenocrysts and in the groundmass. Olivine, normally confined to the phenocryst generation, is an essential constituent of the basanites and may be abundant. *Alkali amphibole* and *biotite* occur to a small extent as phenocrysts in some of the rocks; both show signs of resorption.

Among the *accessory constituents*, little prisms or needles of apatite are nearly always to be found; iron-ore is variable in amount, and perofskite is common in a number of rocks.

Textures and structures

These rocks are nearly always porphyritic and the groundmass is often very fine grained. Some are holocrystalline; in many the groundmass is hypocrystalline, but a wholly vitreous groundmass is rare. Vesicular structure is common.

Varieties, with some examples

True tephrites and basanites are not common rocks. Many that have been described as such have appreciable contents of potash and are the volcanic equivalents of essexites, potash feldspar-bearing teschenites and similar plutonic types (see below).

The *nepheline tephrite* from Garachio, Pico de Teyde, Tenerife, is a very fine-grained aggregate of fluidally arranged laths of plagioclase with grains of augite and iron-ore and interstitial nepheline. It carries sporadic large partially resorbed crystals of brown amphibole. The tephrite from Tarrafal, S. Antaõ, Cape Verde Islands (1), has broken and partially resorbed xenocrysts of brown amphibole and phenocrysts of somewhat sodic augite with some of magnetite, in a groundmass of augite, magnetite, and laths of plagioclase all set in a base of nepheline with some analcite. Accessory apatite and a few flakes of biotite are additional constituents.

Many of the lavas forming the main cone of Fogo (2) are ***nepheline basanites***, usually very fine grained and rich in phenocrysts of titanaugite, olivine and iron-ore lying in a base of poikilitic labradorite, a little nepheline, and glass. Those from the Viana crater, S. Vicente, are similar but have rather more glass.

Nepheline basanites are found among the late products of volcanic evolution in the Hawaiian islands as, for instance, the lava from the Kilea cone, W. Maui. This has scattered phenocrysts of olivine and microphenocrysts of titanaugite in a groundmass of labradorite laths, clinopyroxene, grains of magnetite and a colourless base of low birefringence that is made up, at least in part, of nepheline.

Nepheline and nepheline–analcite basanites occur, also, in the E. Otago volcanic province of New Zealand (3), in New South Wales, in a somewhat altered state as a minor phase among the alkalic igneous rocks of the Balcones province, Texas, and elsewhere. A good example of a nepheline basanite containing only a little plagioclase and transitional to olivine nephelinite occurs on Ponape Island, the largest of the Caroline Islands, in the west Pacific Ocean (4).

Analcite–nepheline basanite forms a dyke at Butterton, Staffordshire (5). This has phenocrysts of olivine in a matrix with abundant grains of titanaugite, iron-ore and laths of labradorite, lying in an apatite-rich base of analcite, some nepheline and a little alkali feldspar.

Analcite basanites (fig. 16.2B) are well displayed among the late Palaeozoic lavas of Ayrshire (6). A typical example near Stair has abundant microphenocrysts of olivine and a groundmass of augite granules, iron-ore, and plagioclase microlites in a base of turbid analcite.

The lavas of Vesuvius (7) are the best-known *leucite tephrites* (fig. 17.1), though they do have accessory olivine. They are rich in leucite and commonly have abundant euhedral phenocrysts of that mineral with a few of pale greenish-brown augite, sometimes a few of labradorite, with or without a few of olivine. The fine-grained groundmass has small laths of labradorite, little subhedral prisms of augite, small rather rounded crystals of leucite with a little residual nepheline or glass, or both, and accessory olivine, iron-ore and apatite.

Sometimes, as in the lava of 1631, the phenocrysts are largely of augite with only a few of leucite and the bulk of the latter mineral is present as abundant microphenocrysts. Again, very small flows and the borders of larger flows may be formed of a porphyritic *leucite hyalo-tephrite* in which numerous phenocrysts of leucite, often skeletal in form, with a few of augite, olivine and iron-ore lie in a brown glass.

Similar leucite tephrites associated with *leucite basanites* occur in the Ringgit–Beser volcanic complex of Java. Some of the latter from Mt. Mouriah on the north coast of that island (8) have minor alkali feldspar and are transitional to leucite trachybasalt.

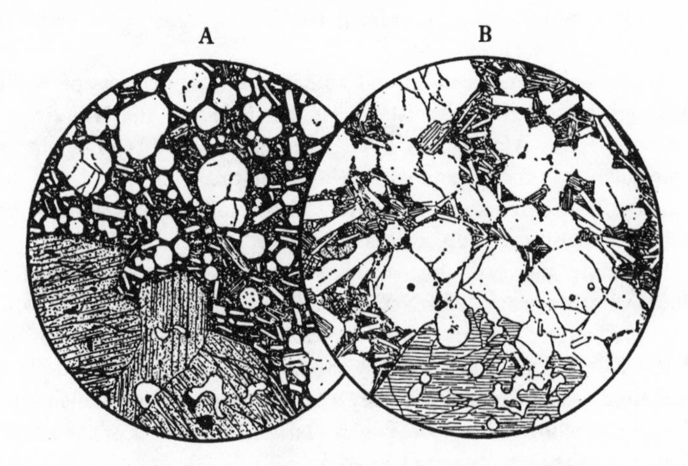

Fig. 17.1. Leucite tephrites, Vesuvius; × 20.

A. Lava of 1813. Glomeroporphyritic augite, small phenocrysts of clear leucite and labradorite, in a fine-grained groundmass with interstitial glass..

B. Atrio del Cavallo. Made up essentially of abundant leucite, augite, and labradorite.

In the leucitic rocks mentioned above, potash is considerably in excess over soda, but in the leucitic tephrites and basanites of the Laacher See district, Germany, the potash content is either less than, or about equal to, that of soda and they are, for the most part, *leucite-nepheline tephrites and basanites.* The phenocrysts here are mainly titanaugite, or titanaugite and olivine, sometimes with some of largely resorbed amphibole and biotite, while leucite is confined to the ground-mass. In a number there is some residual glass which is relatively abundant in the hauyne-bearing rock from Niedermendig, famous as a source of Roman millstones.

Leucite–nepheline basanites and tephrites are found also in the Korath Range, S. Ethiopia (9), where the tephrites have phenocrysts of partially resorbed kaersutite in addition to those of titanaugite, and where, in both the tephrites and the basanites, leucite as well as nepheline is interstitial, as is also the case with leucite basanites from the province of Gerona, Spain.

There are also volcanic equivalents of the essexites, of the nepheline monzonites, and of the potash feldspar-bearing teschenites resembling the tephrites and basanites but characterised by the presence of varying amounts of *sanidine* as a further essential constituent. This may sometimes be found mantling plagioclase phenocrysts, or it may be confined to the groundmass where it may form narrow rims to the plagioclase laths or may occur interstitially. It is frequently difficult to see owing to the fine grain of the groundmass or it may be occult in residual glass, and it is by no means uncommon to find chemical analyses of such rocks with 2 or 3 per cent of potash but with no mention of potash feldspar in the accompanying petrographic descriptions. Here again, therefore, it is frequently necessary to have a chemical analysis and to calculate the norm before the rock can be classified.

These volcanics have received various names in the past. Those with subordinate sanidine were often lumped together with the tephrites and basanites, or were called 'ordanchite' or 'phonolitic tephrite'. Those somewhat richer in sanidine were called 'latite-phonolite' or 'tephritic phonolite' or, more rarely, 'tahitite'. An attempt has been made recently, following an earlier suggestion (10), to bring some order into this chaos. It will be recalled that two series of volcanic rocks were recognised among those equivalent to alkali monzonites, alkali monzodiorites and alkali monzogabbros, namely, a more potassic series trachybasalt–trachyandesite–tristanite, and a more sodic series hawaiite–mugearite–benmoreite (see p. 83). It is proposed that volcanic rocks intermediate between tephrite and phonolite should be classified in a similar manner into a more potassic feldspathoidal trachybasalt–tristanite series and a more sodic feldspathoidal hawaiite–benmoreite series (11). This course is followed here.

Feldspathoidal trachybasalts may be illustrated first by the rock from Klocher Klause, Austria. This is a *nepheline trachybasalt*, dense and blue-grey in hand specimen, with phenocrysts of embayed olivine and of colourless pyroxene with margins of titanaugite. These are set in a groundmass of plagioclase laths, little euhedral prisms of titanaugite, some interstitial sanidine, nepheline, and possibly a little analcite, with accessory iron-ore and apatite. A similar rock from Steinberg, near Feldbach, has a partly glassy groundmass and is of interest because it has irregular schlieren of black glass, light brown in section, having the chemical composition of a phonolite.

Nepheline trachybasalts from Nyassa have abundant phenocrysts of augite, olivine and some of calcic plagioclase that may be rimmed with sanidine, in a groundmass of calcic plagioclase laths, titanaugite, olivine, nepheline, iron-ore and apatite with residual brownish or colourless glass. They are associated with *nepheline trachyandesites* in which more or less resorbed brown alkali amphibole accompanies augite and sparing olivine in the phenocrysts generation, and in which sanidine occurs with plagioclase in the groundmass. The nepheline may appear as anhedral grains or it may be occult in residual glass.

The nepheline trachyandesites of the E. Otago volcanic province, New Zealand (12), are variable rocks. One type has microphenocrysts of olivine, pale green augite that may have a purplish core, and original kaersutite, now replaced by aggregates of augite, iron-ore, feldspar and nepheline. The groundmass is composed of little pale green augite prisms, laths of plagioclase and grains of iron-ore, with interstitial alkali feldspar, nepheline and a little sodalite. Another has prominent phenocrysts of anorthoclase bordered by oligoclase-andesine, and smaller ones of olivine, augite, partially resorbed kaersutite, oligoclase-andesine and nepheline, while in a third variety much of the groundmass consists of pale brown glass.

Analcite trachybasalts and *analcite trachyandesites* are found on the island of Procida, Italy (13), as flows and as blocks in volcanic tuffs. The ferromagnesian constituents are greenish augite, which may build phenocrysts, olivine, and a little biotite.

The best-known feldspathoidal tristanites are the *hauyne tristanites* called 'tahitite' occurring, more especially, in the island of Tahiti and in that of Gran Canaria, Canary Isles (14). The latter have phenocrysts of augite, partially resorbed alkali amphibole, hauyne and sometimes sphene, in a trachytic groundmass of oligoclase and alkali feldspar with hauyne, augite or aegirine-augite and alkali amphiboles, with abundant accessory iron-ore, sphene and apatite. There are also more vitreous types in which the groundmass consists of brown glass crowded with microlites of feldspar, hauyne and clinopyroxene.

These hauyne tristanites from Gran Canaria grade into phonolites and call to mind the 'latite-phonolites' of Cripple Creek, Colorado (15), which are feldspathoidal tristanites lying close to phonolite and having nosean, sodalite and analcite in place of hauyne.

Other rocks belonging here have phenocrysts of hauyne but nepheline in the groundmass (Malagasy), and among these is one with numerous phenocrysts of anorthoclase and some of melanite garnet. Others, *nepheline tristanites*, are devoid of hauyne and have nepheline, with or without a little sodalite or analcite, in the groundmass (Comores, Tahiti).

More rarely, rocks of this group have leucite as their feldspathoid, as is the case with the *leucite trachyandesites* and *leucite tristanites* (16) of Tristan da Cunha (17). The former are holocrystalline and leucite occurs interstitially in the groundmass. The latter may be holocrystalline with leucite as little rounded crystals in the groundmass, or they may be more or less vitreous and then the leucite is occult.

Among the so-called tephrites of the Roman comagmatic region in Italy, there are a number of rocks that differ from the true tephrites of the region in having sanidine as an essential, and sometimes abundant, additional constituent. They are, in general, more leucocratic than the normal tephrites and whereas the differentiation index of the latter is usually somewhere near 50, these rocks have differentiation indices ranging up to 80 in types transitional to leucite trachyte. They thus cover the range from leucite trachybasalt to leucite trachyte but are abnormal in that the normative plagioclase remains calcic throughout and the modal plagioclase is normally labradorite. They are rich in leucite and in those that are also rich in sanidine the potash content may be 10 per cent or more. These unusual rocks are found among the old Somma lavas of Vesuvius (18) and at Monte Vico, Toscanella, Monte Corvallo, etc.

Feldspathoidal representatives of the hawaiite–benmoreite series seem to be rarer, but *nepheline hawaiite* occurs in the E. Otago volcanic province, New Zealand (19). This has small phenocrysts of calcic andesine, with some of titanaugite and olivine, in a very fine-grained groundmass composed of titanaugite, olivine, plagioclase, alkali feldspar and nepheline with iron-ore and a few flakes of biotite. A phonolite from the same province approaches nepheline benmoreite in composition, and it seems likely that other members of the series will be found here.

Nepheline benmoreite from Anjouan Island, Comores, has partially resorbed phenocrysts of brown amphibole with some of augite and of magnetite in a matrix of quite large crystals of andesine, little crystals of green augite and magnetite, microlites of alkali feldspar, nepheline and some analcite.

The alkaline lavas of the Takakusayama district, Japan (20), probably of lower Miocene age, are unfortunately rather altered and oxidised but, among the types present, are *analcite hawaiites* and *analcite mugearites*.

Chemistry

Average nepheline tephrite and basanite (table 16.1, column 5) has a composition similar to average theralite (column 4) but is richer in

MgO and TiO_2 and a little lower in alkalies and Al_2O_3. These differences would probably be lessened if more analyses of theralites were available.

Average leucite tephrite and basanite (column 6) has high K_2O and is unlike any of the plutonic types averaged. Note how, owing to the way of calculating the norm, the bulk of the leucite appears as *or*, while the albite molecule of the plagioclase appears as *ne*.

Average feldspathoidal trachybasalt (differing from average tephrite and basanite in its higher content of K_2O), trachyandesite and tristanite (columns 7, 8, 9) form a progressive series in which SiO_2 increases to a moderate degree together with Al_2O_3 and alkalies while total iron, MgO, CaO and TiO_2 decrease. Normative *or* shows a considerable increase, as does normative *ab*, but normative *an* decreases more slowly, and normative *ne* does not show a great deal of change. These features lead to a steady increase in the differentiation index.

Nepheline hawaiite and benmoreite (columns 10, 11) differ from the above more especially in their lower ratio of potash to soda, and the ratio of normative *or* to normative *ab* + *an* in these is a good deal lower than in the corresponding rocks of the feldspathoidal trachybasalt to tristanite series.

Plutonic equivalents of feldspathoidal trachybasalts, of most feldspathoidal trachyandesites, and of feldspathoidal hawaiites and feldspathoidal benmoreites, are to be found among the essexites and teschenites, while the equivalents of some feldspathoidal trachyandesites and of the feldspathoidal tristanites are found among the nepheline monzonites.

Hypabyssal equivalents

Hypabyssal representatives of this group may be termed microtheralite, microessexite, etc., fine-grained rocks intermediate in grain size between the plutonic types and the lavas. Mineralogically, they resemble the plutonic rocks, and they may be aphyric or porphyritic.

A good example of *microessexite* comes from the Voandrozo valley, Malagasy. This has crystals of zoned plagioclase mantled with orthoclase and crystals of brown alkali amphibole, green at the margins, associated with augite and biotite. The interstices of the rock are filled by nepheline with a little sodalite and analcite. Another, that is either intrusive or forms a segregation vein, from Bambao, Anjouan Island, Comores, has a discontinuous framework of bytownite and titanaugite in relatively large crystals with the interstices occupied by microlites of orthoclase and poikilitic aegirine-augite together with olivine and prisms of apatite, the whole being moulded by nepheline.

The 'essexite porphyry' dykes of Stoffel, Germany (21), are somewhat altered *porphyritic microessexites* with phenocrysts of titanaugite that may be rimmed by aegirine-augite, of calcic plagioclase and subordinate anorthoclase, both of which may have margins of sanidine, and of sparing serpentinised olivine. These lie in a groundmass of sanidine and nepheline, with aegirine-augite, a little alkali amphibole, biotite and zeolites with abundant iron-ore and some needles of apatite.

Among the Scottish teschenites a group has been described as of 'porphyritic or basaltic type' (22). Some of these are fine grained and are *microteschenites.*

References and notes

(1) Part, G. M., 1950. Volcanic rocks from the Cape Verde Islands. *Bull. Brit. Mus. (Nat. Hist.) Miner. Ser.* **1**, no. 2, 48–9.

(2) Part, 1950, p. 49.

(3) Coombs, D. S. & Wilkinson, J. F. C., 1969. Lineages and fractionation trends in unsaturated rocks from the east Otago volcanic province (New Zealand) and related rocks. *J. Petrology* **10**, 451–5.

(4) Yagi, K., 1960. Petrochemistry of the alkali rocks of the Ponape Island, western Pacific Ocean. *International geol. Congr., XXI Session, Norden.* Part **13**, 113.

(5) Scott, A., 1925. *In* The geology of the country around Stoke-upon-Trent. *Mem. geol. Surv. U.K.*, 186–99.

(6) Tyrrell, G. W., 1928. A further contribution to the petrography of the late-Palaeozoic igneous suite of the west of Scotland. *Trans. geol. Soc. Glasgow* **18**, 274–5.

(7) Washington, H. S., 1906. The Roman comagmatic region. *Carnegie Inst. Wash. Publ.* **57**, 103–8, 115–21. Part, G. M., 1944. The Vesuvian lava of the March 1944 eruption. *Geol. Mag.* **81**, 176–80.

(8) Iddings, J. P. & Morley, E. W., 1915. Contributions to the petrography of Java and Celebes. *J. Geol.* **23**, 231–7.

(9) Brown, F. H. & Carmichael, I. S. E., 1969. Quaternary volcanoes of the Lake Rudolf region: 1. The basanite–tephrite series of the Korath Range. *Lithos* **2**, 239–60.

(10) Tilley, C. E. & Muir, I. D., 1964. Intermediate members of the oceanic basalt–trachyte association. *Geol. Fören. Stockholm Förhandl.* **85**, 443.

(11) Coombs & Wilkinson, 1969, pp. 440–501.

(12) Coombs & Wilkinson, 1969, pp. 444–50.

(13) Narici, E., 1932. Contributo alla petrografia chimica della provincia magmatica Campana e del Monte Vulture. *Zeit. Vulk.* **14**, 222–3 and table p. 225.

(14) Hernandez-Pacheco, A., 1970. The tahitites of Gran Canaria and haüynitisation of their inclusions. *Bull. Volc.* **33**, 709–18.

(15) Lindgren, W. & Ransome, F. L., 1906. Geology and gold deposits of the Cripple Creek district, Colorado. *Prof. Pap. U.S. geol. Surv.* **54**, 68–83.

(16) Tilley & Muir, 1964, pp. 441–3.
(17) Baker, P. E., Gass, I. G., Harris, P. G. & Le Maitre, R. W., 1964. The volcanological report of the Royal Society expedition to Tristan da Cunha, 1962. *Phil. Trans. Roy. Soc.*, Ser. A, **256**, 496–7, 551–4.
(18) Savelli, C., 1967. The problem of rock assimilation by Somma–Vesuvius magma. *Contr. Miner. Petrol.* **16**, 328–53.
(19) Coombs & Wilkinson, 1969, pp. 479–82.
(20) Tiba, T., 1966. Petrology of the alkaline rocks of the Takakusayama district, Japan. *Sci. Rept. Tohoku Univ.*, ser. III, **9**, 541–610.
(21) Lehmann, E., 1930. Der Basalt vom Stöffel (Westerwald) und seine essexitisch-theralithischen Differentiate. *Chem. der Erde* **5**, 329–55.
(22) Walker, F., 1923. Notes on the classification of Scottish and Moravian teschenites. *Geol. Mag.* **60**, 242–9.

18

Ultra-alkaline igneous rocks and carbonatites

The *ultra-alkaline plutonic rocks*, of medium to coarse grain, form a small and highly specialised group, frequently associated with nepheline syenites, and consisting essentially of a feldspathoid and ferromagnesian minerals.

The commonest group has nepheline as its feldspathoid and is divided on the basis of colour index into *urtite* (leucocratic), *ijolite* (mesotype), and *melteigite* (melanocratic), the latter passing gradually into one type of alkali pyroxenite (jacupirangite) as nepheline dwindles to the status of an accessory constituent.

Almost the whole series is well displayed in the Fen district, Norway, where the dominant type is a rock hovering on the boundary between ijolite and melteigite. The rocks here are medium grained with a more or less subhedral-granular texture, and the essential constituents are nepheline and aegirine or aegirine-augite, the latter sometimes having a core of titanaugite in the most mafic members. Melanite garnet, commonly poikilitic and relatively late, occurs sporadically throughout and is sometimes abundant; biotite is normally very subordinate except in a special variety, biotite ijolite. Apatite and sphene are common, cancrinite is widespread, and primary calcite makes up several per cent of many of the rocks.

Similar ijolitic rocks occur elsewhere in Scandinavia (Iiwaara, Finland, is the type locality for ijolite), in Africa (Chilwa Island, Malawi; Spitzkop, Transvaal; Budeda and Napak (1), Uganda; Homa Bay, Kenya (2) etc.) and in the Kola Peninsula, U.S.S.R. (Turja; Khibina and Lovozero complexes). Some of those from Khibina have a marked fluidal texture due to the parallel arrangement of elongated pyroxene and sphene crystals, while the Lovozero complex (Lujavr–Urt) contains the type urtite.

Ijolites rich, sometimes very rich, in melanite garnet and grading to urtite and melteigite make up part of the inner core of the Magnet Cove alkali igneous complex, Arkansas (3) (fig. 14.3A). Some have sodic diopside with minor biotite (garnet ijolite), others may have biotite almost to the exclusion of pyroxene (biotite–garnet ijolite). The garnet is partly euhedral, partly as anhedral masses, and the accessory constituents include perofskite, sometimes occurring as residual cores in the garnet, sphene, apatite and primary calcite. These rocks are commonly medium grained but local segregations within the biotite–garnet ijolite are pegmatitic with crystals of biotite as much as 15cm across.

Some ijolites have amphibole as well as pyroxene, as is the case with the ijolite of the Ice River complex, British Columbia (4), where a brown alkali amphibole may be as abundant as clinopyroxene and occurs frequently as large poikilitic crystals.

All the ijolitic rocks mentioned so far have had a sodic diopside, aegirine-augite or aegirine as their clinopyroxene, but there are some characterised by the occurrence of titanaugite instead. Such are those of the Ozernaya Varaka massif, S.W. Kola, U.S.S.R., with poikilitic texture and sometimes rich in melanite garnet or in perofskite; those of Jombo (5), where the titanaugite is bordered by aegirine; and the melanocratic rock found as blocks at Ambaliha, Ampasindava bay, Malagasy, with accessory olivine and biotite and often a little anorthoclase.

Many ijolitic rocks are unaltered, but in many others some alteration, especially of nepheline, is apparent and it is replaced by 'hydronephelite' (natrolite), muscovite, etc. When the hydrothermal alteration is extreme, rocks little resembling the originals are produced. Some of the ijolitic rocks of the Fen district have been affected in this way and end up as muscovite–biotite–calcite, biotite–calcite, chlorite–calcite, or even chlorite–quartz assemblages.

Among the ejected blocks found at some volcanic centres in Italy, such as Mte. Somma (Vesuvius), Mte. Albana and Mte. Cavo (Latium), and Villa Senni (Alban Hills), are subvolcanic rocks of medium to quite coarse grain composed of *leucite* and ferromagnesian minerals and ranging from highly leucocratic to melanocratic. A melanocratic rock of this kind with dominant clinopyroxene, some olivine and less biotite forms a small stock-like mass at the head of Shonkin Creek, Highwood Mountains, Montana (6). Others, at Shonkin Creek and elsewhere, have *pseudoleucite* in place of leucite.

Rarer rock types that fall here include those that have melilite and nepheline; melilite, nepheline and leucite; sodalite; sodalite and analcite; analcite; together with variable amounts of ferromagnesian constituents.

The *ultra-alkaline volcanic rocks* are named according to the feldspathoid, or related mineral, that is present and the most important are *nephelinites* and *leucitites* of one kind or another. Normally mesotype or melanocratic, dense or porous, they are black, dark grey or dark green in hand specimen.

Simple *nephelinites* have typically titanaugite, nepheline and some iron-ore as their main constituents. The more abundant *olivine nephelinites* (often called 'nepheline basalt') have essential olivine in addition.

These rocks are usually, but not always, porphyritic with phenocrysts of titanaugite (which may have a more or less colourless core) and sometimes of nepheline; some have brown amphibole, often partially resorbed, in addition. Olivine nephelinites may have a magnesian olivine as the only phenocrystal mineral or it may be accompanied by titanaugite. Iron-ore may appear as microphenocrysts.

The groundmass is made up of little crystals of titanaugite, which may be margined by green clinopyroxene; iron-ore; and nepheline occurring as little euhedral or as interstitial anhedral grains or, more rarely, as poikilitic plates enclosing the pyroxene and iron-ore. These may be joined by minor amounts of a mineral of the sodalite group, by analcite or, in a few, by leucite or biotite or both. In some, again, there is a varying amount of residual brown glass. Apatite, perofskite, and sphene may be present as further accessory constituents.

Both nephelinites and olivine nephelinites are well displayed at some volcanic centres in Uganda, as at Napak (7) and Mt. Elgon (8), and in Kenya (9), also in the Cape Verde Islands (fig. 18.1B).

Olivine nephelinites are found in the Comores Islands, where some have a glassy groundmass; in the Campos de Calatrava volcanic field, Spain, where the pyroxene may be brownish green or green; in the Eifel district, Germany, where some have accessory leucite and biotite; in the Bohemian Mittelgebirge, Czechoslovakia; in the Balcones igneous province, Texas (10); and elsewhere.

The nephelinite of Oberwiesenthal, Erzgebirge, Germany, is of interest because it contains cognate xenoliths of ijolitic rocks. More unusual leucocratic nephelinites are found in the Selvagen Islands where the phenocrysts are of nepheline and hauyne with a few of aegirine-augite and the groundmass pyroxene is *aegirine*. Here, too, may be noted the olivine nephelinite of Shannon's Tier, Tasmania (11), associated with melilite-bearing rocks, and having *monticellite* as well as olivine among its constituents.

Melilite is present as a minor accessory constituent in certain nephelinites and olivine nephelinites. In some rocks, however, it becomes quite abundant and an essential constituent. In such *melilite nephelinites* and *melilite–olivine nephelinites* the mineral occurs in the groundmass as little rectangular crystals flattened on {001}, but may also build microphenocrysts or small phenocrysts. These rocks are frequently associated with nephelinite and olivine nephelinites as in the Eifel district, the Bohemian Mittelgebirge, the Campos de Calatrava volcanic field, and the Balcones igneous province, all mentioned above. A similar association is seen in the Honolulu volcanic series of Oahu, one of the Hawaiian islands (12).

Analcitites, and the more common *olivine analcitites*, have analcite in place of nepheline. Good examples of olivine analcitite are to be found among the late Palaeozoic lavas and minor intrusions of Scotland. That occurring near Ochiltree (13), for instance, has numerous small phenocrysts of partly altered olivine with one or two of pyroxene set in a matrix consisting of a mass of little prisms of purplish titaniferous augite,

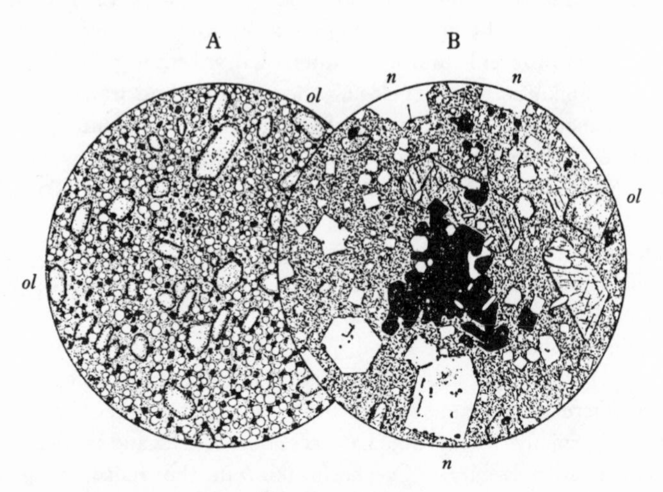

Fig. 18.1. × 20.

A. Olivine leucitite, El Capitan, New South Wales. Composed of olivine (*ol*), largely altered, little crystals of clear leucite, augite, and iron-ore.

B. Olivine nephelinite, Fogo, Cape Verde Islands. Phenocrysts of olivine (*ol*), titaniferous augite, nepheline (*n*) and iron-ore in a fine-grained groundmass of titaniferous augite and nepheline.

sparse grains of iron-ore and some interstitial analcite. A similar rock from Kidlaw quarry, south of Haddington, has a little alkali feldspar as an additional constituent.

The olivine analcitite from The Basin, west of Cripple Creek, Colorado (14), has accessory dark red-brown biotite and feldspar, while in that from Dirnbach, E. Styria, the pyroxene is greenish yellow and a little nepheline accompanies the analcite.

The remarkable rocks called 'blairmorite' from the Crowsnest Volcanics, Alberta (15), are analcitites. One variety has large phenocrysts of analcite in a groundmass of small analcite crystals, euhedral aegirine-augite and nepheline with a little sanidine and melanite garnet, all embedded in an irresolvable matrix. Another variety has smaller analcite phenocrysts, has more potash feldspar and is transitional to analcite trachyte.

When leucite takes the place of nepheline we have the rock types *leucitite* and *olivine leucitite* ('leucite basalt') (fig. 18.1A). These are rather rare rocks occurring at only a few localities.

The best-known *leucitites* are those of the Vulsinian, Latian and Hernican districts of Italy (16). These are dense, dark grey to black

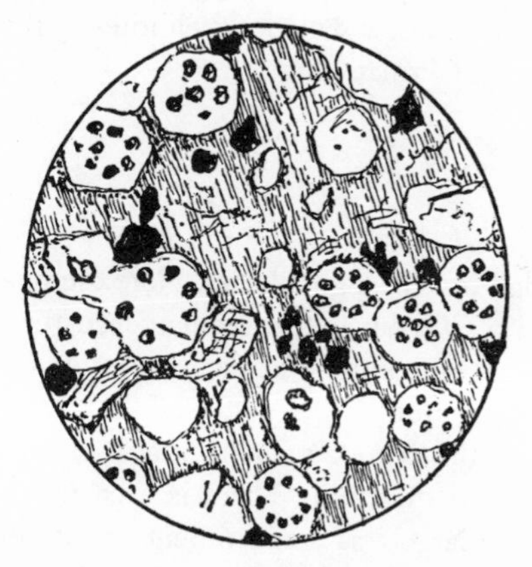

Fig. 18.2. Melilite leucitite, Capo di Bove, near Rome; × 100. Small leucite crystals with zonally arranged inclusions, augite and iron-ore, all enclosed by a large crystal of yellowish striated melilite.

rocks, more or less aphyric apart from rare phenocrysts of leucite or augite or both. They are made up of numerous small rounded leucite crystals, sometimes with a skeletal development in part (Monte Rado), and tiny prisms and grains of pale grey augite. The accessory constituents include small grains of iron-ore in varying amount, needles of apatite, and a small quantity of interstitial base, sometimes of plagioclase or of nepheline or both. Small grains of olivine are present in some, and many have a little interstitial yellowish melilite. The latter occasionally becomes important enough for the name *melilite leucitite* to be applied (Capo di Bove) (fig. 18.2). In others, plagioclase becomes noticeable (Monte Jugo, Pofi), marking a transition to leucite tephrite.

The *olivine leucitites* ('leucite basalts') of the Eifel district, Germany, have phenocrysts of olivine and of augite that is largely pale green to yellow in a fine-grained groundmass of leucite, augite and iron-ore with accessory nepheline, biotite, apatite, and sometimes hauyne. Those of the Laacher See district have phenocrysts of biotite, partly or wholly altered to rhönite, in addition to olivine and augite and are associated with nepheline-bearing leucitites, many of which are rich in hauyne.

Rare volcanic rocks are also known in which melilite becomes dominant over, or occurs to the exclusion of, any feldspathoid. These are the *melilitites* and *olivine melilitites* ('melilite basalts').

Melilitites may be exemplified by those occurring at the active volcano of Mt. Nyiragongo, Congo (17), where they are associated with leucitites and nephelinites, and olivine melilitites by those of the Swabian Alps, Württemberg, Germany.

Chemistry

Urtite, ijolite and melteigite (table 18.1, columns 1 to 3). All show very low SiO_2 and high to very high alkalies. Note the relatively high P_2O_5; also the presence of CO_2 due to the frequent occurrence in these rocks of primary calcite and cancrinite. The decrease of Al_2O_3 and alkalies, and the increase in total iron, MgO and CaO on passing from urtite to melteigite reflects the increasing percentage of ferromagnesian constituents in the same direction. There is a corresponding increase in TiO_2, present here largely as a constituent of iron-ore, perofskite and melanite garnet. The K_2O present in these rocks is almost entirely in the nepheline.

The ultra-alkaline volcanic rocks (table 18.1, columns 4 to 9). These again are characterised by very low SiO_2. All are rich in lime and also, in general, in MgO and, relatively, in TiO_2 and P_2O_5. Note that although the nephelinites and olivine nephelinites are the volcanic representatives of the urtite–melteigite group, they are not the equivalents of most of the latter and, indeed, for all these ultra-alkaline lavas it is difficult to find exact plutonic equivalents. Observe how the presence of melilite leads to the appearance of *cs* in the norm (columns 6 and 7). The higher H_2O content of most of these lavas, especially noticeable in columns 6 and 7, is due largely to the ready alteration of melilite and nepheline, making it difficult to obtain perfectly fresh samples.

TABLE 18.1. *Average chemical compositions of some ultra-alkaline [igneous rocks]*

		1	2	3	4	5	6	7	8	9
SiO_2		42.44	42.13	41.11	40.54	40.56	37.97	37.08	47.11	43.64
TiO_2		0.31	1.69	2.76	3.41	2.90	3.30	3.31	1.25	2.54
Al_2O_3		27.11	18.58	12.78	13.55	11.08	10.83	8.08	15.74	10.28
Fe_2O_3		2.43	4.27	6.41	8.16	4.79	4.61	5.12	4.54	5.11
FeO		1.70	4.12	5.01	6.08	7.64	7.21	7.23	4.54	5.89
MnO		0.10	0.18	0.23	0.25	0.22	0.29	0.18	0.27	0.15
MgO		0.62	3.32	5.91	5.52	13.54	13.79	16.19	5.24	13.86
CaO		4.90	11.59	15.60	13.02	12.92	15.02	16.30	11.01	10.66
Na_2O		13.80	8.97	4.72	4.89	3.15	3.35	2.30	2.02	2.16
K_2O		4.12	2.63	2.69	2.30	1.34	1.25	1.36	6.72	4.09
H_2O^+		0.62	0.93	0.80	1.57	1.10	1.45	1.89	0.87	0.72
P_2O_5		0.42	1.20	1.10	0.70	0.75	1.01	0.96	0.44	0.63
CO_2		1.43	0.39	0.88	–	–	–	–	–	–
or		6.4	7.2	4.7	11.1	–	–	–	15.3	6.9
ab		–	–	–	–	–	–	–	–	–
an		–	2.5	5.6	8.1	12.2	10.8	7.5	14.2	6.1
ne		63.3	41.2	21.6	22.4	14.5	15.3	10.5	9.1	9.9
lc		13.7	6.5	8.9	1.7	6.1	5.7	6.5	19.0	13.8
di	*wo*	2.4	11.0	17.2	16.0	18.1	12.0	10.7	15.7	17.8
	en	1.5	8.3	14.8	13.8	13.8	9.4	8.6	11.4	14.5
	fs	0.8	1.6	–	–	2.4	1.2	0.8	2.8	1.1
ol	*fo*	–	–	–	–	14.0	17.6	22.3	1.2	14.1
	fa	–	–	–	–	2.5	2.5	2.5	0.3	1.2
wo		2.8	7.7	7.4	5.6	–	–	–	–	–
cs		–	–	–	–	1.0	8.8	12.8	–	–
mt		3.5	6.3	8.6	10.7	7.0	6.7	7.4	6.5	7.4
il		0.6	3.2	5.3	6.5	5.5	6.2	6.2	2.3	4.9
ap		1.0	2.8	2.7	1.7	1.8	2.4	2.3	1.0	1.5
cc		3.3	0.9	2.0*	– †	–	–	–	–	–
DI		83.4	54.9	35.2	35.2	20.6	21.0	17.0	43.4	30.6

1. Urtite (average of 7).
2. Ijolite (average of 16).
3. Melteigite (average of 14).
4. Nephelinite (average of 10).
5. Olivine nephelinite (average of 25).
6. Melilite–olivine nephelinite (average of 12).
7. Olivine melilitite (average of 10) (*Bull. geol. Soc. Am.* **65**, p. 1029, no. XV).
8. Leucitite (average of 7) (*Ibid.*, p. 1031, no. X).
9. Olivine leucitite (average of 11) (*Ibid.*, p. 1031, no. XI).

*Also *hm* 0.5 †Also *hm* 0.8

Hypabyssal equivalents

The hypabyssal representatives of the ultra-alkaline plutonic and volcanic rocks differ essentially from the former only in their finer grain. Most that have been described have nepheline as their feldspathoid and, in this connection, the *microijolites* and *micromelteigites* of the Fen district, Norway, and of the Turja Peninsula, Kola, U.S.S.R., may be mentioned. African occurrences include microijolite at Spitzkop, Transvaal, and micromelteigite at Jombo, Kenya. They are found either as dykes or as a marginal facies of ijolitic intrusions and, though commonly aphyric, are sometimes porphyritic with phenocrysts of clinopyroxene (Magnet Cove, Arkansas), or of clinopyroxene and nepheline (Fen).

Carbonatites

The *carbonatites* are interesting rocks found in association with ultra-alkaline, mafic feldspathoidal and alkali ultramafic igneous rocks (18) in a number of localities. Well-known localities are those of Alnö in Sweden (19) and Fen in Norway, but some of the best examples are found in eastern and southern Africa (20), while Magnet Cove, Arkansas, Iron Hill, Colorado (21) and Nemegos in Canada (22) may be cited from among those in the north American continent.

Carbonatites may occur as dykes, as cone sheets, or as plugs often enclosed in a ring complex of feldspathoidal rocks. They are most commonly composed of calcite or dolomite, sometimes of both, and the calcite crystals may have exsolution blebs of dolomite. The carbonate is normally accompanied by minor amounts of other minerals and, of these, magnetite, apatite, pyrochlore, manganiferous biotite, with or without other silicates, and complex rare-earth bearing carbonates are the most important.

Sr, Ba, Nb, rare earths, P and Ti have a relatively high concentration in these rocks, a very different minor element pattern to that of sedimentary carbonate rocks. There is, in fact, no doubt that these carbonatites are magmatic, and evidence in favour of this view is provided both by experimental work and by the occurrence of *carbonatite lavas*. An interesting example of the latter has been recorded from the Oldoinyo Lengai volcano, Tanzania (23). This is made up of stumpy crystals of a complex sodium-calcium-potassium carbonate in a turbid matrix which is mainly hydrated sodium carbonate with minute crystals of fluorite. The lava is black when first extruded on to the floor of the crater but turns white within three days.

There is evidence that the common calcitic and dolomitic carbonatites have crystallised from magmas that were originally alkali-rich but have lost their alkalies to the surrounding rocks (24).

References and notes

(1) King, B. C., 1949. The Napak area of southern Karamoja, Uganda. *Mem. geol. Surv. Uganda* **5**, 37–9.
(2) Pulfrey, W., 1950. Ijolitic rocks near Homa Bay, western Kenya. *Q. Jl. geol. Soc. Lond.* **105**, 425–54.
(3) Erickson, R. L. & Blade, L. V., 1963. Geochemistry and petrology of the alkalic igneous complex at Magnet cove, Arkansas. *Prof. Pap. U.S. geol. Surv.* **425**, 28–30.
(4) Allan, J. A., 1914. Geology of the Field map area, B.C. and Alberta. *Mem. geol. Surv. Canada* **55**, 147–52.
(5) Baker, B. H., 1953. *In* Geology of the Mombasa–Kwale area. *Rept. Geol. Surv. Kenya* **24**, 37–8.
(6) Burgess, C. H., 1941. Igneous rocks of the Highwood Mountains, Montana. Part IV. The Stocks. *Bull. geol. Soc. Am.* **52**, 1814–15.
(7) King, 1949, pp. 26–33.
(8) Davies, K. A., 1952. The building of Mt. Elgon (East Africa). *Mem. geol. Surv. Uganda* **7**.
(9) Smith, W. Campbell, 1938. Petrographic description of volcanic rocks from Turkana, Kenya Colony, with notes on their field occurrence from the manuscript of Mr A. M. Champion. *Q. Jl. geol. Soc. Lond.* **94**, 533–4.
(10) Spencer, A. B., 1969. Alkalic igneous rocks of the Balcones province, Texas. *J. Petrology* **10**, 274.
(11) Tilley, C. E., 1928. A monticellite–nepheline–basalt from Tasmania: a correction to mineral data. *Geol. Mag.* **65**, 29–30.
(12) Winchell, H., 1947. Honolulu series, Hawaii. *Bull. geol. Soc. Am.* **58**, 1–48. Jackson, E. D. & Wright, T. L., 1970. Xenoliths in the Honolulu volcanic series, Hawaii. *J. Petrology* **11**, 405–30.
(13) Tyrrell, G. W., 1928. A further contribution to the petrography of the late-Palaeozoic igneous suite of the west of Scotland. *Trans. geol. Soc. Glasgow* **18**, 276.
(14) Cross, W., 1897. An analcite-basalt from Colorado. *J. Geol.* **5**, 684–93.
(15) Mackenzie, J. D., 1914. The Crowsnest volcanics. *Geol. Surv. Canada Mus. Bull.* **4**, 19–28.
(16) Washington, H. S., 1906. The Roman comagmatic region. *Carnegie Inst. Wash. Publ.* **57**, 121–6, 130–41.
(17) Sahama, Th. G., 1962. Petrology of Mt. Nyiragongo. *Trans. geol. Soc. Edinb.* **19**, 1–28.
(18) Including kimberlite.
(19) Eckermann, H. von, 1948. The alkaline district of Alnö Island. *Sver. Geol. Undersok.*, Ser. Ca, no. **36**.
(20) Smith, W. Campbell, 1956. A review of some problems of African carbonatites. *Q. Jl. geol. Soc. Lond.* **112**, 189–219. King, B. C. & Sutherland, D. S., 1960. Alkaline rocks of eastern and southern Africa. Part I. Distribution, ages and structures. *Sci. Prog.* **48**, 298–321. King, B. C., Le Bas, M. J. & Sutherland, D. S., 1972. The history of

the alkaline volcanoes and intrusive complexes of eastern Uganda and western Kenya. *Q. Jl. geol. Soc. Lond.* **128**, 173–205.

(21) Larsen, E. S., 1942. Alkalic rocks of Iron Hill, Gunnison County, Colorado. *Prof. Pap. U.S. geol. Surv.* **197**-A.

(22) Hodder, R. W., 1961. Alkaline rocks and niobium deposits near Nemegos, Ontario. *Bull. geol. Surv. Canada* **70**.

(23) Dawson, J. B., Bowden, P. & Clark, G. C., 1967. Activity of the carbonatite volcano Oldoinyo Lengai, 1966. *Geol. Runds.* **57**, 865–79.

(24) Gittins, J., Allen, C. R. & Cooper, A. F., 1975. Phlogopitisation of pyroxenite; its bearing on the composition of carbonatite magmas. *Geol. Mag.* **112**, 503–7.

19
Lamprophyres

The lamprophyres form a rather peculiar group of fine-grained hypabyssal rocks, occurring typically in thin dykes or sills. They differ from normal types containing the same essential minerals by the marked abundance of their ferromagnesian minerals and by their characteristic texture. Another striking feature is the very common presence of alteration products, especially calcite and chlorite.

Some, the *calc-alkali lamprophyres*, are associated with intrusions of calc-alkali igneous rocks such as diorite, granodiorite and granite; others, the *alkali lamprophyres*, are associated with intrusions of alkali igneous rocks, especially nepheline syenites.

Mineralogy

One of the commonest ferromagnesian constituents is a magnesian *biotite* in more or less euhedral tablets. It frequently shows colour variation, being commonly darker at the borders than in the core of the crystal. It may be found partly or wholly altered to chlorite.

Amphibole is a constituent of some, building relatively long prismatic crystals. It is a common green or brown hornblende in calc-alkali lamprophyres but a deep brown alkali variety in the alkali lamprophyres.

Clinopyroxene is present in many. It is a colourless to pale green augite in calc-alkali lamprophyres, and this is often pseudomorphed by a mixture of chlorite, calcite and iron oxide. In the alkali lamprophyres it is titanaugite.

Orthopyroxene is found occasionally accompanying augite, but is rare.

Magnesian olivine is found in euhedral crystals or more rounded grains in a number but, especially in calc-alkali lamprophyres, it is commonly pseudomorphed by serpentine (or chlorite) and carbonate with or without magnetite. At other times it is replaced by fibrous amphibole, giving rise to the so-called 'pilite' pseudomorphs. *Monticellite* is present in certain alkali lamprophyres.

The ferromagnesian constituents of a lamprophyre may occur both as phenocrysts, usually small, and in the groundmass. The light constituents, on the other hand, are virtually confined to the groundmass. The *feldspars* that may be present include both lime-bearing plagioclase and potash feldspar. While at least partially euhedral, they not infrequently have a divergent or sheaf-like arrangement. Some of the more siliceous calc-alkali lamprophyres carry a little interstitial *quartz*, which may be intergrown with the feldspar. Some alkali lamprophyres have interstitial *analcite* and, in some extreme types, this mineral occurs to the exclusion of feldspar. In another extreme type, *melilite* becomes important.

Among the *accessory constituents* apatite needles are common, and often abundant, though not always easily seen; iron-ore is variable in amount; pyrite is found in many calc-alkali lamprophyres, and perofskite in some alkali lamprophyres.

Texture and structure

The texture of these rocks is governed by the euhedral character of the constituent minerals, giving rise to what is known as *panidiomorphic texture*. A number also show *porphyritic texture*, with phenocrysts confined to the ferromagnesian minerals, as already mentioned. A common feature is *ocellar structure*, the ocelli being small rounded bodies giving circular or oval outlines in section. They may consist, for instance, of an outer margin of plagioclase crystals projecting into the interior of the ocellus which is filled with large plates of calcite; or calcite may form the centre, surrounded by a zone of chlorite and this, in turn, margined by a zone of quartz. In almost all cases, laths of biotite or some other ferromagnesian mineral are arranged tangentially round the outside of the bodies.

Some varieties with examples

1. Calc-alkali lamprophyres

These may be classed as mica lamprophyres or as hornblende lamprophyres. *Mica lamprophyres* having potash feldspar as their dominant feldspar have been termed *minettes*, while those in which lime-bearing plagioclase is the dominant feldspar have been termed *kersantites*. However, the altered nature of the feldspar in some of these rocks renders determination difficult, so that these names cannot always be applied.

Mica lamprophyres are well displayed in the southern part of the Lake District and elsewhere in the north of England (1) (fig. 19.1A, B). They include both minettes and kersantites and many contain subordinate augite, often altered. Some had olivine originally, now pseudomorphed. They are also found in connection with the dyke phase of Lower Old Red Sandstone igneous activity in Scotland. Those of Wigtownshire, for instance, are kersantites (2) with pseudomorphs after pyroxene and olivine, and occasionally show good ocellar structure.

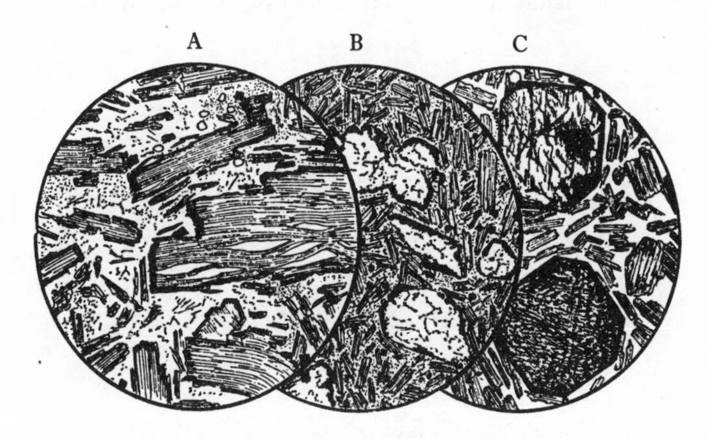

Fig. 19.1. Mica lamprophyres; × 18.

A. Helm Gill, near Dent, Yorkshire. The mica flakes have a dark border and a bleached interior.

B. Rawthey Bridge, near Sedbergh, Yorkshire. Shows original olivine now pseudomorphed by carbonate with a border of iron oxide.

C. St. Helier, Jersey. With octagonal cross-sections of augite, largely replaced by secondary products.

Further examples, largely augite-bearing minettes and usually much altered, are to be seen in Cornwall (near Falmouth, Newquay, etc.) (3) and in the Channel Islands, especially Jersey (4) (fig. 19.1C).

Both minettes and kersantites occur among the lamprophyres of the Odenwald, the former, in particular, having minor interstitial quartz sometimes intergrown with the feldspar. Some of the rocks contain hornblende, commonly as phenocrysts, and are transitional to hornblende lamprophyres. The same is true of some of the kersantites associated with the Lausitz and Riesengebirge's 'granites'.

A feature of many mica lamprophyres is the occurrence of xenocrysts of rounded quartz and feldspar, the former surrounded by a narrow rim of augite grains (or their alteration products) due to reaction with the magma. Xenocrysts of plagioclase may, similarly, have a narrow shell of potash feldspar.

Hornblende lamprophyres with potash feldspar dominant have been termed *vogesite*, those with dominant lime-bearing plagioclase, *spessartite*. Both occur in various parts of the Highlands of Scotland. Relatively fresh vogesite and spessartite dykes of Devonian age with green hornblende, variable amounts of biotite and some interstitial quartz cut rocks of the Strontian complex, Argyllshire (5), while somewhat altered examples are found in the Assynt district (6) (fig. 19.2A). These have

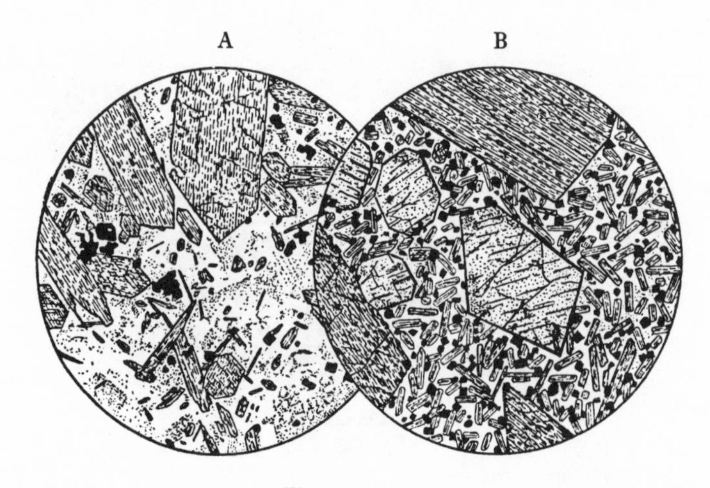

Fig. 19.2. × 20.

A. Hornblende lamprophyre (vogesite), Loyne Bridge, Assynt, Scotland. Euhedral green hornblende, potash feldspar, a little plagioclase and accessory iron-ore.

B. Camptonite, Maena, near Gran, S. Norway. A porphyritic type with conspicuous phenocrysts of kaersutite, titaniferous augite and rare altered olivine (not shown), in a groundmass of kaersutite, plagioclase and abundant accessory iron-ore.

phenocrysts of hornblende, sometimes colour zoned, in a base of hornblende, feldspar and a little interstitial quartz. Some have phenocrysts of augite, or its pseudomorphs, in addition. The feldspar is sericitised with development of chlorite and epidote but is, apparently, albite-oligoclase with variable orthoclase.

The vogesites of the Odenwald have a rather pale brown hornblende and include types with more or less biotite transitional to minette and others with relatively abundant plagioclase as well as biotite, transitional to hornblende-rich kersantite.

Some of the spessartites associated with the Lausitz 'granite' are aphyric with olive brown hornblende, narrowly margined by green, as acicular or stumpy prisms. Others have phenocrysts of brown hornblende with green margins and the groundmass hornblende is green. Augite is common, sometimes as phenocrysts, sometimes as little prisms in the groundmass.

2. *Alkali lamprophyres*

The most important alkali lamprophyres are camptonite, monchiquite, fourchite and alnoite.

Camptonites have essential deep brown kaersutite and lime-bearing plagioclase. Many, perhaps most, have titanaugite, with or without olivine, in addition and in some rocks the pyroxene is present in excess of amphibole. A number have a small quantity of interstitial analcite, emphasising their alkaline nature. They may be aphyric or porphyritic.

They are known from numerous localities, occurring especially as dykes associated with alkali dolerites or with major intrusions of alkali rocks (Oslo district, Ditro, Monteregian Hills, Kola Peninsula, etc.), a particularly well-known example being the porphyritic camptonite from Maena, Gran, in the Oslo district with conspicuous phenocrysts of kaersutite and titaniferous augite and rarer ones of altered olivine (fig. 19.2B).

Typical camptonites occur in Ross-shire; the Ross of Mull (7), with interstitial analcite; and the Orkney Islands (8). Those of Carboniferous to Permian age in Argyllshire (9) show transitions to alkali dolerite as do those in the northern Highlands of Scotland (10) and those of the Skaergaard Peninsula, E. Greenland (11).

Monchiquites are characterised by having abundant ferromagnesian minerals, among which olivine is essential, set in a more or less colourless isotropic base. This base is commonly identified as analcite, but in some rocks that have been described as monchiquite it is, apparently, made up of glass. The common ferromagnesian constituents, other than olivine, are titanaugite and kaersutite. Less commonly, biotite is present.

Fourchite is a name that has been given to the same kind of rock without olivine (fig. 19.3).

Monchiquites and fourchites are associated with camptonites at many of the localities already mentioned. The monchiquites of the Orkney Islands have small phenocrysts of olivine, usually pseudomorphed and often abundant, commonly accompanied by some of titanaugite and sometimes kaersutite. The groundmass has numerous little prisms of pyroxene and grains of iron-ore, set in a mesostasis of analcite or of brownish glass (12). In most cases, however, this mesostasis has been partly, or even wholly, replaced by an aggregate of carbonate, zeolites and serpentinous material. Small flakes of biotite are usually present and in some rocks ragged poikilitic plates of this mineral become abundant. Other rocks here have some nepheline accompanying analcite. Many are amygdaloidal, the amygdales being filled with analcite and calcite.

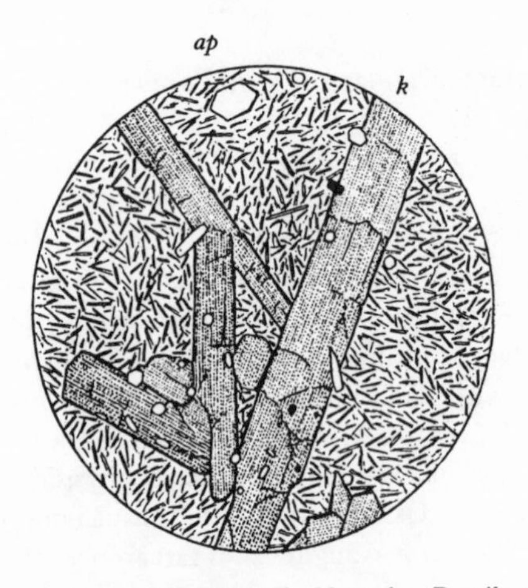

Fig. 19.3. Fourchite, Fernando Noronha, Brazil; × 20.

Phenocrysts of kaersutite (*k*) and apatite (*ap*), in a groundmass of analcite crowded with little acicular crystals of kaersutite.

Fourchites, as well as monchiquites, have been recorded from the Ross of Mull and from Colonsay (13).

Alnoite represents a more extreme type composed essentially of olivine, biotite and melilite, though most contain some pyroxenes and many have monticellite as well.

Good examples have been described from Isle Cadieux (14) and elsewhere (15) in Quebec. The main type at Isle Cadieux is a fine-grained rock with large poikilitic plates of biotite up to 1 cm across full of corroded grains of magnesian olivine and minor brownish to greenish augite. Monticellite, in smaller grains, also encloses the olivine and augite, and there is a variable amount of melilite in small tabular crystals that may be zoned. Apatite in relatively large grains is very abundant, perofskite quite abundant. Carbonate, often crowded with apatite needles, is said to be present mainly as an alteration product of melilite and monticellite, but in rocks from elsewhere the carbonate is thought to be largely primary (Ste. Monique, Como). Melanite garnet appears as an additional constituent in the alnoite at Husereau Farm, St. Joseph du Lac.

Some alnoites are pyroxene-free as is that forming a small pipe-like mass, near Avon, Missouri (16), which has olivine, mostly serpentinised, as large phenocrysts as well as in the groundmass. In others, hauyne or nosean appears as an additional constituent with or without a little nepheline, as in the alnoite found west of Winnett, Fergus Co., Montana (17) and in the so-called 'polzenites' of the Zeughaus dykes, Lausitz, and of the Polzen district, N. Bohemia.

Chemistry

Average chemical compositions of some lamprophyres are given in table 19.1. Average calc-alkali lamprophyres (table 19.1, columns 1–4) have about 50 per cent SiO_2 with moderate Al_2O_3 and with MgO about equal to CaO, and moderate amounts of iron. Alkalies vary considerably in abundance according to the type of lamprophyre, being highest in average minette and lowest in average spessartite, as might be expected. K_2O is strongly dominant over Na_2O in minette, where both biotite and potash feldspar are important, but about the same as Na_2O in kersantite, where potash feldspar is very subordinate or absent. It is slightly dominant over Na_2O in vogesite, where potash feldspar accompanies hornblende, but is dominated by Na_2O in spessartite, where plagioclase is the essential feldspar.

As already remarked, most lamprophyres are altered to some extent, and this accounts for the presence of CO_2 and for the relatively high content of H_2O^+ even in those rocks that do not contain mica.

Average chemical compositions of some alkali lamprophyres are given in table 19.1 columns 5–8. Average camptonite, monchiquite and fourchite have low SiO_2 with moderate Al_2O_3. They have roughly the same MgO contents as average calc-alkali lamprophyres but this is associated with higher CaO and they are also richer in iron. They have higher Na_2O than calc-alkali lamprophyres, and, in general, lower K_2O.

Average alnoite is characterised by very low SiO_2, low to very low Al_2O_3, very high MgO and CaO and rather low alkalies with K_2O in excess of Na_2O.

Note that TiO_2 is more abundant in alkali than calc-alkali lamprophyres as also, in general, is P_2O_5. Note further the general presence of *hy* and absence of *ne* in the norms of the calc-alkali lamprophyres, the reverse being the case with the alkali lamprophyres.

TABLE 19.1. *Average chemical compositions of lamprophyres*

		1	2	3	4	5	6	7	8
SiO_2		50.78	51.00	53.27	50.16	44.40	41.56	42.47	31.58
TiO_2		1.46	1.12	1.21	1.18	3.16	3.63	2.41	2.54
Al_2O_3		12.92	15.33	14.91	15.97	14.31	13.49	15.29	8.88
Fe_2O_3		3.20	3.46	2.99	3.03	3.95	5.37	5.17	5.12
FeO		4.51	5.53	5.38	6.68	7.39	7.78	5.67	6.16
MnO		0.15	0.05	0.08	0.18	0.24	0.17	0.17	0.22
MgO		8.09	7.53	6.48	7.25	7.08	8.09	6.45	18.67
CaO		7.78	7.20	6.08	8.09	10.55	11.02	11.84	17.64
Na_2O		2.09	2.90	3.02	2.92	3.21	3.87	3.79	1.30
K_2O		5.48	3.19	3.45	1.97	1.66	1.20	2.27	2.12
H_2O^+		1.82	1.89	2.32	1.99	1.87	2.57	2.91	1.68
P_2O_5		0.91	0.42	0.35	0.32	0.59	0.92	0.60	1.51
CO_2		0.81	0.38	0.46	0.26	1.59	0.33	0.96	2.51
qz		–	–	1.7	–	–	–	–	–
or		32.8	18.9	20.0	11.7	10.0	7.2	13.3	–
ab		17.8	24.6	25.1	24.6	23.1	12.6	7.3	–
an		9.2	19.2	17.2	24.7	19.5	15.9	18.1	12.2
ne		–	–	–	–	2.3	10.8	13.3	6.0
lc		–	–	–	–	–	–	–	3.5
kp		–	–	–	–	–	–	–	4.4
di	*wo*	7.8	4.8	3.3	4.9	8.0	12.6	12.9	–
	en	5.9	3.4	2.2	3.1	5.6	9.4	10.0	–
	fs	1.1	0.9	0.8	1.5	1.7	2.0	1.5	–
hy	*en*	1.3	8.1	14.0	9.9	–	–	–	–
	fs	0.3	2.4	4.8	4.4	–	–	–	–
ol	*fo*	9.1	5.1	–	3.6	8.5	7.6	4.3	32.7
	fa	1.6	1.7	–	1.7	3.0	1.6	0.7	2.6
cs		–	–	–	–	–	–	–	15.2
mt		4.6	5.1	4.4	4.4	5.6	7.9	7.7	7.4
il		2.9	2.1	2.3	2.3	6.1	6.8	4.6	4.9
ap		2.1	1.0	0.8	0.7	1.3	2.2	1.4	3.7
cc		1.8	0.9	1.0	0.6	3.6	0.8	2.2	5.7
DI		50.6	43.5	46.8	36.3	35.4	30.6	33.9	13.9

1. Minette (average of 23).
2. Kersantite (average of 12). } Mica lamprophyres
3. Vogesite (average of 6).
4. Spessartite (average of 7). } Hornblende lamprophyres
5. Camptonite (average of 14).
6. Monchiquite (average of 8).
7. Fourchite (average of 9).
8. Alnoite (average of 6).

The strongly undersaturated nature of alnoite leads to the presence of *lc* and *kp* as well as *ne* in the norm with abundant *cs*, due to modal melilite.

Note, finally, the lower differentiation indices of average alkali lamprophyres when compared with those of average calc-alkali lamprophyres.

References and notes

(1) Harker, A., 1892. Lamprophyres of the north of England. *Geol. Mag.* **9**, 199–206.

(2) Read, H. H., 1926. The mica-lamprophyres of Wigtownshire. *Ibid.* **63**, 422–9.

(3) Flett, J. S., 1906. *In* Geology of the country near Newquay. *Mem. geol. Surv. U.K.*, 58–61. Smith, H. G., 1929. Some features of Cornish lamprophyres. *Proc. Geol. Ass. Lond.* **40**, 260–8.

(4) Smith, H. G., 1936. New lamprophyres and monchiquites from Jersey. *Q. Jl. geol. Soc. Lond.* **92**, 365–81.

(5) Gallagher, M. J., 1963. Lamprophyre dykes from Argyll. *Miner. Mag.* **33**, 419–20.

(6) Sabine, P. A., 1953. The petrography and geological significance of the post-Cambrian minor intrusions of Assynt and the adjoining districts of North-West Scotland. *Q. Jl. geol. Soc. Lond.* **109**, 146.

(7) Flett, J. S., 1911. *In* The geology of Colonsay and Oronsay with part of the Ross of Mull. *Mem. geol. Surv. U.K.*, 90.

(8) Flett, J. S., 1935. *In* The geology of the Orkneys. *Mem. geol. Surv. U.K.*, 174–7.

(9) Gallagher, 1963, pp. 422–7.

(10) Ramsay, J. G., 1955. A camptonitic dyke suite at Monar, Ross-shire and Inverness-shire. *Geol. Mag.* **92**, 297–309.

(11) Vincent, E. A., 1953. Hornblende-lamprophyre dykes of basaltic parentage from the Skaergaard area, East Greenland. *Q. Jl. geol. Soc. Lond.* **109**, 21–47.

(12) Flett, 1935, p. 181.

(13) Flett, 1911, pp. 43–6, 90–1.

(14) Bowen, N. L., 1922. Genetic features of alnoitic rocks at Isle Cadieux, Quebec. *Am. J. Sci.* (5) **3**, 1–34.

(15) Stansfield, J., 1923. Extensions of the Monteregian petrographical province to the west and north-west. *Geol. Mag.* **60**, 433–53.

(16) Singewald, J. T. & Milton, C., 1930. An alnoite pipe, its contact phenomena, and ore deposition near Avon, Missouri. *J. Geol.* **38**, 54–66.

(17) Ross, C. S., 1926. Nephelite-hauynite alnoite from Winnett, Montana. *Am. J. Sci.* (5) **11**, 218–27.

20

Pyroclastic rocks (1)

Volcanism may give rise not only to extrusion of lava, but also to the products of explosive activity, the *pyroclastic rocks* (2). Lava, already solidified in the crater, is disintegrated into fragments and dust, while some of the liquid lava in the throat of the volcano may be blown out to solidify rapidly in the air as rounded volcanic bombs, the smaller lapilli, or fragments of glass. This fragmental matter settles to form beds of volcanic agglomerate, composed dominantly of coarse fragments, or beds of volcanic ash, composed dominantly of very fine material.

The deposition of ash from the air, in this way, constitutes an *ash-fall* or, when indurated, an *ash-fall tuff*. This may be composed of a mixture of tiny rock fragments, crystals, and glass shards with their characteristic cusp-like outlines but, not infrequently, one component is dominant and the tuff is then described, according to its *character*, as a lithic tuff, a crystal tuff, or a vitric tuff. Alternatively, the tuff may be described according to its *composition* as a rhyolite tuff, a trachyte tuff, a basalt tuff, and so on.

Ash-fall tuffs are generally well stratified and types of varying character may be found alternating with one another.

An *ash-flow* is formed when a fluidised mixture of gas and ash is ejected explosively at high temperature from a vent or fissure and travels rapidly down the slopes of a volcano or along the ground surface. *Ash-flow tuffs* are dominantly vitric in character, the violent explosion of viscous magma, commonly of rhyolitic to rhyodacitic, or dacitic composition, resulting in the formation of glass shards. Crystals ('phenocrysts') of felsic minerals such as plagioclase, sanidine and quartz, and of mafic minerals such as pyroxenes or hornblende and biotite are usually present as well but in widely varying amounts in different occurrences. Lithic fragments may be found also and, in particular, fragments of pumice (often in a collapsed state) are almost universal.

The glass shards may remain plastic enough to become partially or wholly welded together. Such *welded ash-flow tuffs* have been termed ignimbrites (3) (fig. 20.1).

Ash-flow tuffs form units some 10–200m in thickness and are unsorted and unstratified. Another feature, not found normally in either ash-fall tuffs or lavas of similar composition, is the uniformity of individual units over very wide areas.

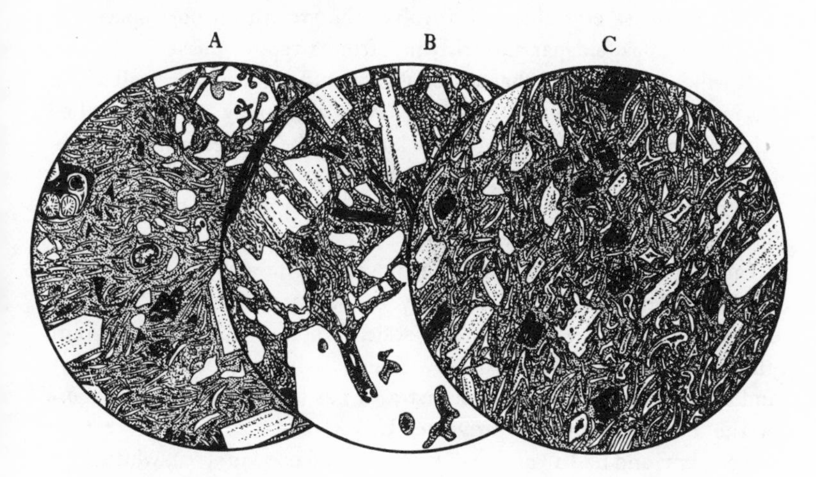

Fig. 20.1. Welded ash-flow tuffs (ignimbrites), New South Wales; × 15. A. From Westbrook; B. From Paterson-Lamb's Creek area; C. From Martin's Creek.

In general, if a flow unit is thick enough, an upper zone of no welding will merge downwards into a zone of partial welding which will be succeeded by a zone of complete welding, separating these upper zones from lower zones of partial and no welding. Where sufficient lateral thinning of such a flow occurs, the zone of complete welding will be absent and the upper and lower zones of partial welding will merge (4).

A number of ash-flow tuff units show vertical compositional zoning (5).

Crystallisation takes place in many ash-flow tuffs and may include both *devitrification* and *vapour phase crystallisation*.

One of the commonest effects of devitrification is the replacement of glass shards by a very fine-grained parallel intergrowth consisting, in the more recent ash-flow tuffs, of feldspar and cristobalite. The two minerals begin to crystallise at the margins of a shard and move inwards to meet at a central discontinuity, marked by a dark line. This is the axiolitic texture (6), observed only in ash-flow, and not in ash-fall, tuffs.

Devitrification may, however, be more extensive with feldspar–cristobalite intergrowths or aggregates showing continuity over considerable areas of tuff structure, and then the original shards may be obscured to a varying degree. The development of *spherulites* with their radial disposition of feldspar and cristobalite, a feature that is by no means uncommon, usually results in the destruction of the tuff structure.

Vapour phase crystallisation involves the growth, in pore spaces, of crystals such as feldspar and tridymite from a vapour phase.

It will be appreciated that some ash-flow tuffs in which crystallisation has occurred may closely resemble lavas, and many rocks described originally as rhyolites, etc., are now known to be welded ash-flow tuffs.

In pre-Tertiary ash-flow tuffs, quartz takes the place of cristobalite and tridymite. These minerals were presumably present originally but have inverted in the course of time.

Small deposits of ash-flow tuff may come from a single volcano (7); the larger deposits from a series of vents or from fissures (8) and some of these are on a very extensive scale. Much of the great 'rhyolite plateau' of the North Island of New Zealand is built up of such rocks (9), and they cover large areas in Sumatra. It has been estimated that those of the Basin and Range province, U.S.A., once covered more than 200 000 km^2 and had a volume of 65 000 to 100 000 km^3 (10), while those in the Andes between latitudes 16° and 27°S. cover an area of some 150 000 km^2 (11).

All those mentioned have been of Tertiary or more recent age, but ash-flow tuffs are known from earlier geological periods. Thus the welded tuffs north of Helena, Montana, U.S.A., are of Cretaceous age (12), while in the British Isles, for instance, examples can be found of Old Red Sandstone age in Scotland (13), of Ordovician age in the Lake District (14), of Ordovician and Precambrian ages in north Wales (15), and of Precambrian age in Shropshire (16).

References and notes

(1) Much of this chapter is based on an excellent paper by Ross, C. S. & Smith, R. L., 1961. Ash-flow tuffs: their origin, geologic relations and identification. *Prof. Pap. U.S. geol. Surv.* **366**, 81 pp.

(2) An interesting classification scheme for explosive volcanic eruptions, based on the area of dispersal and degree of fragmentation of the resulting pyroclastic deposits, is given in Walker, G. P. L., 1973. Explosive volcanic eruptions – a new classification scheme. *Geol. Runds.* **62**, 431–46.

(3) Marshall, P., 1935. Acid rocks of the Taupo–Rotorua volcanic district. *Trans. Roy. Soc. N.Z.* **64**, 1–44.

(4) Smith, R. L., 1960. Zones and zonal variations in welded ash flows. *Prof. Pap. U.S. geol. Surv.* **354**-F, 149–59.
(5) Lipman, P. W., Christiansen, R. L. & O'Connor, J. T., 1966. A compositionally zoned ash-flow sheet in southern Nevada. *Ibid.* **524**-F, 1–47.
(6) Ross & Smith, 1961, figs. 71–74, p. 71; fig. 95, p. 77.
(7) E.g. the Cimino volcano, central Italy. See Mittempergher, M. & Tedesco, C., 1963. Some observations on the ignimbrites, lava domes and lava flows of M. Cimino (central Italy). *Bull. Volc.*, ser. II, **25**, 343–58.
(8) Good examples of plugs and dykes, filled with welded tuff, from which some of the surrounding tuff-flows were probably discharged, can be found in: Cook, H. E., 1968. Ignimbrite flows, plugs and dikes in the southern part of the Hot Creek Range, Nye County, Nevada. *Mem. geol. Soc. Am.* **116**, 107–52.
(9) These have been compared with some American examples in Martin, R. C., 1959. Some field and petrographic features of American and New Zealand ignimbrites. *N.Z. J. Geol. Geophys.* **2**, 394–411.
(10) See Cook, 1968, pp. 139–40.
(11) Zeil, W. & Pichler, H., 1967. Die känozoische Rhyolith-Formation im mittleren Abschnitt der Anden. *Geol. Runds.* **57**, 48–81.
(12) Barksdale, J. D., 1951. Cretaceous glassy welded tuffs – Lewis and Clark County, Montana. *Am. J. Sci.* **249**, 439–43.
(13) Roberts, J. L., 1966. Ignimbrite eruptions in the volcanic history of the Glencoe cauldron subsidence. *Geol. J.* **5**, 173–84.
(14) Oliver, R. L., 1954. Welded tuffs in the Borrowdale volcanic series, English Lake District, with a note on similar rocks in Wales. *Geol. Mag.* **91**, 472–83.
(15) Fitch, F. J., 1967. Ignimbrite volcanism in north Wales. *Bull. Volc.* ser. II, **30**, 199–219.
(16) Dearnley, R., 1966. Ignimbrites from the Uriconian and Arvonian. *Bull. geol. Surv. U.K.* **24**, 1–6.

II: Sedimentary rocks

21

Sedimentary rocks – introduction

The physical, chemical, and biological processes of weathering cause disintegration of primary rock fabrics and lead to the redistribution of the rock components, often in a physically and chemically modified form. The products of disintegration include (a) chemically unaltered mineral or rock fragments (resistates), (b) particulate material derived from the alteration of the parent rock constituents, and (c) dissolved material derived from solution or alteration of primary minerals. The mechanical accumulation and chemical precipitation of these components leads to the formation of sediments, and their fractionation by both mechanical and chemical processes leads to the development of the several end members of the sedimentary rock series (1).

Terminology

A number of terms are used in connection with the broad classification of sedimentary deposits, including 'detrital', 'clastic', 'fragmental', 'chemical', and 'precipitated'. Because of their variable usage it is necessary to define these terms for the purposes of this text.

The term *detrital* is applied to solid products of weathering which have been removed from their site of origin. Particles produced by mechanical fragmentation of pre-existing fabrics are termed *clastic.* The terms 'detrital' and 'clastic' are also used for rocks largely composed of detrital or clastic grains, and they are sometimes applied to chemical or organic rocks in order to emphasise the transported or fragmented nature of their constituents. The terms *chemical* or *biochemical* are applied to constituents formed by physicochemical or biochemical precipitation respectively and also to sediments largely composed of such constituents. The term *terrigenous* is used to denote constituents that are the products of subaerial erosion.

The supply of sedimentary constituents is augmented by explosive volcanic activity which produces rock fragments, crystals, lapilli, and volcanic ash. These accumulate to form purely pyroclastic deposits or are deposited in admixture with constituents of detrital origin. Hydrothermal activity releases dissolved components which may give rise to local precipitates.

Deposition and diagenesis

The solid products of weathering may accumulate more or less in place as *residual deposits* but usually they are removed by the agencies of ice, wind, and water to be subsequently deposited when the transporting medium no longer has the power to sustain particle motion. Dissolved components are carried by rivers to lakes or to the sea, where they are added to the general reserve of dissolved salts. These salts are selectively precipitated when and where local conditions are favourable.

Diagenesis may be broadly defined as including the processes of physical and chemical change which take place within a sediment after its deposition and before the onset of either metamorphism or weathering. The earliest diagenetic changes take place at the sediment surface often through inherent disequilibrium between sedimentary particles and the depositional medium. Such processes (which are sometimes included under the term *halmyrolysis*) include solution and chemical alteration and are enhanced by low rates of sedimentation.

Most diagenetic changes take place beneath the surface where connection with the depositional medium is restricted or severed, allowing the development of local chemical environments within the pore waters. Changes in pore water chemistry result partly from chemical decomposition of sedimentary components but the dominant influence in newly buried sediments is that of bacterial activity. Bacteria derive energy by oxidation of organic matter, for which they extract oxygen both from pore solutions and from more readily reduced ions and molecules. The resulting depletion in oxygen and release of carbon dioxide are dominant factors in the early diagenesis of most aqueous deposits.

The processes so far referred to may be included in *chemical diagenesis*. They are accompanied to varying degrees by *physical diagenesis*, involving the reorganisation and compaction of primary fabrics and resulting in reduction in interparticle pore space. In subaqueous deposits pore space reduction causes expulsion and migration of pore fluids, which may become involved in chemical diagenesis.

With increasing depth of burial, sediments are subjected to conditions of high temperature and pressure. These favour large-scale intergranular solution and recrystallisation together with the formation of new minerals with stability fields considerably different from those of primary or early diagenetic origin. Under such conditions diagenesis grades into metamorphism; no precise textural or mineralogical limits can be placed on this extreme of diagenesis. The effects of circulating groundwaters depend on the mineralogy of the rocks concerned; for example sandstones may undergo little change until they physically disintegrate in the weathering process whereas more chemically reactive rocks such as saline deposits may display a variety of textural and mineralogical changes.

The following terms are used to denote the various processes of chemical diagenesis. *Solution* involves the total removal of a mineral without its substitution by another. *Cementation* involves the precipitation of mineral matter in void spaces of either depositional or diagenetic origin. The modification of mineral textures during diagenesis is termed *neomorphism*. Where evidence permits, neomorphic changes may be ascribed to *recrystallisation*, involving textural modification without change in composition, or to *replacement*, involving 'molecule-by-molecule' change in mineral composition. The precipitation of mineral matter during diagenesis is known as *authigenesis*, and the mineral constituents of a sedimentary rock may be classed as *primary* (i.e. present at the time of deposition) or *authigenic*.

Classification and description

The sedimentary rocks may be arbitrarily classed according to their mineralogical, and to a lesser degree physical, constitution. Further subdivision is possible, but the more detailed classifications (which are often designed for specialised studies) are not considered here. The principal sediment classes in this text are as follows:

Detrital silicate deposits
- Argillaceous (clay and silt grade)
- Arenaceous (sand grade)
- Rudaceous (pebble grade)

Residual deposits

Chemical deposits
- Calcareous
- Dolomitic
- Siliceous
- Phosphatic
- Ferruginous
- Manganiferous
- Carbonaceous
- Saline

The processes of mechanical transport and deposition allow the coming together of sedimentary constituents of widely differing mineralogy and origin; very few constituents are restricted to a single type of deposit and many constituents, particularly those of terrigenous and skeletal origin, are common to almost all sedimentary groups. Furthermore the textures of grain accumulation and primary and diagenetic crystallisation are largely independent of mineralogy and environment and thus transgress even the most basic lithological groupings.

For these reasons it should be appreciated that while it is convenient to class sediments into groups, these groups only serve to illustrate extremes of an infinite variety of mineralogical and textural assemblages. This section is accordingly designed to allow the analysis of lithologies not specifically covered in the text. All of the main sedimentary constituents are individually described and discussed in the following chapters and it is hoped that this information, combined with knowledge of the general principles of textural interpretation (next chapter), will allow insight into most sedimentary lithologies encountered in thin section.

References and notes

(1) For more detailed and critical accounts of the classification of sedimentary rocks and of the broader aspects of sedimentology, the following texts are recommended: Blatt, H., Middleton, G. & Murray, R., 1972. *Origin of sedimentary rocks*. New Jersey: Prentice-Hall. Garrels, R. M. & Mackenzie, F. T., 1971. *Evolution of sedimentary rocks*. New York: Norton. Hatch, F. H., Rastall, R. H. & Greensmith, J. T., 1971. *Petrology of the sedimentary rocks*. London: Murby. Pettijohn, F. J., 1975. *Sedimentary rocks*. New York: Harper & Row (3rd ed.). Texts on procedural and descriptive sedimentary petrography include the following: Carozzi, A., 1960. *Microscopic sedimentary petrography*. New York: Wiley. Carver, R. E., ed., 1971. *Procedures in sedimentary petrology*. New York: Wiley–Interscience. Milner, H. B., 1962. *Sedimentary petrography : vol. 1 Methods in sedimentary petrography*; *vol. 2 Principles and applications*. New York: Macmillan.

22

Sedimentary textures

The physical make-up of a sedimentary rock, relating to the size, shape, distribution, and mutual relationships of its components, constitutes the rock texture or fabric. The contrasting modes of accumulation of sedimentary components result in the development of several basic textures which may be grouped as 'particulate' and 'non-particulate'.

Particulate texture results from the accumulation of discrete particles; it is characterised in its unaltered state by point contacts between particles and by the associated intergranular spaces. Particulate texture is characteristic of detrital sediments but also occurs in non-detrital rocks where crystals or organic tests have accumulated through gravity settling.

Non-particulate textures are dominated by the *crystalline texture* which results from the crystallisation within or at the surface of a sediment and is characterised by an interlocking crystal mosaic. Crystalline texture is mostly diagenetic in origin. Other non-particulate textures include those produced by organic binding of sedimentary particles.

In addition to the above rock-forming textures many sediments display a variety of other textures resulting largely from post-depositional modification; these include solution textures and textures produced by deformation.

Particulate texture

Under the microscope particulate texture is directly visible only in relation to particles of silt grade or larger; in finer-grained sediments a 'microparticulate' texture can sometimes be inferred from other criteria. In the more coarsely-grained rocks there is a tendency for the particles to fall into two groups based on size and distribution. These groups are respectively termed the *grains*, or relatively coarse particles, and the *matrix*, consisting of relatively fine particles which appear to be packed around the grains. In many sediments this visual grouping actually represents the presence of two distinct size classes within the particle assemblage (fig. 22.1A); in others (fig. 22.1B) the distinction between grains and matrix is relatively arbitrary but in practice still serves as a useful basis for the description of rock texture and composition.

Where all of the particles are of closely similar size a matrix as such does not exist, and the rock consists wholly of grains. The interstitial

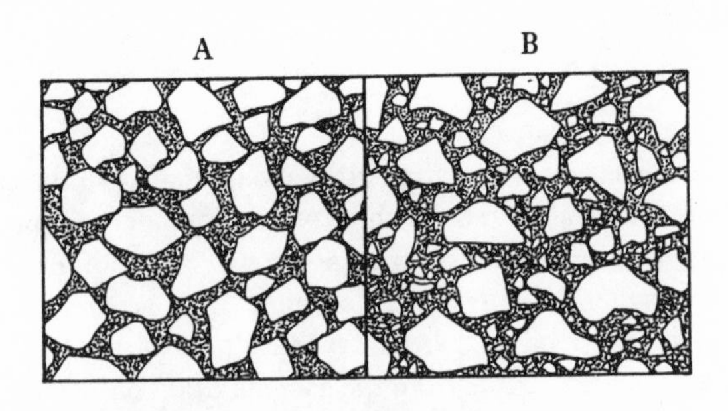

Fig. 22.1. Grains and matrix.

A. Grains and matrix clearly differentiated.
B. Particle assemblage lacking true grain–matrix distinction.

sites may remain open or may become filled during diagenesis by a *cement*, which usually consists of an interlocking crystal mosaic (figs. 22.6 A, B). The original pore space may also be reduced during diagenesis by *intergranular solution* in which adjacent grains undergo solution at their points of contact (fig. 25.2A). This solution increases the surface area of contact and reduces pore space. Ultimately the process may lead to the total loss of pore space, and the resulting aggregate of 'welded' grains may bear a strong resemblance to a crystalline aggregate.

In order to describe grain assemblages a number of properties are taken into account: grain size, grain sorting, grain form, grain orientation, and grain packing. These are dealt with individually in the following sections.

Grain size

The most widely used grain size classification (1) is based on a logarithmic metric scale (table 22.1). Unconsolidated or disaggregated sediments are normally analysed by sieving (for coarser particles) and by settling techniques (for finer particles). In thin section it is customary to measure the apparent long diameter of particles, in contrast to the intermediate

TABLE 22.1. *Grade scales and terminology of sedimentary grains and lithified grain aggregates*

Diam. mm				Diam. ϕ units
		BOULDER	Boulder conglomerate	
256				−8
128		COBBLE	Cobble conglomerate	−7
64				−6
	vc			
32				−5
	c			
16			Pebble conglomerate	−4
	m	PEBBLE		
8				−3
	f			
4				−2
	vf	(GRANULE)		
2				−1
	vc			
1				0
	c			
$\frac{1}{2}$				1
	m	SAND	Sandstone	
$\frac{1}{4}$				2
	f			
$\frac{1}{8}$				3
	vf			
$\frac{1}{16}$				4
	c			
$\frac{1}{32}$			Siltstone	5
	m			
$\frac{1}{64}$		SILT	- - - - - - - - -	6
	f			
$\frac{1}{128}$				7
	vf		Mudstone/shale	
$\frac{1}{256}$				8
		CLAY		

The relationship between the millimetre and ϕ scales is expressed by: $\phi = -\log_2 \text{diam. (mm)}$; vc = very coarse; c = coarse; m = medium; f = fine; vf = very fine.

diameter measured in sieving. Thin section measurements differ more seriously from sieve measurements because of the random sectioning effect (fig. 22.2A). Nevertheless for general purposes size analysis in thin section provides useful information, particularly where comparisons are to be made with other thin section data.

Although the classification of particulate sediments according to grain size is essentially descriptive it also serves as a useful basis for genetic interpretation as the grain size of a sediment is related to the nature and power of the transporting medium, and hence indirectly to the environment of deposition. In certain circumstances, however, the grain size distribution may also reflect limited grain size availability in rocks of the source area.

Grain sorting

The sorting of a grain population expresses the degree to which the grains approach a uniform size. For loose grains the sorting may be expressed quantitatively by means of analysis of the sieve size distribution. In thin section the same drawbacks affect the assessment of sorting as affect the assessment of grain size. For general purposes sand-grade sediments may be classed as well sorted, moderately sorted, and poorly sorted (fig. 22.2). Where the grain population is very distinct from the

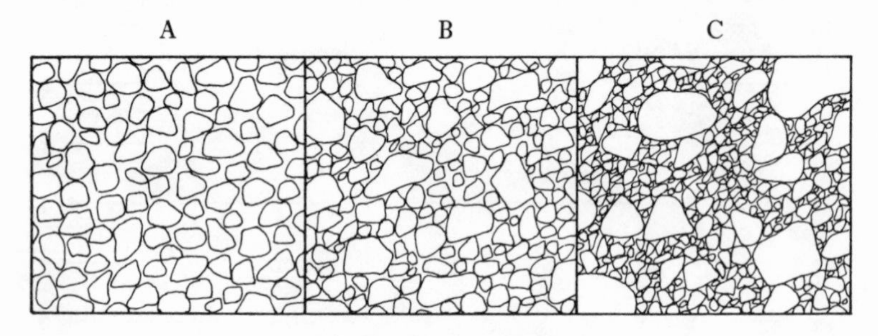

Fig. 22.2. Grain sorting.
A. Well sorted. B. Moderately sorted. C. Poorly sorted.

matrix it may be desirable to assess the sorting of the grains independently. Where two distinct modes are expressed by the grains themselves the grain size distribution is termed *bimodal*.

The sorting of a sediment is ultimately controlled by grain size availability and by processes of deposition. At the site of deposition the important factors are the rate of deposition and the variability of the depositing current. The highest degree of sorting is commonly associated with steady continuous currents or prolonged wave action. The lowest degree of sorting is associated with deposition by mass movement or turbid flow, or with deposition from the melting of debris-laden ice. Where a muddy matrix occurs in relation to a well-sorted grain assemblage the low overall sorting is open to several interpretations: post-depositional downward filtering of mud, bioturbation, or authigenic clay cementation. A high degree of sorting is also open to more than one interpretation owing to the possibility of its being inherent in the grain population of a sandstone source rock.

Grain form

The *form* of a grain may be defined as the expression of its external morphology. It is normally described in terms of shape, sphericity, roundness and surface texture.

The *shape* of a grain is a measure of the relation between its three axial dimensions. For general descriptive purposes four classes may be distinguished: equant, disc-shaped, blade-shaped, and rod-shaped. In thin section the determination of shape is not possible on individual grains, but it may be possible with an assemblage of grains of distinctive and relatively uniform shape, such as mica flakes and many skeletal components. In order to eliminate the effects of preferred orientation sectioning in more than one direction is required.

The *sphericity* of a grain is a measure of its approach to the shape of a sphere. Particles of different shape (e.g. a rod and a disc) may have the same sphericity. In practice sphericity is determined numerically from the measurements of the long, short, and intermediate dimensions of a grain. For thin section purposes a 'projection sphericity' may be used; this is defined as the diameter of the largest inscribed circle of a grain divided by the diameter of the largest circumscribed circle. A more simple measure, which is accurate for most comparative purposes, is that of the ratio of length to width. In either case the usual caution must be exercised in relation to preferred orientation. A simple division into low and high sphericity may be carried out by reference to standard images (2).

The *roundness* of a grain is a measure of the degree of angularity of surface projections, and may be defined as the average radius of curvature of the corners of the grain divided by the radius of the maximum inscribed circle. The two-dimensional aspect of sectioning presents little problem in the accurate determination of roundness but the procedure is too tedious for normal purposes and roundness is normally estimated by reference to a visual standard (3).

Shape and roundness are imparted to a grain prior to its deposition. The degree of roundness is generally equated with the degree of abrasion although it may also be inherited by grain recycling. In comparing roundness of different grain assemblages several factors must be taken into account. The relative hardness of the grains is an obvious factor. More important is the effect of grain size; with decreasing particle size the cushioning effect of the medium becomes more effective, so that rounding is barely developed in grains of silt grade. Comparisons of roundness are therefore best made using sediments of similar grain size. As discussed on p. 250, grain shape can sometimes be used to determine the source rock of terrigenous silicate grains.

Surface texture is a feature not normally distinguishable under the microscope. Textures resulting from surface diagenesis may, however, be visible and usually take the form of pitting due to solution or replacement. Their recognition as diagenetic is essential as the associated angularity would otherwise be interpreted as indicating minimal transport and abrasion.

Grain orientation

The most common expression of this is the bedding orientation of flat or elongate grains. Current flow may cause parallel alignment of elongate particles within the plane of the bedding. Disc or blade-shaped particles, on the other hand, tend to be *imbricated* by current flow (i.e. stacked horizontally with an upstream tilt). Unless particles are of markedly low sphericity, these preferred orientations are detectable only by measurement of a large number of grains. Preferred orientation is important in its effect of imparting directional permeability in reservoir rocks.

Preferred orientation is not normally visible in fine-grained sediments, but where the shape of the particles is related to internal crystallographical arrangement, as in the clay minerals and micas, the orientation will be reflected in aggregate parallel extinction.

Grain packing

Packing involves the mutual arrangement of grains in a three-dimensional framework. In most sediments the term *packing* is used to express the distributional relationship between grains and matrix. Where the grains appear to constitute a three-dimensional framework by virtue of intergranular contact, the sediment is said to be *grain-supported*. Where the grains are dispersed within a matrix to such an extent that three-dimensional continuity does not exist, the rock is described as *matrix-supported*.

A more comprehensive classification in terms of packing has been developed in relation to limestones (4), and seems equally applicable to other rock types. Grain-supported sediments are subdivided into *grainstones* (fig. 22.3A), closely packed aggregates devoid of interstitial support, and *packstones* (fig. 22.3B), relatively loosely packed aggregates in which the proportion of grain contacts is reduced by the presence of an interstitial matrix. The matrix-supported sediments are divided into *wackestones* (fig. 22.3C), with more than 10 per cent grains and *mudstones* with fewer than 10 per cent grains. It is important not to confuse 'mudstone' the texture with 'mudstone' the rock (p. 238).

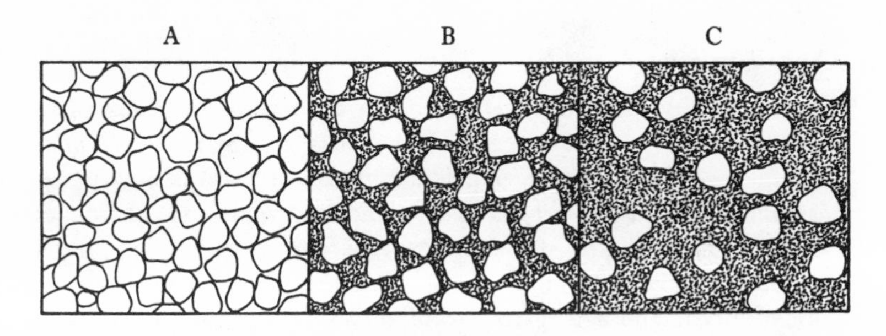

Fig. 22.3. Packing textures.

A. Grainstone texture; grain-supported, without matrix.
B. Packstone texture; grain-supported, with interstitial matrix.
C. Wackestone texture; matrix-supported, with 'floating' grains.

The degree of packing is significant in terms of the deposition of a granular aggregate. Rapid deposition of grains and matrix particles favours the development of wackestones; mild agitation and reworking of the deposited sediment will organise the deposit into a packstone. Deposition under continuously agitated conditions on the other hand favours the development of matrix-free grainstones, with the finer sediment remaining in suspension.

Non-particulate textures

These textures result from the accumulation of constituents by direct adhesion to previously deposited constituents. This adhesion is related to crystal interlocking in the case of precipitated components and to the presence of an organic matrix in the case of agglutinated components.

Accretionary textures

These are displayed in shells and other bodies of direct organic origin (see pages 270–4) and also in accretionary bodies of less certain origin. The latter are typified by grains which show an internal structure of concentric shells; these include *ooliths*, ranging up to 2mm diameter, and *pisoliths*, with diameters greater than 2mm (fig. 25.3A). These consist of a concentrically layered *envelope* surrounding a *nucleus*. Other accretionary structures are those of algal origin which may form continuous layers and beds; distinctive examples are the *stromatolites* which consist of a series of pillars with domed internal lamination.

Crystalline textures

These may first be divided according to the size of the constituent crystals: *macrocrystalline* or *sparry* (greater than 20 μm), *microcrystalline* (less than 20 μm) in which individual crystals are visible only under magnification, and *cryptocrystalline* in which individual crystals cannot be distinguished under the microscope. Cryptocrystalline aggregates may be truly amorphous but are often very finely crystalline.

Some crystalline aggregates are of relatively uniform crystal size; these are described as *equicrystalline* (fig. 22.4A). *Inequicrystalline* textures include the *porphyroblastic texture* (fig. 22.4B), in which a few crystals are distinctly larger than those of the groundmass, and *poikiloblastic texture* (fig. 22.4C), in which one crystal encloses smaller

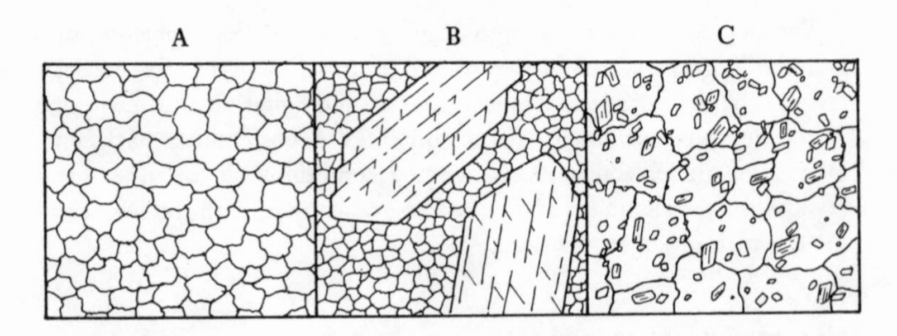

Fig. 22.4. Crystalline textures.
A. Equicrystalline mosaic. B. Porphyroblastic texture. C. Poikiloblastic texture.

crystals or grains of another (5). Other distinctive textures include *fibrous texture* in which crystals are arranged in subparallel fashion to form fringes (fig. 22.5A) or as radiating aggregates to form *radial fibrous texture* (fig. 22.5B), the complete expression of which is *spherulitic texture* (fig. 22.5C).

Most crystalline textures are of diagenetic origin and many primary crystalline textures are extensively modified or destroyed during diagenesis. In order to interpret diagenetic crystal textures it is necessary to distinguish between those which have originated by cementation and those which have resulted from neomorphic alteration of primary sediment. The textures associated with these two processes are not always mutually exclusive, but some of the more distinctive features are described in the following sections.

Crystalline textures associated with cementation. Cementation is the process of chemical precipitation of a crystalline aggregate within pre-existing pore spaces. The texture of the cement depends on factors such as mineralogy, rate of crystallisation, and influence on seeding of the surfaces of crystallisation. Most surfaces of nucleation differ in mineralogy from the cement and in consequence nucleation takes place at numerous closely spaced sites, leading to the formation of a finely crystalline cement layer. Progressive competitive growth of adjacent crystals causes a reduction in the number of growing crystals and a consequent increase in size. The resulting texture of coarsening away from nucleation surfaces (fig. 22.6A) is characteristic of cements, but does not alone constitute an unequivocal distinction from neomorphic textures. One feature that may provide such a distinction is the presence of more than 30 per cent of enfacial junctions among the observed inter-crystal triple junctions (6). An *enfacial junction* is one in which two

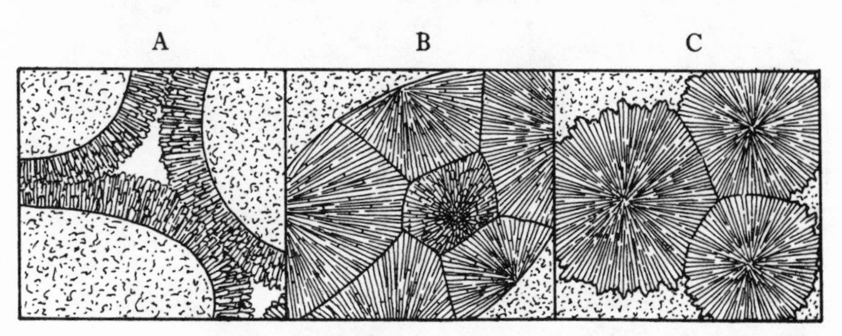

Fig. 22.5. Crystalline fibrous textures.
A. Normal fibrous texture. B. Radial fibrous texture. C. Spherulitic texture.

crystals about a third along a planar surface (fig. 22.6A). This relationship is most likely to occur in cements, where free-growing faces develop without competition from adjacent crystals.

Cements may be incomplete and merely form a *fringe* around the grains; often an initial phase of fringe cementation is followed by a later phase of more extensive and more coarsely crystalline cement (fig. 22.6B). Sharply defined fringes are restricted to cements and thus when present provide definite evidence of cementation as opposed to neomorphism. *Syntaxial rims* or overgrowths (fig. 24.2A) are developed where cement has been precipitated in optical continuity with the crystals of grains of the same mineral composition. This feature is most commonly developed in relation to single crystal grains such as detrital quartz grains and echinoderm fragments. Syntaxial rims normally result from cementation but are rarely associated with neomorphism of a primary matrix (p. 279).

Crystalline textures associated with neomorphism. These include textures resulting from replacement or recrystallisation of an original crystalline, amorphous, or particulate fabric. The products of crystallisation of amorphous matter are varied, but generally display a relatively uniform crystalline texture in which spherulitic or radial elements are sometimes present. Further recrystallisation leads to the formation of progressively coarser crystal mosaics. Similarly homogeneous textures can develop from the neomorphism of primary crystalline or particulate textures, but spherulitic, porphyroblastic, and poikiloblastic textures also develop. Recrystallisation of matrix material may be controlled to some

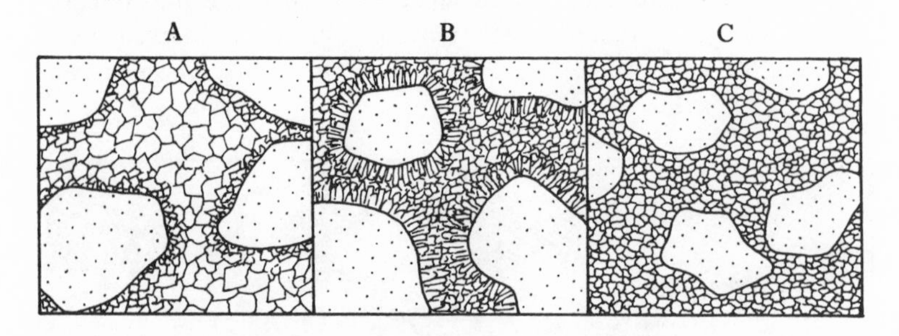

Fig. 22.6. Crystalline textures.

A. Cement texture showing increase in crystal size away from surfaces of seeding; the crystal junctions are enfacial in some cases.

B. Two generations of cement, with early fringe cement followed by equicrystalline cement.

C. Equicrystalline mosaic.

extent by the surfaces of included grains, as where syntaxial rims of neomorphic origin have developed.

Where neomorphism is incomplete the interpretation of crystal textures is relatively straightforward, but in many rocks the neomorphic origin of the textures has to be established from the textures themselves. Useful features in this respect are the presence of dispersed inclusions of the primary deposit – either corroded relics of primary chemical constituents or dispersed inclusions of clay, organic matter, or diagenetic minerals (e.g. pyrite). Such inclusions are unlikely to occur in cements. The packing of grains in a sediment may also give some indication of the presence of a primary mud matrix, especially in the mudstones and wackestones. In the case of packstones, however, the distinction between neomorphic and cement textures is less readily made (fig. 22.6C); evaluation of the proportion of enfacial junctions (see preceding section) is perhaps the best means of distinction as in neomorphic aggregates the proportion ranges up to only 5 per cent.

Neomorphism may also affect cement textures, sometimes by recrystallisation or by replacement, as in the transformation of aragonite to calcite.

Deformation textures

Physical deformation due to compactional stresses is a striking feature of sediments possessing inherent structural weakness; in most sediments, however, it is one of the least conspicuous of the diagenetic effects. Deformation takes place either by squeezing of a matrix between compacting grains (fig. 22.7A) or by distortion of the grains themselves.

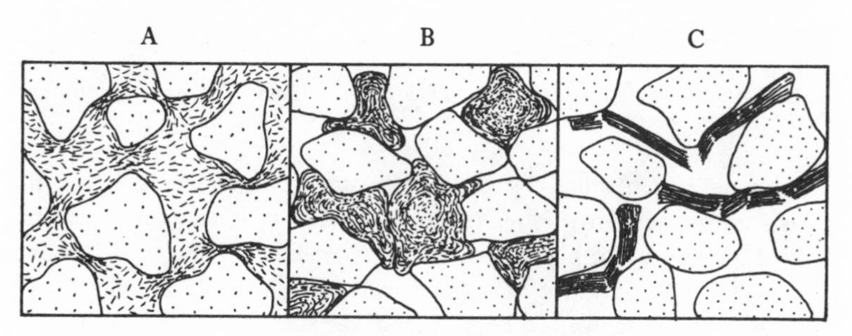

Fig. 22.7. Deformation textures.

A. Compaction of matrix mud around included grains, with local reorientation of clay mineral flakes.

B. Plastic deformation of incompetent grains in a grain-supported framework.

C. Brittle fracture of weakly structured grains in a grain-supported framework.

Grain weakness may be related to shape and packing, to structural weakness, or to plasticity (figs. 22.7B, C). Plastic deformation of resistant crystals is sometimes revealed by the development of strained extinction, and that of clay particles is often indicated by alignment of the clay flakes causing local zones of aggregate parallel extinction.

Solution textures

Solution by purely chemical action on a sedimentary component may be termed *passive solution*; it typically leads to the formation of a cavity within a rock which may become the site for cementation or which may be compressed during compaction in the absence of prior cementation (fig. 23.2B). In other instances solution appears to have resulted from (or at least to have been related to) directed pressure; this is known as *pressure solution.* Pressure solution commonly takes the form of intergranular solution in which original point contacts are reduced to surface contacts by mutual or one-sided solution (fig. 22.8A). Sometimes the

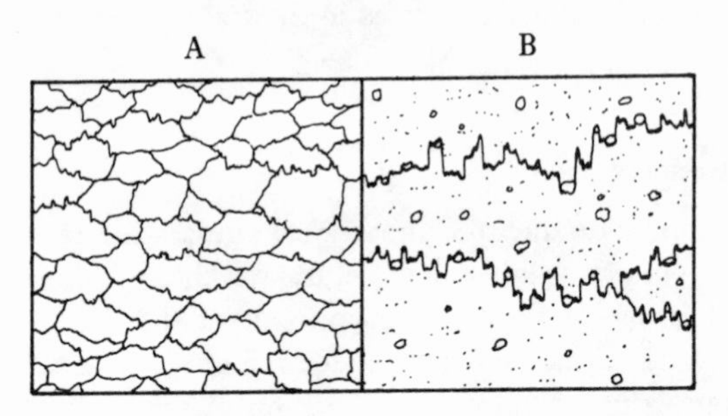

Fig. 22.8. Solution textures.

A. Intergranular solution, with the development of irregular and locally microstylolitic sutures.
B. Stylolite, showing concentration of insoluble material along the solution surface.

solution is accompanied by reprecipitation in immediately adjacent areas – a process known as *solution transfer.* The contacts between such welded grains may be planar, curved, or indented.

In rocks of fine-grained homogeneous texture, the solution effects of directed pressure may affect large areas of the rock fabric. Such solution is expressed in the form of irregular layers along which insoluble residue (clay, organic matter, etc.) is concentrated. The solution surfaces may be more or less planar but more commonly display a structure of interlocking columns attached to the opposing bodies of the rock (fig. 22.8B);

these columnar surfaces are known as *stylolites*. Evidence for solution lies partly in the concentration of insoluble matter but also in the truncation of structures within the rock on either side.

Solution surfaces may develop in relation to local inhomogeneities and particularly bedding planes; in this way they may accentuate lithological contrasts by eliminating gradational contacts. Solution surfaces which have developed along bedding planes are probably related to overburden pressure; those developed at an angle to the bedding may be of tectonic origin. Although it is not possible to give an accurate figure of total volume loss along solution surfaces, the amplitude of opposing projections of stylolitic surfaces gives a minimum figure for thickness loss (7).

Shrinkage textures

These include the internal shrinkage cracks of septarian nodules (p. 239) and shrinkage associated with the carbonation of wood fragments. Such cracks are often filled with cement.

Geopetal textures

Geopetal textures are those which permit determination of the 'way-up' of a sedimentary unit. In this text only those of post-depositional origin will be considered. The most conspicuous of these result from accumulation of downward filtering mud in cavities within the deposited sediment. The cavities are usually primary but exceptionally have developed during very early diagenesis by solution or shrinkage. The former must be interpreted with caution as internal shell sedimentation may take place before final deposition of the shell; accordingly way-up should be deduced from more than one internal fill showing similar orientation (fig. 22.9A). Other way-up structures are produced by solution or collapse (fig. 22.9B); these are generally reliable although they may possibly be of late origin and develop in relation to tectonically tilted beds.

Displacive crystallisation textures

This process is comparable to ice-wedging. A relatively common occurrence is one in which carbonate or sulphide crystallisation within mica crystals causes disruption and expansion of the original crystals (fig. 22.9C). Cone-in-cone texture (p. 244) also appears to be of displacive origin.

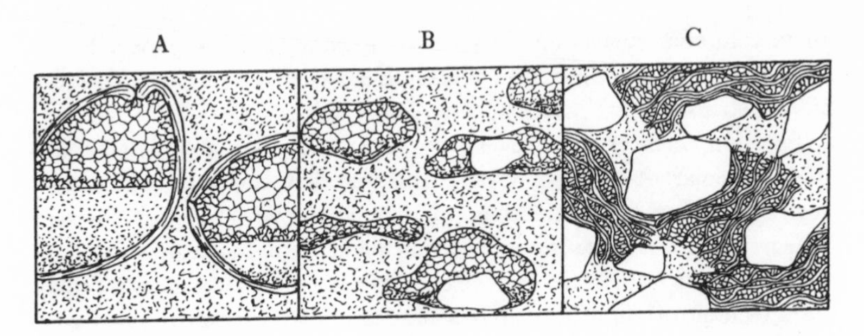

Fig. 22.9.

A. Geopetal texture; mud has accumulated within the shell cavities after the final accumulation of the shells.
B. Geopetal texture; oolith nuclei and thin residual films have collapsed under gravity after solution of the oolitic envelope.
C. Displacive crystallisation within a mica flake, causing disruption and expansion of the crystal normal to the basal cleavage.

Establishing a diagenetic sequence

By examining the mutual relationships of diagenetic textures it is often possible to establish a sequence in the diagenetic history of a sediment. Four major processes are involved: compaction, solution, cementation, and neomorphism, and it is by studying the interaction of these processes that the main sequence of diagenetic events can be established. The sequence is often complicated by the multiphase nature of one or more of the main processes (e.g. two generations of cement). Of the four processes, solution, cementation, and neomorphism are usually genetically connected.

The time relationship between chemical diagenesis and compaction provides a useful starting point in determining a diagenetic sequence. Where solution occurs before compaction in a matrix-supported deposit the dissolved component may be completely obliterated or be represented by a compressed film of insoluble matter (fig. 23.2B). Where compaction has imparted rigidity to the matrix, solution of a component produces a cavity which may later undergo cementation. Cavities may also be produced by relatively late solution which follows cementation.

In grain-supported rocks compaction is represented by fracture or distortion of grains and by intergranular solution. Where these features are displayed in a fully cemented rock it is clear that cementation has followed compaction. At the other extreme, total cementation prior to compactional pressure would prevent any actual deformation. Where cementation is in the form of thin fringes the effects of compaction may be evident in the form of dislocation, buckling, and fracturing of the fringes (fig. 25.1A); in such cases the compaction phase is generally followed by wholesale cementation.

The relative timing of compaction and replacement can sometimes be determined, particularly where replacement has caused local modification in rigidity (fig. 29.2B).

A significant event in the diagenesis of many marine sediments is the transformation of the metastable aragonite to its stable polymorph calcite. Possible relationships between this transformation and other diagenetic events are covered in chapters 25 and 27.

References and notes

(1) Except that the 'granule' class is included with the pebble class, this scale is essentially that proposed by Wentworth, C. K., 1922. A scale of grade and class terms for clastic sediments. *J. Geol.* **30**, 377–92.

(2) A two-fold division into low and high sphericity is included in the standard images of Powers, M. C., 1953. A new roundness scale for sedimentary particles. *J. sedim. Petrol.* **23**, 117–19.

(3) Six classes of roundness are recognised in the standard images of Powers (see note 2); a five-class scheme is proposed by Pettijohn, F. J., 1975. *Sedimentary rocks.* New York: Harper & Row (3rd ed.), fig. 3–24.

(4) Dunham, R. J., 1962. Classification of carbonate rocks according to depositional texture, *in* Ham. W. E., ed., *Classification of carbonate rocks.* Mem. Am. Ass. Petrol. Geol. 1, pp. 108–21.

(5) A terminology specific to sedimentary rocks has also been devised: Friedman, G. M., 1965. Terminology of crystallisation textures and fabrics in sedimentary rocks. *J. sedim. Petrol.* **35**, 643–55.

(6) Bathurst, R. G. C., 1971. *Carbonate sediments and their diagenesis*, Developments in sedimentology 12, Amsterdam: Elsevier, pp. 423–5.

(7) A discussion of the literature on stylolites can be found in Bathurst, 1971, pp. 468–73.

23
Argillaceous rocks

Introduction

Traditionally the argillaceous rocks have been divided into two size classes: *siltstone* (composed largely of coarse silt grains) and *shale* or *mudstone* (distinguished by the presence or absence of fissility respectively). This scheme is used in the following text, but for detailed comparative work a scheme of three grades (*silt*, *mud*, and *clay*) may be preferred (1).

Argillaceous rocks are mostly detrital in origin but it is customary to include also the fine-grained residual rocks and clay beds derived by the alteration of volcanic ash.

Origin

Depending on the nature of the weathering, argillaceous constituents consist of varying proportions of resistate minerals (quartz, feldspar, and mica) and clay minerals and hydroxides derived from the alteration of source rock minerals. Temperate weathering permits partial retention of the more mobile cations within the clay structure. In potassium-rich rocks the clay mineral illite is formed; magnesium-rich rocks produce montmorillonite. Tropical weathering causes greater leaching of metal ions which favours the formation of kaolinite. Extreme leaching removes even the silica, so that hydroxides of aluminium and iron are formed in place of clay minerals. Deposits of these hydroxides are termed *laterites*. Chlorite is abundant in low-grade metamorphic pelites but is not an abundant detrital mineral because of its low resistance to chemical weathering.

The composition of a detrital assemblage may be altered before sedimentation by the process of differential settling, which favours early precipitation of more coarsely crystalline clays (such as kaolinite) and those which form floccules in the presence of salt water. Fundamental chemical modification of clays during transport is probably limited, although some clays are capable of rapid ion exchange (e.g. enrichment in potassium on contact with sea-water). Under conditions of very slow sedimentation, true structural transformation may take place, and is probably enhanced by high water temperature.

Neomorphism of clays may take place in diagenesis but is difficult to detect in thin section because of the small scale of the textural modifications. In the long term illite and chlorite develop at the expense of kaolinite and montmorillonite, but burial depths of over a thousand metres are usually required. In the shorter term both kaolinite and montmorillonite are known to form in early diagenesis.

Authigenic clays are associated with the alteration of volcanic ash deposits. Montmorillonite is the principal constituent of *fuller's earth* and *bentonite* deposits (2) and kaolinite is the principal constituent of *tonsteins*, some of which are of volcanic origin (3).

Recently deposited muds possess a highly porous microparticulate texture but their relatively low permeability inhibits extensive cementation. Clean-washed silts, however, more closely resemble sands in their physical properties and are therefore commonly cemented by silica and carbonates. In mudstone and shale cementation is usually limited to the formation of local concretions and where an intimate association of clay and calcite occurs throughout a unit it is usually of depositional rather than diagenetic origin. Authigenic pyrite is widely distributed among argillaceous sediments; it forms under reducing conditions and is particularly characteristic of marine deposits because of the greater availability of sulphate ions in solution.

Concretions initiated during early diagenesis develop by cementation of primary interparticle pores and thus locally preserve the pre-compactional fabric of the host sediment. Such concretions are usually augmented by later diagenetic replacement of the enclosed particles. More rarely concretions are wholly of late diagenetic origin and are dominantly replacive. Recrystallisation of initial concretionary aggregates may be accompanied by inward expulsion of water, which collects in internal shrinkage cracks or *septaria*. These cracks usually act as sites of later cementation.

The depositional environment largely controls the amount of included organic matter. Where bottom conditions are oxidising, organic matter will be incorporated only where moderate or high rates of sedimentation prevail. Reducing conditions favour a greater retention of organic matter, especially where sedimentation rate is low. Early diagenesis of organic matter is greatly influenced by the activity of bacteria which remove oxygen and nitrogen from proteins and carbohydrates and release carbon dioxide, methane, and ammonia. The residue is rich in fatty acids and amino acids. Further diagenesis involves the largely inorganic conversion of the organic residues into hydrocarbons. Burial and heating may cause distillation of the remaining hydrogen, leaving purely carbonaceous residues.

Petrography of argillaceous constituents

The clay minerals are difficult to identify under the microscope on account of their very fine grain size. *Illite* possesses properties similar to those of muscovite; it is typically colourless and displays high relief, high birefringence, and parallel extinction. *Montmorillonite* also possesses high relief and birefringence but its extremely small particle size is often distinctive; it often shows its full polarisation colours only where compactional orientation has occurred. In contrast to illite and montmorillonite, *kaolinite* possesses low birefringence; it is distinguished in the presence of finely crystalline quartz by its high relief and by the common development of groups of parallel-oriented flakes in the form of 'books' or 'worms'. *Chlorite* is usually pale green and displays very low birefringence. Chamosite (p. 315) and finely divided glauconite (p. 322) may be confused with chlorite, but well-crystallised material displays higher birefringence.

The resistate constituents consist principally of quartz and feldspar which, because of their low birefringence and relief, are often underestimated in thin section. Fine silt-sized quartz particles are highly angular and usually elongate; many such particles may result from abrasion of larger grains. Detrital feldspar is usually less abundant than quartz and typically occurs as cleavage fragments which are too small to display diagnostic twinning patterns. The third common resistate mineral is mica which, as a result of hydraulic sorting, may occur as relatively large grains.

Also encountered are fossil remains, pyroclastic grains, and terrigenous rock fragments, together with bituminous or carbonaceous organic matter. Common authigenic minerals are pyrite, calcite, siderite, dolomite, and phosphates.

Petrography of argillaceous rocks

Siltstones are the product of bed-load transport and deposition and possess the textural and mineralogical character of sandstones; accordingly they will not be treated in detail here. The silt grains consist essentially of quartz and feldspar with variable amounts of mica and chlorite. The quartz and feldspar grains are angular and commonly display low sphericity (fig. 23.1A). Siltstones with grainstone texture may undergo extensive cementation or intergranular solution.

Mudstones and shales usually consist of a mixture of clay and silt-sized particles. These may show a more or less homogeneous distribution (fig. 23.1B) and in the absence of extensive bioturbation this homogeneity represents deposition under exceptionally stable environmental conditions or rapid deposition from turbidity flows. Undermelting of ice sheets may produce similarly unsorted argillaceous deposits.

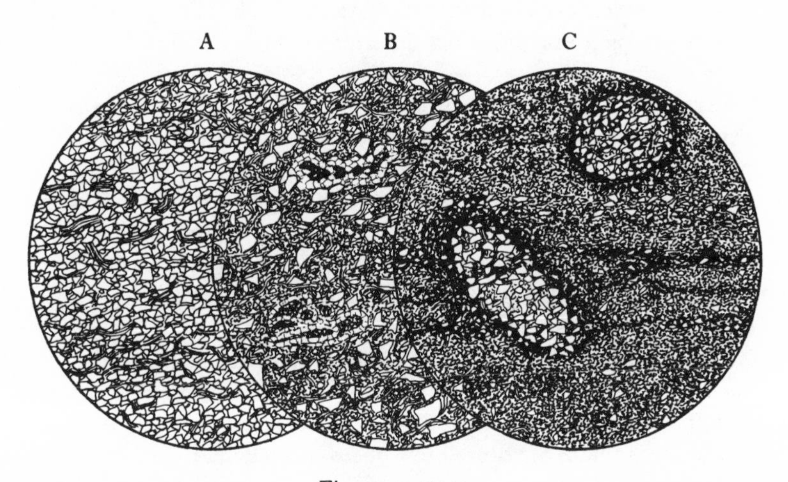

Fig. 23.1. × 20.

A. Siltstone; well-sorted coarse silt with abundant muscovite flakes. Old Red Sandstone, Caithness.
B. Mudstone, silty with arenaceous foraminiferans showing compactional deformation. Tertiary, Kent.
C. Mudstone, with burrows revealed by contrast in silt content and pyritised margins. Tertiary, Kent.

More commonly argillaceous rocks display segregation of coarse silt particles in the form of laminae. Sharply defined silt and sand laminae indicate intermittent deposition, with phases of surface winnowing or bed-load transport. Diffuse lamination may reflect continuous but variable current action, but it may also result from the disturbance of sharper laminae by small burrowing organisms.

A more ordered arrangement of the grain size is represented by graded bedding, which normally consists of coarse particles resting on a sharply defined base, and passing up into progressively finer grained material. Sometimes the sharp basal surface may reveal signs of erosion, even on a microscopic scale. Some graded units are interpreted as annual varves; these are usually only a few millimetres thick and display regular repetition. Graded beds deposited by waning turbidity flows range from a few centimetres up to several tens of centimetres in thickness. Varves may display lamination and good sorting because of their relatively slow deposition. Varves of glacial origin are characterised by highly angular grains, unstable minerals, and often by the absence of true clay minerals.

Varves may also be represented by alternation of detrital mud with chemically precipitated constituents. Annual peaks of physicochemical or biochemical precipitation may interrupt steady detrital sedimentation or *vice versa*.

Fissility may be present in both laminated and unlaminated rocks. Perfect fissility is found in clay-grade rocks whose component clay particles show bedding-plane orientation. Fissility is also enhanced where abundant organic matter is present as thin shreds strung out parallel to the bedding (fig. 23.2A). The organic matter appears yellow to brown in thin section except where it has been carbonised. Kaolinitic deposits rarely display good fissility owing to the edge-to-face attraction of the clay particles.

Shales which display little bedding-plane orientation of the clay particles may nevertheless part easily along planes marked by enrichment in clay, organic matter, or mica. These planes probably represent intermittent phases of reduced rate of sedimentation.

Burrowing on a large scale causes a general destruction of depositional fabric but smaller-scale activity is often represented by burrows which appear in thin section as discrete tubular structures. These burrows are distinguished by the interruption of primary structure, by contrast in grain size of the fill and matrix sediment, and often by concentrations of pyrite around the burrow margin (fig. 23.1C). The absence of burrowing in marine deposits is often indicative of deposition in deoxygenated environments; in such cases the sediment is usually rich in organic matter and pyrite. Non-marine deposits are commonly unaffected by burrowing and rarely contain more than scattered tubules.

Rootlet structures can usually be recognised by the presence of a marginal carbonaceous residue, but where the organic matter has been totally decomposed they are preserved as sediment-filled tubes which are easily mistaken for burrows in thin section.

Accessory matrix components include chemically precipitated muds (such as aragonite, chamosite, and glauconite) which may be intimately associated with the detrital clay. Among accessory grains the most common are organic constituents, detrital sand grains, glauconite grains, and faecal pellets. The organic constituents include the more common varieties (p. 270) and forms which are less often seen in thin section such as diatoms and agglutinated foraminiferans. Diatoms are originally siliceous but in argillaceous rocks they are commonly preserved in pyrite. Agglutinated foraminiferans consist of detrital grains set in a matrix of organically precipitated silica, calcite, or indeterminate isotropic material. In thin section they are likely to be overlooked as sand patches or burrow fills except where the chambers have been preserved by early diagenetic authigenesis (fig. 23.1B).

Compaction usually causes illite, montmorillonite, and chlorite to show some degree of bedding-plane orientation. This is reflected in aggregate polarisation, with extinction parallel to the bedding. The aggregate birefringence colours are useful in the identification of the clay mineral involved. In the absence of bedding-plane orientation the colours may be detachable where clay particles have been compressed around resistant grains.

Chemical diagenesis is largely represented by solution and cementation; replacement usually plays a minor role. Solution particularly affects organic hard parts composed of aragonite, magnesian calcite, or opaline silica, the final product depending on the timing of solution relative to compaction (fig. 23.2B). Precipitation of carbonates and phosphate is often concretionary and tends to preserve relatively unstable constituents such as shells or volcanic glass particles which are deformed or dissolved in the surrounding mudstone or shale. Within the concretion shells may remain unaltered or may be represented by open moulds, cement casts, or compressed films. Often the relative stability of calcitic over aragonitic shells is emphasised (fig. 23.2B).

Solution voids and septarian cracks are commonly sites of cementation; the cements are derived from compaction of the surrounding muds and include calcite, dolomite, siderite, and kaolinite. Minor elements may be concentrated in the form of sulphides: chalcopyrite, galena, and sphalerite.

In fresh samples the margins of concretions are seen to be gradational; this gradation is usually rapid and is readily obscured by weathering. A crude concentric zoning may be displayed and is related either to variation in the proportion of inclusions or to variation in concretion mineralogy (e.g. a gradation from siderite into calcite). Concretions in red mudstones are characterised by bleaching due to carbonation and reduction of the iron oxides.

Late diagenetic concretions develop largely by replacement or displacement of the matrix. Chert nodules developed in argillaceous deposits are often of replacive type. An example of displacive concretion development is provided by cone-in-cone concretions (fig. 23.2C). Crystal growth separates layers of the host sediment and often rearranges them along crystal boundaries. Cone-in-cone structure is believed to represent crystallisation in a stress field, possibly self-generated.

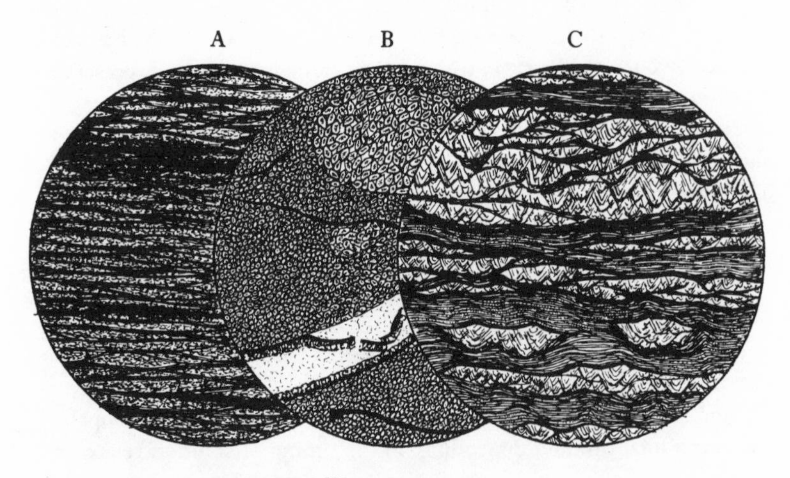

Fig. 23.2. × 20.

A. Bituminous shale. Jurassic, Yorkshire.

B. Siderite concretion. Originally aragonitic shells have been dissolved prior to compaction and are represented by pyrite films. A calcite shell has been dissolved out after the major phase of compaction; the cavity has been cemented by an early siderite rim (showing local dislocation) and later kaolinite. Burrow sections are revealed by coarser siderite crystals. Jurassic, Yorkshire.

C. Cone-in-cone concretionary clacite occurring as thin bands within a dark organic shale. Jurassic, Skye.

Pigmentation of argillaceous sediments is related to the mineralogy of iron-bearing constituents and to the abundance of organic derivatives (4); it is therefore greatly influenced by diagenesis. A red or purplish pigmentation is produced by haematite; this may be of primary origin but commonly forms during diagenesis in oxidising terrestrial environments either by dehydration of hydroxides or by direct diagenetic precipitation. A strong green colour is usually attributable to glauconite or chamosite, but chlorite and iron-rich mica or illite may impart varying degrees of green colour. The grey colour of many argillaceous deposits is due to the combined effect of pyrite and finely divided organic derivatives of various types. Concentrations of organic material, whether bituminous or carbonaceous, may impart a very dark grey or black colour to argillaceous rocks and render them almost opaque in thin section.

Laterite. This rock type includes a range from ferruginous laterite to bauxitic laterite. All types are characterised by a great variety of textures, some of which are primary (e.g. relict igneous textures, fig. 23.3A), and

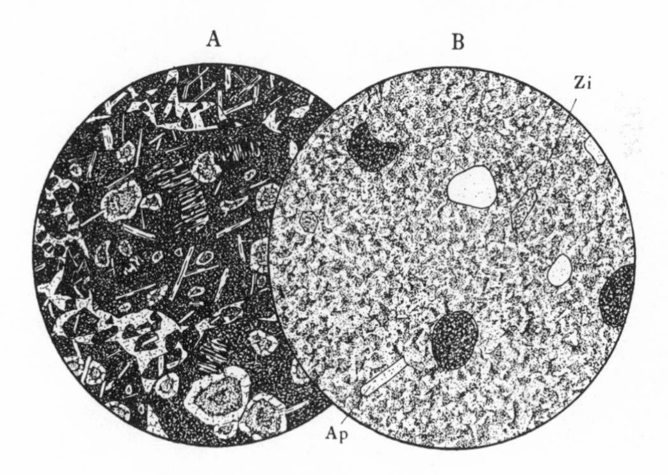

Fig. 23.3. × 20.

A. Red bole; conglomerate of lateritised basalt fragments displaying altered olivine crystals and incipient pisolitic structure. Fragments cemented by analcite. Tertiary, Skye.

B. Fuller's earth; a groundmass of montmorillonite with scattered quartz grains and glauconite (stippled). The igneous accessory minerals apatite (Ap) and zircon (Zi) are present in euhedral form. Cretaceous, Bedfordshire.

others, such as concretionary and pisolitic textures, which are diagenetic. Although laterites are typically residual deposits, the lateritic components may be reworked to form detrital deposits, often in admixture with components of non-residual origin.

Bentonite. This term is applied to montmorillonitic clays which display a high degree of water adsorption and base exchange. The rocks consist primarily of montmorillonite but may contain other clay minerals together with grains of feldspar, biotite, magnetite, and zircon. The latter are often euhedral and indicate derivation of the deposits by the alteration of volcanic ash. Sometimes this is confirmed by the presence of relict pumice fragments or glass shards. Fuller's earth (fig. 23.3B) is a variety of bentonite in which the montmorillonite contains varying amounts of detrital sand-grade inclusions, commonly quartz, glauconite, and organic grains. These deposits probably result from the alteration of reworked tuffs which accumulated as clean washed sand-grade deposits.

References and notes

(1) Ingram, R. L., 1953. Fissility of mudrocks. *Bull. geol. Soc. Am.* **64**, 869–78.
(2) Jeans, C. V., Merriman, R. J. & Mitchell, J. G., 1977. Origin of Middle Jurassic and Lower Cretaceous fuller's earths in England. *Geol. Mag.* **12**, 11–44.
(3) Spears, D. A., 1970. A kaolinite mudstone (tonstein) in the British Coal Measures. *J. sedim. Petrol.* **40**, 386–94.
(4) McBride, E. F., 1974. Significance of color in red, green, purple, olive, brown, and gray beds of Difunta Group, north-eastern Mexico. *J. sedim. Petrol.* **44**, 760–73.

24
Arenaceous rocks

Introduction

The limits of sandstone against coarser and finer sedimentary rock types are arbitrary (1). Some sediments show such poor sorting that a complete gradation in particle size is present from pebble to clay, with no well-defined mode. Such rocks, which include many greywackes, may not be true sandstones at all, but are usually treated as such for purposes of discussion. The many sandstone classifications that have been proposed fall into two main categories according to whether emphasis is placed on grain composition or rock texture.

Classification according to grain composition is achieved by distinguishing three types of grain: (a) quartz, quartzite, and chert grains (2); (b) feldspar grains; (c) 'unstable' or *labile* rock fragments. Quartz dominates over other grain types in most varieties of sandstone, and accordingly quartz-rich or *quartzose sandstone* is defined by a very high percentage of quartz – usually 90 or 95 per cent. The quartz-poor sandstones are divided into *feldspathic sandstone* or *lithic sandstone* according to whether feldspar or rock fragments predominate. In genetic terms such classification emphasises both grain provenance and the effects of weathering and transport.

Classification according to texture is concerned with the proportion of grains to matrix; two main categories are recognised: the matrix-poor sandstones or *arenites* and the matrix-rich sandstones or *wackes*. Although the original dividing line was taken at 10 per cent, the figure of 15 per cent is perhaps more often used. Another view is that a distinction between grain-support and matrix-support is of more significance than a rigid percentage figure. Problems associated with this classification include the difficulty of estimating the proportion of matrix and sometimes of even defining the matrix. The emphasis is nevertheless of value to the economic geologist, whose concern with sandstone usually involves intergranular porosity and reservoir properties.

This text attempts to cover both aspects of classification by regarding sandstones both in terms of their primary grain composition and in terms of their 'muddiness' and its effect on diagenesis. The distinction between 'arenite' and 'wacke' is here based on grain-support or matrix-support respectively. The arenites themselves are regarded as being 'clean', 'muddy', or 'mud-packed' according to the amount of intergranular matrix.

Despite the setting up of rigorous sandstone classifications a number of traditional terms are still in common use. *Orthoquartzite*, as generally used, is equivalent to quartzose sandstone as defined here. An *arkose* is a feldspathic sandstone with a high proportion of alkali feldspar; 25 per cent is the lower limit most commonly used. The term *greywacke* is best used in its original sense as a field term for dark and generally poorly sorted sandy rocks which typically form thick and monotonous sequences of synorogenic association.

Origin

Much literature is readily available on this subject and only those aspects directly related to thin section study are discussed here.

The nature of a sandstone is determined by the nature of the source rock and the type of weathering that it experiences, together with the modes of transport and deposition. Chemical weathering causes preferential decomposition of those minerals which are chemically unstable under surface conditions, particularly those of high-temperature origin and those rich in ferrous iron. Among the ferromagnesian minerals olivine is the least stable, followed by pyroxene and then amphibole. Of the feldspars anorthite is the least stable, followed by the sodic plagioclases and finally the potash feldspars. The composition of the mineral assemblage is thus largely determined during weathering of the source rock, and subsequent chemical modification is minimal until after deposition. The transport stage is dominated more by the physical degradation of soft and easily cleaved minerals. The lack of cleavage and the hardness of quartz is partly responsible for its dominance in detrital sands, and the roundness of quartz grains provides a useful indication of the degree of mechanical attrition experienced by a grain population.

The elimination of unstable minerals by chemical and mechanical weathering leads to the formation of sands of dominantly quartzose composition, often referred to as 'mineralogically mature' sands. Because these retain very little information as to the mineralogy of their source rocks, such information has been sought by studies on the nature of the quartz grains themselves (p. 250). Any such study may be complicated by recycling of the more resistant grains, and it is usually possible to determine only the ultimate source of the grains rather than their immediate source (which may have been pre-existing sediments). Recycling is sometimes revealed by the association of angular and rounded grains of the same size and by the presence of abraded authigenic overgrowths in quartz grains.

Inferences about weathering of source rocks are similarly subject to limitations. The presence of unstable minerals favours a relatively dry (hot or cold) climate but a stable grain assemblage may be inherited from a pre-existing sandstone irrespective of weathering conditions. Also climatic conditions conducive to chemical weathering may be ineffective in areas of high relief due to the rapid stripping off of only partially decomposed material. Transport mechanisms and their duration may be reflected in grain rounding, but once again the possibility of grain recycling must be considered. Also any comparison of roundness must take into account the dependency of rounding on grain size (p. 228).

The environment of deposition of a sandstone is best inferred from large-scale features, such as sedimentary structures and facies associations, but petrographical analysis can provide additional information on the mechanics of deposition by determination of sorting and packing. Thin section studies may also reveal the presence of diagnostic grain types not evident in hand specimen (microfossils, ooliths, glauconite grains, etc.).

Petrography of sandstone constituents

Quartz grains

These may consist of a single crystal (monocrystalline grains) or several (polycrystalline grains). *Monocrystalline quartz grains*, when examined in detail, are seen to display a number of distinctive features and attempts have been made to relate these to the primary source rock. *Undulose extinction* is a common feature but is of little genetic importance as it occurs in crystals of both plutonic igneous and metamorphic origins (3). Quartz crystals displaying uniform extinction are typical of volcanic and hypabyssal rocks and constitute only about 20 per cent of the crystals in plutonic rocks. Despite the abundance of strained quartz in source rocks, unstrained quartz commonly predominates in sediments, apparently because lattice dislocation within strained crystals renders them more susceptible to both chemical and mechanical breakdown.

The proportion of strained quartz is thus a function of mineralogical maturity as well as source.

Grain shape has been used to distinguish between quartz crystals of metamorphic and igneous origin, the former supposedly having a lower sphericity than the latter. In most cases, however, there seems to be little in the way of a reliable relationship, besides which sphericity is not well represented in thin section. Within polycrystalline quartz grains variation in sphericity may be more significant (see below).

Inclusions are found in many quartz grains, although they are commonly too small to be identified optically. Many inclusions consist of fluid-filled vacuoles, often arranged in subparallel lines; these occur in quartz from metamorphic rocks (chiefly coarsely crystalline gneisses) and from plutonic rocks, but appear to be most abundant in hydrothermal quartz. Mineral inclusions include feldspar, mica, zircon, tourmaline, rutile, apatite, kyanite, and sillimanite. Of these, acicular rutile appears to be characteristic of granitic quartz, and kyanite and sillimanite of metamorphic quartz. Quartz from volcanic or hypabyssal rocks can sometimes be recognised by the presence of glassy inclusions.

Polycrystalline quartz grains may be derived from gneisses, schists, quartzites, granites, quartzitic sandstones, and recrystallised cherts. Those from plutonic igneous rocks are mostly restricted to the coarse sand grade on account of the size of the individual component crystals; the crystals tend to be uniform in size and may be simply or irregularly sutured. Grains derived from gneisses and schists are composed of somewhat elongate crystals displaying subparallel alignment; highly sutured contacts are common. Grains derived from quartzites are equicrystalline but may show a wide range in other characters. Grain flattening and irregular suturing are typical of metamorphic quartzites but both features may develop during the diagenesis of unmetamorphosed quartzose sandstone. Wholesale undulose extinction of the quartz crystals indicates a metamorphic origin, and quartzitic sandstones may display cement characters such as enfacial junctions. Unaltered chert grains are finely crystalline and often brownish; they may contain microfossils and other sedimentary inclusions. With increasing recrystallisation the chert grades into a quartzite texture.

Alteration of quartz grains

Monocrystalline grains frequently display solution and authigenic overgrowth. Solution is evident at grain contacts where sutured surfaces develop in place of point contacts. Authigenic overgrowths develop in optical continuity with the host grains, even to the extent of adopting undulose extinction, but are often separated from the host by mineral or fluid inclusions which outline the original grain surface (fig. 24.2A). The overgrowths are often clearer than their host grains.

Total replacement of quartz grains is rare but where authigenic carbonates and clays are present in the matrix the grains may undergo marginal replacement (fig. 24.4B). The latter is often difficult to distinguish from the effects of grain–cement overlap in the plane of the section. More definite evidence of replacement is afforded by carbonate rhombs which penetrate deep into the grains.

Polycrystalline grains are generally less stable than monocrystalline grains; where the two meet at solution contacts the former are preferentially dissolved. They also show a greater degree of internal replacement as the replacing minerals may penetrate along crystal boundaries, often causing isolation of the component crystals. Authigenic overgrowths may develop on polycrystalline grains and may themselves be polycrystalline (4).

Feldspar grains

Because of their greater stability the alkali feldspars greatly exceed the calcic feldspars as sandstone constituents. Plagioclases are the most distinctive varieties on account of their twinning. Fragmentation generally eliminates the simple twinning of orthoclase and small grains may be mistaken for quartz where cleavage is not displayed; distinguishing features are cloudiness (due to vacuoles), incipient alteration to clay minerals, the somewhat lower birefringence, and the lower R.I. which imparts a slight brownish tint to the crystals. Accurate measurement of feldspar proportions is normally carried out on stained slides.

Mineral inclusions in feldspars are mostly composed of quartz. Fluid-filled vacuoles may also be present and, as in quartz, are most common in crystals of non-volcanic origin. Most of the vacuoles in feldspars, however, appear to result from weathering rather than from primary entrapment.

The mineralogy of the feldspar itself provides the best indication of source rock composition, particularly where less stable varieties are present. A high proportion of plagioclase implies proximity to source and dominance of calcic plagioclase may reflect a direct volcanic contribution. Zoned crystals are also indicative of volcanic origin. A low proportion of plagioclase may reflect source rock composition, selective destruction during weathering, or both. Recycling of feldspars is of minor importance except where erosion and deposition have taken place under dry conditions.

Alteration of feldspar crystals

The effects of diagenetic alteration are easily confused with those of hydrothermal alteration and weathering. All three may cause clouding of crystals by the development of fluid inclusions. Alteration to finely divided kaolinite or micaceous minerals is also common to all three processes. Only total replacement of grains can definitely be ascribed to diagenesis as such degraded grains would have been in too weak a state to survive transport. Replacement by carbonate follows grain margins and internal cleavage planes; the latter process may reduce a crystal to a number of isolated relics whose original unity is revealed only by their optical continuity (fig. 24.4B). Authigenic chlorite may similarly invade feldspar crystals.

Feldspar is affected by intergranular solution and where in contact with quartz usually shows the greater volume loss. Pressure may also result in plastic deformation, as revealed by distortion of twin lamellae. Passive solution of feldspar may be detected in uncemented sandstone, especially by the use of stained impregnation; it usually results in etching of grains along their cleavages. In cemented sands early passive solution would be taken for replacement.

Authigenic feldspar rarely attains volumetric importance as a cement. It usually develops as authigenic overgrowths on detrital grains, and is accordingly most abundant in feldspathic sandstones. The overgrowth is usually clearer than the host and is separated from it by a layer of inclusions. Continuity of twinning is displayed by authigenic plagioclase although the sharpness of the twin lamellae may decrease outwards; microcline normally develops untwinned overgrowths or sometimes overgrowths with patchy irregular extinction.

Micaceous grains

Muscovite is the dominant mineral because chlorite and biotite are relatively unstable and are abundant only close to their source or where weathering is dominantly mechanical. The buoyancy of mica flakes prevents much abrasion during transport but during diagenesis they are less physically resistant, with compaction causing flexing and fracture of muscovite and plastic deformation of biotite flakes. Carbonate or sulphide minerals sometimes cause disruption of crystals by internal crystallisation along cleavage planes, and carbonate cementation is often accompanied by replacement of included micas, especially biotite.

Lithic grains

Igneous lithic grains normally consist of finely crystalline volcanic rock types. Basic varieties, consisting largely of plagioclase feldspar or basaltic glass, are restricted to immature sandstones; acidic varieties, with a greater quartz content, are more widely distributed.

Sand-grade fragments of sedimentary origin are mostly quartzose but grains of limestone and mudstone occur locally. During diagenesis limestone grains are liable to solution and mudstone grains to compactional distortion against more resistant grains (fig. 24.4A).

Metamorphic rock fragments include finely crystalline gneisses, schists, and slates. Generally these are recognised by their fabric, but because of their fineness of grain the slate texture may be difficult to distinguish from that of laminated shale.

Heavy minerals

Most sandstones contain a small proportion of accessory minerals of terrigenous origin (5). They are mostly denser than quartz and feldspar

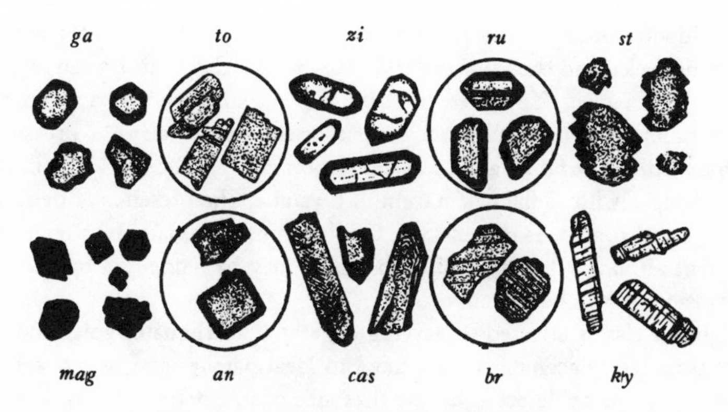

Fig. 24.1. Accessory minerals of sandstones.

The crystals within the small circles are magnified 80 to 100 times, the others about 20 times. *ga*, garnet, worn dodecahedra and fragments. *to*, tourmaline. *zi*, zircon. *ru*, rutile. *st*, staurolite, displaying post-depositional etching. *mag*, magnetite, octahedra and worn grains. *an*, anatase, pyramidal and tabular habits. *cas*, cassiterite, showing zones of growth. *br*, brookite. *ky*, kyanite, cleavage fragments.

and are studied after concentration by means of heavy liquids. These *heavy minerals* include garnet, zircon, tourmaline, rutile, apatite, kyanite, staurolite, and magnetite (fig. 24.1), together with many less common varieties.

The scarcity and small size of heavy minerals renders them inconspicuous in thin section but garnet and tourmaline can often be distinguished on account of their distinctive optical properties and their occurrence as relatively large grains. Zircon may also be conspicuous by its abundance and very high relief.

Diagenesis of heavy minerals mostly involves solution, with the formation of crystals with jagged etched outlines, but authigenic overgrowths have been observed on tourmaline grains, and wholly authigenic crystals of anatase and brookite are occasionally abundant.

Other grains

Grains of non-terrigenous origin include shell fragments, glauconite pellets, phosphate pellets, and pyroclastic particles. These are rarely sufficiently durable to be reworked. Shell fragments, in particular, are liable to solution in sandstones, and may in this way contribute to calcite cements.

Detrital clay matrix

The identification of clay minerals is discussed in relation to the argillaceous rocks, and the same criteria apply where clay occurs as an intergranular matrix. Unaltered matrix clay usually shows a random orientation but compaction may cause orientation of clays at pressure points within the fabric and recrystallisation may produce poorly defined clay fringes with radial orientation of crystals. The presence of detrital quartz silt grains usually serves to distinguish primary matrix from cement although detection of such grains may be hampered by quartz corrosion or by chertification.

Matrix clay is affected to varying degrees by carbonate replacement, but usually replacement is restricted to local patches and, where more complete, can be detected by the presence of fine detrital or diagenetic inclusions.

Clay cement

Cements can usually be distinguished from primary matrix by the textures displayed, although these differ considerably with mineralogy

(6). Chlorite cements form well-defined fringes of fibrous crystals oriented normal to the grain boundaries; sometimes these rims may coalesce to form complete pore fillings but in many cases the chlorite fringes enclose central masses of later cement minerals such as quartz, carbonate, or other clay minerals. Illite may form fringing cements but is also precipitated as fine equicrystalline aggregates. Kaolinite is by far the most common cementing clay mineral; it does not form fringes but occurs as unoriented crystal aggregates. The aggregates are usually equicrystalline, whereas neomorphic kaolinite tends to display small-scale porphyroblastic texture. Montmorillonite may form crystal aggregates but is most common as pellicles displaying crude tangential orientation of the constituent crystals.

Fringes and pellicles often display superficial replacement of the enclosed grains, leading to diffuse grain outlines.

Iron oxide and hydroxide cements

The pigmentation of many red sandstones is related to the presence of haematite in the form of pellicles. These are usually too thin to cause substantial pore filling. They are opaque or slightly translucent and consist of finely divided clay with a deceptively small proportion of included haematite – usually not more than a few per cent. Such coatings have been ascribed both to crystallisation of original amorphous coatings and to diagenetic redistribution of iron derived from the decomposition of iron-bearing detrital minerals (7).

Both haematite and limonite may result from the oxidation of siderite or ferrous silicates; the primary textures are usually destroyed but where oxidation is only local a gradation from the primary to the secondary texture may be displayed (fig. 24.4C).

Silica cement

Authigenic silica most commonly occurs as syntaxial overgrowths (p. 232) of relatively uniform thickness (fig. 24.2A). Less commonly syntaxial quartz develops as isolated prismatic crystals projecting from a limited area of the host grains. It is uncertain whether this type of cementation reflects unusual grain surface texture or whether it is controlled by pore water chemistry and rate of crystallisation. Non-syntaxial quartz cements usually show a tendency for development of chalcedony or microquartz textures (p. 294); rapid crystallisation may account for this less controlled type of crystallisation. In other cases lack of syntaxial overgrowth is related to the presence of unusually thick clay pellicles.

Opal and chalcedony cements are present in relatively young sandstones of unusual facies, such as those occurring in silica-rich desert soil horizons, and in sandstones with unstable siliceous components, such as volcanic glass particles or siliceous organic remains.

Carbonate cements

Calcite typically forms anhedral crystal mosaics ranging from microcrystalline to coarsely sparry and poikiloblastic in texture. Apparent grain separation may result from marginal replacement of detrital silicate grains; more rarely true grain separation is associated with finely crystalline cements of partly displacive character. Where carbonate shell fragments were present in the original deposit their solution has sometimes led to the formation of relatively early fringe cements.

Dolomite and siderite more commonly occur as isolated or clustered rhombs; complete cementation is rare.

Other cements

Other minerals which may act as cements to sandstones include feldspars (p. 252), anhydrite, gypsum, halite, barite, and pyrite. Anhydrite occurs both as finely crystalline and dense intergranular aggregates and as coarsely crystalline poikilitic crystals; it is frequently associated with dolomite (fig. 24.2C). Gypsum may similarly be finely or coarsely crystalline; it generally results from the hydration of anhydrite under surface conditions. Halite and barite are both unusual as cements; they typically occur as coarse crystal aggregates.

Petrography of sandstones

The fundamental properties of a sandstone are the nature of the grains, the presence or absence of a detrital matrix, and the nature of the diagenetic modifications. All are to some extent interrelated, and the last two particularly so. Accordingly the sandstones are here described in two complementary sections, one emphasising their primary grain composition and the other their diagenesis in relation to depositional texture.

Quartzose sandstones

These are necessarily restricted in mineral variety; quartz grains are accompanied by alkali feldspar and sometimes chert grains. Abrasion, sometimes over more than one cycle, causes relatively rapid size reduction of feldspars with the result that they are relatively depleted in the coarser grades. Heavy minerals include the most stable varieties such as zircon and tourmaline. The mineralogical maturity of quartzose sandstones is also reflected in the nature of the quartz grains themselves, in that the ratio of strained to unstrained quartz is low.

Textural variation is also limited and for the most part the sandstones are clean-washed and free from detrital clay matrix. Because of their multicycle origin most quartzose sandstones exhibit at least moderate rounding of the grains and, especially where transport has taken place under desert conditions, rounding may be extreme. Muddy quartz arenites usually result from depositional mixing of mud and clean sand and consequently do not necessarily differ in grain rounding or sorting from their clean-washed counterparts.

Because they generally possess a grainstone texture, quartzose sandstones are often lithified by cementation or intergranular welding. Calcite cement may occur, particularly where fossils were present in the original deposit, but quartz is the dominant cementing mineral (fig. 24.2A). Grain welding (fig. 24.2B) is often accompanied by quartz cementation (solution transfer, p. 234).

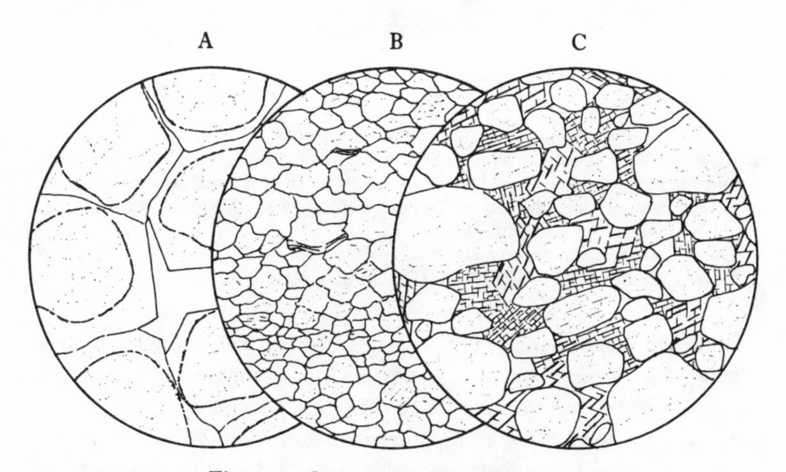

Fig. 24.2. Quartzose sandstones; × 20.

A. Cementation by syntaxial overgrowth on quartz grains; haematite pellicles reveal the original grain outlines. Permian, Cumbria.
B. Induration by vertically-directed pressure solution. Carboniferous, Northumberland.
C. Cementation by dolomite (rhombic cleavage) and later anhydrite (rectangular cleavage). Restricted quartz overgrowths are also present. Permian, North Sea.

Feldspathic sandstones

Unlike the quartzose sandstones, the feldspathic sandstones include sediments of widely differing mineralogical maturity. The proportion of feldspar itself varies considerably and it is customary to distinguish a subdivision of highly feldspathic sandstone or arkose (fig. 24.3). Also feldspathic sandstones of all degrees grade into lithic sandstones (fig. 24.4A) as the proportions of feldspar and lithic grains approach equality.

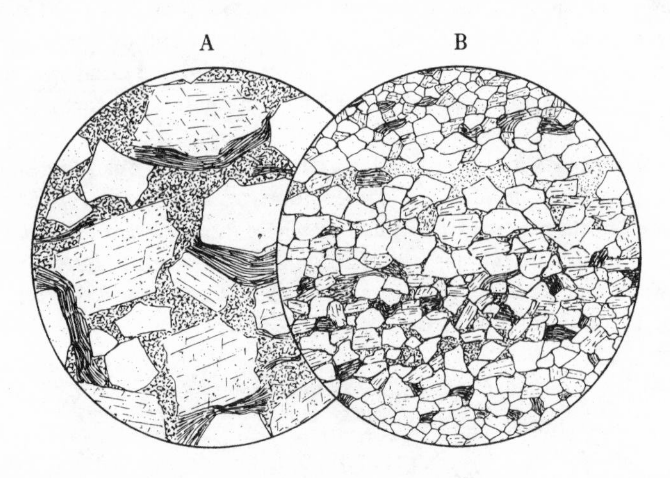

Fig. 24.3. Arkose; × 20.

A. Coarse arkose. The poor sorting of the constituent feldspar, quartz, and mica grains is suggestive of a residual arkose, but the siltstone matrix indicates that the constituents are detrital although probably very close to their source. Old Red Sandstone, Caithness.

B. Detrital arkose. The grains show relatively good sorting and the quartz and feldspar grains are of similar size. Torridonian, Ross & Cromarty.

Feldspathic sandstones rich in quartz display many of the features associated with quartzose sandstones, including the development of a quartzite-like texture through diagenesis. With decreasing quartz content, however, the hallmark of mature sandstone become less apparent; grain rounding tends to decrease and in general the sandstones become more poorly sorted and more muddy in nature. Many exceptions, related both to grain assemblages and to depositional conditions, can be found to these general trends. More consistent trends are shown by grain mineralogy: as the feldspar content increases the proportion of polycrystalline and strained quartz increases, as does the proportion of incipiently altered feldspars. These trends reflect the lessening effects of weathering and transport in the elimination of less resistant grains. At the same time the heavy mineral suite becomes less restricted, with garnet often being conspicuous.

Arkoses are usually the product of deposition close to acid plutonic source rocks. Typical characteristics are angularity of grains, abundant altered feldspars (often closely associated with fresh feldspars), a high proportion of strained quartz, and abundant mica; the grains are often poorly sorted and lie in a muddy matrix. The extreme case is found in

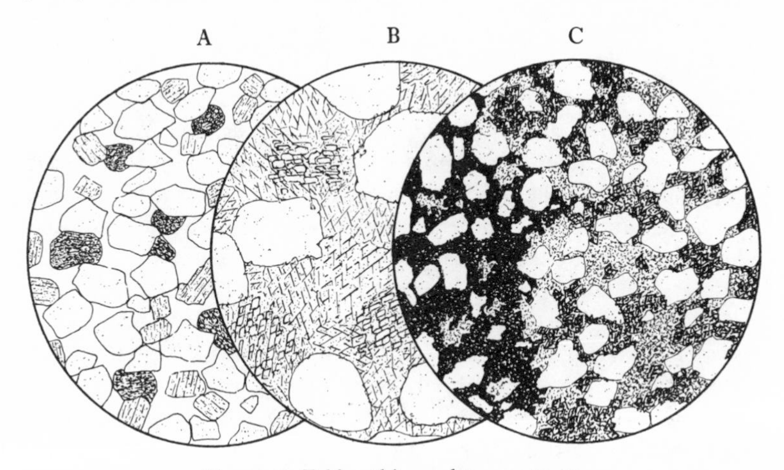

Fig. 24.4. Feldspathic sandstones; × 20.

A. Uncemented sandstone, showing compactional distortion of lithic grains (glassy volcanic particles). Tertiary, Skye.
B. Calcite-cemented sandstone, showing marginal replacement of quartz grains and extensive replacement of feldspar grains, with the reduction of the latter to separate relics with common optical orientation. Jurassic, Yorkshire.
C. Sand with cement of siderite and kaolinite weathered to limonite (left) in which rhombic outlines can locally be distinguished. Jurassic, Yorkshire.

the residual arkoses which lie more or less directly on their source rock; they tend to be very coarsely grained and the feldspars are typically larger than the associated quartz grains (fig. 24.3A). Feldspathic sandstones derived directly from volcanic sources are relatively rare, and differ from other types in the rarity of quartz and the dominance of calcic plagioclase.

Diagenesis in quartz-rich types resembles that of the quartzose sandstones but feldspathic sandstones typically show a greater range of cement types, with clay minerals being of particular significance. The variety of cements probably reflects a greater range in facies association as much as the composition of the primary deposit. For example the siderite–kaolinite cement association of many non-marine fluviodeltaic sandstones (fig. 24.4C) is probably related to the expulsion of pore waters from surrounding muds.

Residual arkoses possess little cement owing to the presence of a mud matrix; where present it is usually calcareous. In these and other feldspathic sandstones the presence of cementing or replacive calcite in the groundmass is associated with replacement of included feldspar grains (fig. 24.4B).

Lithic sandstones

These usually result from rapid erosion, transport, and deposition; in consequence they are often composed of grains derived from a single source area. Weathering determines the type of lithic grain assemblage, with chemical weathering favouring dominance of the more resistant types such as rhyolites and cherts. It is rare for lithic grains to occur alone, and in fact quartz is usually the dominant grain type. Feldspars are not abundant because the main source rocks (sediments and low-grade metamorphics) are themselves poor in sand-grade feldspar. Micas are more common, especially where the source rocks are metamorphic. Lithic sandstones are occasionally clean-washed, with a grainstone texture, but because of their usual association with rapid sedimentation they are generally mud-packed or mud-supported (fig. 24.5A).

A distinctive feature of lithic sandstones is the tendency for the grains to be deformed under compactional pressure; this may be displayed in both grainstone and packstone textures. Where the grains are highly mobile, deformation may completely obliterate the primary texture so that it is difficult to make out original grain outlines. More commonly a small proportion of the grains may undergo intense deformation, such that they become squeezed into intergranular sites and form a ***pseudo-matrix***. This loss of definition of the weaker grains may be enhanced by diagenetic recrystallisation and replacement.

In grainstones grain deformation may be prevented by early cementation. Calcite and clay minerals are common cements although syntaxial overgrowths may develop in quartz-rich varieties. Volcanic sands commonly possess chlorite fringe cements.

Hybrid sandstones

These include sandstones which contain significant proportions of non-terrigenous material (fig. 24.5B). They normally result from sediment mixing within the basin of deposition. The non-terrigenous material commonly has a profound effect on diagenesis especially where it occurs as a matrix (e.g. calcitic, phosphatic, or chamositic mud). Lime mud matrices are liable to recrystallisation, chertification, dolomitisation, and other modifications found in limestones. Where the non-terrigenous components are in granular form (glauconite and phosphate pellets,

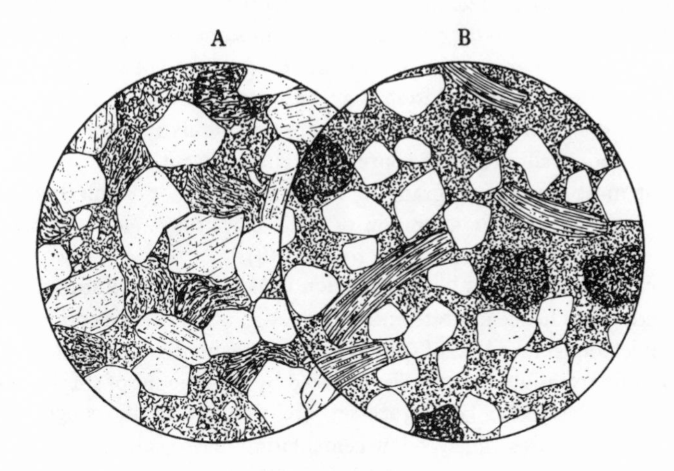

Fig. 24.5. × 20.

A. Lithic sandstone; mud-packed, showing plastic deformation of lithic grains.
B. Hybrid sandstone; sand grains are set in a matrix of carbonate mud.

ooliths, shell fragments), the effect on sandstone diagenesis is usually less, but solution of calcareous components may provide material for later cementation.

Diagenesis of clean arenites

Sandstones in which interstitial pore space is clay-free or only partly filled by clay are usually affected by cementation or by intergranular solution or deformation. Deformation is largely restricted to lithic grains which yield plastically under compactional stresses; where lithic grains are abundant the primary pore space may be virtually eliminated by the formation of pseudomatrix (p. 260). Alternatively grains may yield by fracture, as often observed in shell fragments, glauconitic pellets or other brittle grains. Even feldspar or quartz grains may undergo fracture in this way. Deformation within individual grains may be revealed by distortion of structure or of optical continuity.

Intergranular solution is a common process of pore space reduction in uncemented sandstones (fig. 24.2B). The surfaces of contact may be roughly planar, curved, sinuous, or microstylolitic. This variation may partly reflect the intensity of intergranular pressure, but the microstylolitic sutures in particular seem to be caused by the presence of clay impurities on the grain surfaces.

The occurrence of intergranular solution is erratic; its intensity is sometimes related to depth of burial, but in some deeply buried formations intergranular solution is minimal despite the absence of cement. Lack of intergranular solution may be related to the displacement of pore waters by hydrocarbons, but more commonly it is a reflection of the stability of silicates in the pore water environment. Alkaline solutions would most favour intergranular solution and it is possible that on a local scale at least the alteration of illitic clays may release potassium hydroxide and thereby promote solution (8).

Pore space reduction by cementation usually follows deformation but precedes, and hence inhibits, intergranular solution. An exception to the latter relationship is that of solution transfer, where the amount of silica lost by intergranular solution is balanced by the amount of silica cement on which basis the two processes are thought to be concomitant. Tectonic pressures which follow cementation may result in intercrystalline solution within the cement fabric itself.

Cementation itself may involve a complex series of events and it may be possible to establish an order of precipitation (9). A clear example of this is the association of haematite-clay pellicles with quartz overgrowths, in which the pellicles are the earlier as they are related to the surface of the detrital grain (fig. 24.2A). Similarly inclusions of calcite or other minerals in quartz overgrowths can usually be interpreted as preceding the overgrowths, but this is certain only where the included mineral is clearly moulded around the detrital nucleus.

Where authigenic components are not directly related to the grains the time relationships are even less easy to prove beyond doubt, but it is generally assumed that enclosed euhedral crystals are the earlier (fig. 24.2C). Confirmation of this may be provided by the relative distribution of the minerals as the first-formed mineral has the opportunity to fill small interstices, causing later minerals to have a more restricted distribution.

Less equivocal relationships are found where fringing cements are present. For example pores fringed with fibrous chlorite may possess quartz or feldspar at their centres; in such cases there is no doubt that the chlorite is the earlier mineral.

Although it is not possible to establish a universal order of authigenic mineral formation, some of the more common minerals do show definite trends in many sandstones. Already mentioned is the relationship between haematitic pellicles and quartz overgrowths. The wide distribution of syntaxial quartz overgrowths provides a useful diagenetic marker against which other authigenic events can be timed. Calcite and other carbonates appear to post-date quartz precipitation in most cases but chlorite cementation displays a more variable relationship, with fringes being developed on detrital quartz nuclei in some sandstones and on the authigenic overgrowths in others. Authigenic feldspar is probably precipitated at the same time as the quartz.

It is not uncommon to find three major components in a sandstone cement. For example sandstones containing quartz and dolomite cements often possess minor amounts of anhydrite also (fig. 24.2C); in such cases the anhydrite usually appears to have been the last to crystallise.

Diagenesis of mud-packed arenites

The inclusion of large amounts of porous clay renders these sandstones liable to profound textural reorganisation during compaction (fig. 24.6A). Compression of the clay matrix may be sufficient to transform an original

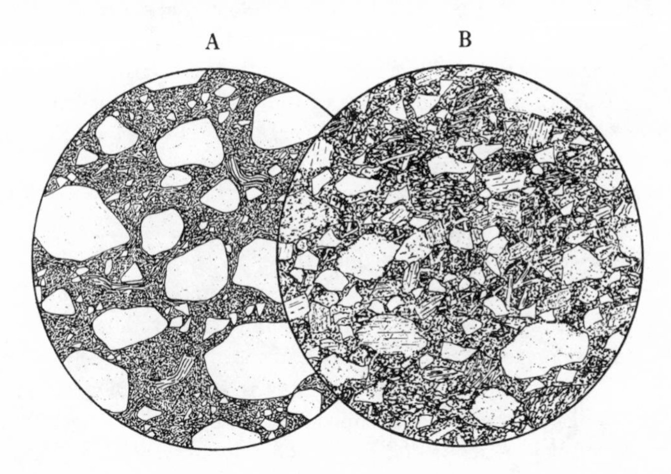

Fig. 24.6. × 20.

A. Sandstone with wackestone texture; moderately sorted sand floating in a muddy matrix.

B. Greywacke; angular and poorly sorted grains show poor definition from the matrix due to decomposition and marginal corrosion. Ordovician, Dumfriesshire.

wackestone texture to a packstone texture. In more mud-rich deposits it may also cause redistribution of grains and matrix with the formation of grain clusters; these clusters possess the texture of arenites and may undergo intergranular solution as in larger sand aggregates. Compaction may also cause reorientation of clay particles so that they become moulded around resistant grains and display distinctly anisotropic fabric under crossed polars. Larger flaky grains, such as mica grains, may also display distortion around more resistant grains.

The absence of large pores and the corresponding low permeability of the deposits prevent the large-scale introduction of minerals during diagenesis but mineral transformations may nevertheless take place where the primary constituents themselves are inherently unstable. Of particular importance is the alteration of clay minerals within the matrix; deep burial increases temperature and pressure to the point where kaolinite and montmorillonite become reconstituted to form illite and chlorite, and even illitic clay becomes recrystallised to more coarsely crystalline mica. The development of chlorite often begins at grain margins by the formation of poorly defined fringes; these fringes develop essentially at the expense of the matrix but minor replacement of the grains also takes place causing a general loss of definition of the primary texture. Recrystallisation of silica in the matrix may further blur the textures. As diagenesis grades into metamorphism the micas and chlorite become more coarsely recrystallised and tend to cluster around grain surfaces and along the cleavage planes of feldspar crystals. In sandstones of more reactive composition some of the changes described may take place in earlier diagenesis; in particular, sandstones with a high proportion of pyroclastic material show early development of chlorite and cherty silica, often accompanied by zeolites.

Grain overgrowth is generally suppressed in muddy sandstones but where they are developed they differ from those of cement origin by possessing diffuse outer margins and by containing fine inclusions of clay and pyrite. Feldspar alteration is common and in some cases albite is seen to develop authigenically in place of anorthite or as freely growing crystals. These changes are particularly characteristic of greywackes, which also display the total decomposition of calcic plagioclase with the formation of clay minerals and micas. This decomposition and that of unstable rock fragments combines with recrystallisation of the matrix to produce a rock with an apparently very high proportion of relatively homogeneously textured matrix (10) (fig. 24.6B). It is probable that the wackestone texture of many greywackes has originated in this way, and that the original sediments may have had a packstone texture. Instances have been cited in which the primary low-matrix texture has been preserved by early concretion formation (11).

Less drastic modifications of texture are produced by the earlier authigenic precipitation of pyrite and the carbonates. Siderite and dolomite tend to occur as fine crystal aggregates; calcite may show more extensive replacement of the matrix and may also invade grain margins. In rocks of greywacke type the calcite may in part be derived from the decomposition of calcic plagioclase.

Rudaceous rocks

These include conglomerates (rounded clasts) and breccias (angular clasts). They may be classified on size of clasts, composition of clasts, or composition of matrix or cement.

Most rudaceous deposits are of fluvial or shoreline origin and typically consist of mud-free gravels or their lithified equivalents. Even well-sorted gravels of this type commonly possess a matrix of downward filtered sand, which gives the deposits a distinctly bimodal grain size distribution. The greater mass of the primary constituents of rudaceous rocks allows much more rapid abrasion and rounding to take place so that roundness is of very different significance to that of sandstones. Angularity in deposits of aqueous origin is a clear indication of minimal transport.

Rudaceous rocks also result from mass sediment flow and from the melting of debris-laden ice. These deposits are characterised by extremely poor sorting and those of glacial origin may show preservation of chemically unstable minerals and rock types.

Because of their large size the essential constituents of rudaceous rocks are lithic clasts rather than monocrystalline grains. These clasts are of great value in provenance studies and so much of the petrographical study of rudaceous rocks concerns the identification and analysis of igneous, sedimentary, and metamorphic lithologies described elsewhere in the text. The matrix of these deposits is composed of mud or sand which can be studied independently as argillaceous or arenaceous lithologies respectively.

References and notes

(1) A detailed account of the definition, classification, petrology, and genesis of sandstones is given by Pettijohn, F. J., Potter, P. E. & Siever, R., 1972. *Sand and sandstone*. New York: Springer.

(2) Chert is sometimes included with the unstable rock fragments on

account of its low resistance to chemical and physical breakdown compared with the other silica grain types.

(3) Blatt, H. & Christie, J. M., 1963. Undulatory extinction in quartz of igneous and metamorphic rocks and its significance in provenance studies of sedimentary rocks. *J. sedim. Petrol.* **33**, 559–79.

(4) Illustrations of this and other types of syntaxial overgrowth in sandstone are to be found in Waugh, B., 1970. Petrology, provenance and silica diagenesis of the Penrith Sandstone (Lower Permian) of north-west England. *J. sedim. Petrol.* **40**, 1226–40.

(5) The diagnostic properties of heavy minerals are given in Milner, H. B., 1962. *Sedimentary petrography*: vol. 1, *Principles and applications*. London: Allen & Unwin. Chapter 1.

(6) Carrigy, M. A. & Mellon, G. B., 1964. Authigenic clay mineral cements in Cretaceous and Tertiary sandstones of Alberta. *J. sedim. Petrol.* **34**, 461–72.

(7) Van Houten, F. B., 1968. Iron oxides in red beds. *Bull. geol. Soc. Am.* **79**, 399–416. Walker, T. R., 1974. Formation of red beds in moist tropical climates: a hypothesis. *Bull. geol. Soc. Am.* **85**, 633–8.

(8) Thompson, A., 1959. Pressure solution and porosity, *in* Ireland, H. A. ed., *Silica in sediments*. Sp. Publs Soc. econ. Paleont. Miner., Tulsa, 7, 92–110.

(9) Waldschmidt, W. A., 1941. Cementing materials in sandstones and their influence on the migration of oil. *Bull. Am. Ass. Petrol. Geol.* **25**, 1839–79. (Describes sequences of cementation involving quartz, calcite, dolomite, and anhydrite.) Gilbert, C. M., 1949. Cementation in some Californian Tertiary reservoir sands. *J. Geol.* **57**, 1–17. (Describes sequences involving dolomite, calcite, and quartz.) Taylor, J. M., 1950. Pore-space reduction in sandstones. *Bull. Am. Ass. Petrol. Geol.* **34**, 701–16.

(10) Cummings, W. A., 1962. The greywacke problem. *Lpool Manchr geol. J.* **3**, 51–72.

(11) Brenchley, P. J., 1969. Origin of matrix in Ordovician greywackes, Berwyn Hills, North Wales. *J. sedim. Petrol.* **39**, 1297–1301.

25
Calcareous rocks

Introduction

Limestones (1) are rocks whose primary constituents are composed dominantly of calcium carbonate. These constituents include precipitated carbonate mud, skeletal fragments, ooliths, intraclasts, and occasionally terrigenous carbonate grains. Calcareous rocks which are not true limestones in the above sense are those with a primary calcium carbonate content of less than 50 per cent but whose total carbonate content exceeds 50 per cent because of diagenetic enrichment.

In purely textural terms limestones may be considered as being composed of grains and matrix, and classed as *grainstone*, *packstone*, *wackestone*, *mudstone*, and *boundstone* (p. 229). Where the nature of the individual grain components is to be emphasised, limestones may be divided into oolitic, shelly, pelletal, and other varieties. Because of the interest in limestones as fluid reservoirs, a further classification is used which places primary emphasis on the nature of the intergranular material, which is divided into microcrystalline calcite or *micrite* and sparry calcite or *spar*. The original dividing point was taken at 4 μm, but slightly higher figures are regarded by some as being more meaningful. On the basis of matrix crystal size the limestones are divided into *micrites* and *sparites*; subdivision is made by grain type, for example, biosparite (shelly), oomicrite, pelsparite (pelletal), intramicrite (intraclastic), and so on (2). Detrital silicates are indicated by the prefixes 'sandy', 'silty', and 'muddy'.

A cruder classification that is useful in the field is based purely on grain size; the grades *calcirudite*, *calcarenite*, *calcisiltite*, and *calcilutite* correspond to pebble, sand, silt, and clay grades respectively. Other field terms in common use are *bioclastic limestone* (composed essentially of skeletal fragments) and *calcite mudstone.*

Origin

Present-day carbonate sediments are found in shallow-water shelf areas, sea-floors, evaporating basins, freshwater lakes, and certain types of soil. A similar range of environments is represented in ancient deposits but most limestones are of warm-water carbonate shelf facies.

Present-day shallow marine carbonate deposits can be related to three main facies: a reef facies, a carbonate sand facies, and a lagoonal mud facies. Skeletal material is abundant throughout but is an essential constituent in the reef facies. Carbonate mud is deposited as minute needles of aragonite which appear to be precipitated both inorganically and by the activity of algae. Aragonite also occurs in the form of ooliths; these dominate many carbonate sand facies and appear to form by largely inorganic precipitation in near-shore zones of maximum wave and current activity.

Such shallow-water components are usually reliable indicators of environment but sometimes they undergo displacement to deeper environments by traction or turbidity currents. Limestones deposited in this way usually show interbedding with detrital silicate muds or sands and may also display graded bedding. The limestone constituents often show a great variety due to mixing of constituents from different shelf facies.

Carbonate deposition in cool shelf environments is restricted to the accumulation of biogenic skeletal particles, especially fragments of bivalves, gastropods, and algae. In deep-water environments the tests of calcareous microplankton may accumulate to form biogenic muds or oozes.

Limestones associated with evaporites and lake deposits display a variety of lithologies ranging from unfossiliferous mudstones to oolitic or shelly carbonate sands. In evaporite sequences limestones mark phases of lowered salinity, usually associated with freer connection with the sea.

Primary accumulations of carbonate material are typically affected by profound textural modification during diagenesis, and considerable enrichment in carbonate may take place by cementation. Cementing carbonate may be derived directly from the overlying sea-water, from the solution of metastable carbonate components (see next section), from intergranular solution, and from migrating pore waters. A few limestones are of wholly diagenetic origin, including those formed in the soil zone of semi-arid areas (*caliche* or *cornstone*).

The existence of present-day analogues greatly assists in the interpretation of ancient limestones, but difficulties arise from the extensive diagenetic imprint encountered in the latter. Solution or alteration of metastable carbonates causes profound textural changes which may completely obliterate the primary mineralogy and fabric.

Mineralogy

The mineralogy of most ancient limestones is dominated by calcite, with the other carbonates dolomite and siderite as authigenic accessories. In recent deposits the mineralogy is more complex because calcite may occur as normal (low magnesium) calcite, *magnesian calcite* (Mg up to 18 per cent), and aragonite. Both magnesian calcite and aragonite are liable to be transformed to low-Mg calcite, especially when the composition of the pore water changes from that of the depositional medium. The mineral transformations do not seem to represent true inversion, but a process of molecule-for-molecule solution and reprecipitation; this is indicated by changes in ionic substitution, with loss of magnesium and often enrichment in iron to form *ferroan calcite*.

The variable iron content of diagenetic calcites and dolomites is often used as an aid to thin section studies by means of staining (3). Differential staining may greatly emphasise limestone fabrics and helps to distinguish phases of cementation and neomorphism.

Petrography of limestone constituents

Ooliths. Calcareous ooliths typically consist of a roughly spherical envelope enclosing a nucleus which may be a carbonate grain (shell fragment, pellet, etc.) or a detrital silicate grain (figs. 25.2A, 26.1B). Diameters usually lie in the range 0.15 to 1.0 mm. Irregularities in the shape of the nucleus are progressively eliminated by the oolitic layers, which always display a trend towards increasing sphericity. Ooliths which possess a relatively large nucleus and a very thin envelope are known as *superficial ooliths*.

Unaltered marine ooliths possess aragonitic envelopes made up of delicate concentric laminae (*c*. 1–3 μm thick) of three types: (a) with unoriented crystals, (b) with tangentially oriented crystals (c-axes radial), and (c) a rare type with radially oriented crystals. The unoriented laminae may be discontinuous and lenticular and they are relatively rich in organic matter. It is possible that algal or other surface colonies are involved in the formation of these layers.

In ancient rocks the ooliths have undergone alteration to calcite; the primary concentric structure is partly obscured by a crude radial texture (fig. 25.2A). This radial texture is produced by the rearrangement of opaque organic or pyritic inclusions and the associated calcite crystals show a rough radial orientation of their c-axes.

Ooliths may also undergo solution, leaving behind the nucleus (if not itself aragonitic) and sometimes collapsed films representing the residue of the organic-rich layers (fig. 22.9B). The resulting cavity may be subsequently cemented by calcite.

Pisoliths. These often resemble ooliths in shape and structure, in which case they are found in close association with oolitic limestones. In other cases the primary limination is diffuse and irregular, and the envelope may contain detrital inclusions; such bodies are regarded as being of algal accretionary origin and are termed *algal pisoliths* or *oncoliths* (fig. 25.3A). They commonly occur in micrites with mudstone or wackestone texture, representing quieter conditions than those of true oolitic precipitation.

Pellets. Sand-sized grains of structureless micrite are termed pellets (2). Those that possess smooth elliptical outlines and which (in any one deposit) fall into one or more distinct size groups are probably of faecal origin – *faecal pellets* (fig. 25.1B). Other pellets may be of clastic origin or may result from the micritisation of skeletal grains (p. 274).

Intraclasts. Fragments of lithified sediment derived by erosion within the basin of deposition are termed *intraclasts*. They possess mudstone and packstone textures and may include a variety of grain types.

Bivalves. Suitable sections of articulated bivalve shells show the typical bilateral symmetry across the hinge-line. These and many other sections show the pronounced thickening of the shell along the hinge-line (4).

Many bivalves are composed of aragonite or a combination of aragonite and calcite. In diagenesis the aragonite typically alters to calcite, with destruction of the primary shell fabric and its replacement by an irregular neomorphic crystal mosaic (fig. 25.1A). Some bivalves possess a two-fold structure consisting of an inner layer of aragonite (represented by neomorphic calcite after diagenesis) and an outer layer of laminated calcite which survives diagenesis.

A few groups of bivalves are wholly calcitic; the ostreids and pectenids possess a two-layered structure consisting of an outer prismatic layer and an inner lamellar layer in which the lamellae are arranged at an angle to the shell surface. In the shell of *Inoceramus* the prismatic layer is particularly prominent (fig. 25.3B); fracture along the prism boundaries may reduce the shell to small rectangular fragments or even individual prisms. These shells may also show compactional dislocation along the prism boundaries.

Aragonite may survive diagenesis under exceptional conditions, but the primary structures are too varied to be covered in detail here, and reference should be made to the comprehensive works already cited. They include prismatic and lamellar structures, vestiges of which may be retained in calcitised shells as minute pyritic or organic inclusions. Sometimes aragonitic shells are replaced by *pseudopleochroic calcite* in which the preferred orientation of included organic particles causes the calcite to change from colourless to brownish with rotation of the stage (5).

Gastropods. Sections of gastropods range from nearly circular when cut perpendicular to the spire to squat or elongate triangular when cut parallel to the spire (fig. 25.4A). The latter sections display various aspects of the coiled living tube, depending on the position of the section relative to the coil axis.

The microstructure of gastropod shells is similar to that of bivalves. Aragonite is the usual mineral component but a two-layered structure of aragonite (inner layer) and calcite (outer layer) is present in some forms.

Ammonoids and nautiloids. These are distinguished by the internal septa and generally by their bilaterally symmetrical planispiral coiling. Some forms possess straight chambered shells.

The shells appear to have been primarily aragonitic and are thus usually preserved in neomorphic calcite. Where aragonite has survived the structure consists of an inner laminated layer covered on both sides by a prismatic layer.

Belemnites. These are usually represented by the stout cylindrical guard which is composed of fibres or prisms of calcite which radiate from an off-centred longitudinal axis (fig. 25.3B).

Brachiopods. Articulated brachiopods generally display a difference in the size of the two valves and in some forms traces of internal spiralia may be present. Apart from a few forms with phosphatic shells (p. 306) the brachiopods have calcitic shells; fossil brachiopods therefore lack the neomorphic calcite textures of most bivalves. A two-layered structure is normally displayed consisting of an outer layer of fine fibres or prisms arranged perpendicularly to the shell margin and an inner layer composed of long fibres or bundles of fibres disposed at low angles to the inner shell margin. In both of the layers the optic axis of the crystals is perpendicular to the length of the shell. The outer layer is usually thin but in the terebratulids and rhynchonellids the outer prismatic layer dominates the shell.

Punctuate brachiopods display small tubes or punctae in the inner layers of the shell; the punctae are arranged perpendicular to or at a steep angle to the shell surface but do not reach the surface as they break up into fine tubules in the outermost layers. The punctae are commonly filled with calcite cement, but sometimes primary micrite has filtered into the pores. Pseudopunctae shells resemble the punctate types but contain small calcite rods in place of punctae. Many brachiopods, being impunctate, show neither feature.

Some Palaeozoic brachiopods possess spines which often become detached from their host shells and act as discrete particles. They appear in section as hollow rods with a fibrous concentric structure (fig. 29.1B). Associated shells display surface protuberances corresponding to spine bases.

Ostracodes. Ostracode valves rarely measure more than a few millimetres across. They show a strong inward curvature along the free margins and where articulated show slight overlap of one valve by the other (fig. 27.2C). The shell consists of a number of layers in which a fine prismatic structure is usually the most prominent.

Trilobites. These are composed of a number of segments whose hard parts usually occur in a disarticulated state. Individual segments display complex curvature, and the lower margin in particular is strongly recurved. The carapace is usually preserved in finely crystalline calcite with a delicate prismatic structure. A crude lamination is sometimes produced by inclusions of organic matter or pyrite. Narrow pores may be present which run more or less perpendicularly to the surface; they are most apparent when mud-filled.

Echinoderms. In shallow marine environments echinoderm skeletons usually break down into their component plates. These plates are distinctive in being composed of single calcite crystals. Where plates or rods are curved the crystals' orientation swings round with the curvature.

A second distinctive feature is the system of micropores that perforates each of the crystal plates. The pores mostly appear circular or elliptical in section but may coalesce to form a reticulate pattern. In echinoid spines elongate pores radiate from the axis. The outer surface of a plate often possesses a dense system of relatively small pores. Crinoid ossicles characteristically possess a central canal (fig. 25.1B), a feature also shown by some types of echinoid spine.

Echinoderm fragments play a distinctive role in limestone diagenesis. Echinodermal calcite is of the magnesian variety and is therefore subject to transformation to low-Mg calcite. In grainstones this transformation is often accompanied by obliteration of the pore structure through cementation by clear sparry calcite in optical continuity with the host grain. Internal cementation may take place on the sea floor but more commonly takes place within the sediment where it is usually accompanied by syntaxial rim cementation (fig. 25.1B). Syntaxial overgrowth is inhibited by the presence of a micrite envelope (p. 274) and to some extent also by the plugging of surface pores by micrite. Embedding in a micrite matrix usually inhibits overgrowth but poorly defined syntaxial rims occasionally develop by neomorphism of the adjacent matrix.

Pore structure is most clearly preserved in packstones, wackestones, and mudstones where the matrix has permeated the pore system. Such preservation sometimes reveals considerable enlargement of the pores by solution, which probably takes place on the sea floor. The highly porous nature of these grains makes them particularly susceptible to replacement by silica, dolomite, or phosphate.

Foraminiferans. Calcareous foraminiferans display a wide variety of shell form although most are multichambered and either straight or regularly coiled (fig. 25.3B); they are mostly less than 1mm in diameter and are composed of low-Mg calcite. The appearance of individual fossils depends to a large degree on the orientation of the section.

The internal structure of the shell is too varied to be fully described here, but the size, morphology, and chambered structure are sufficiently distinctive to permit ready identification.

Corals. The corals display variation both in the internal structure of the individual corallites and in the manner of their compounding in colonial forms. Rugose corals are composed of primary calcite fibres oriented at various angles to the growth surfaces; a central layer rich in organic inclusions is commonly present (fig. 26.1A). Scleractinian corals are primarily aragonitic and are therefore preserved in turbid neomorphic calcite (or clear cement where early solution has taken place). Coral structures can usually be distinguished from those of bryozoans by the greater diameter of the corallites (usually over 1 mm) and by the presence of septa.

Bryozoans. These basically consist of a series of parallel zooecial tubes which may be crossed by transverse plates. The tubes usually have diameters less than 0.5 mm and display circular to polygonal cross-sections (fig. 29.1B). Present-day bryozoans possess calcitic, aragonitic, or combined skeletons but earlier forms are believed to have been calcitic. The internal wall structure is variable but dominated by a delicate lamellar structure parallel to the tube wall surfaces; in some forms, however, the lamellar structure is more complex and in others a fibrous or prismatic structure is developed perpendicular to the surface.

Calcareous algae. These occur as encrusting nodular or sheet-like masses of very finely cellular calcite. The cells are arranged in a lamellar or radial pattern or take the form of intertwined tubules. Two distinct sizes of cell are distinguishable in some varieties. In the absence of strong recrystallisation the skeleton consists of finely crystalline calcite with abundant organic or pyritic inclusions.

Micrite envelopes. In many recent and ancient limestones skeletal grains possess thin pellicles or envelopes of micritic material which lack the characteristic lamination of oolitic accretions. Under high magnification the inner margins of these envelopes are seen to be rather diffuse and delicate tubules (generally *c.* 5 μm diameter) may be distinguishable penetrating the outer margin of the skeletal carbonate. These ***micrite envelopes*** result from the infilling by micrite of densely superimposed superficial perforations of algal and fungal origin (6). The envelopes possess considerable resilience as they may survive solution of the host grains with little or no deformation; this is indicated by undeformed micrite envelopes which enclose cement casts (fig. 25.2C). This relative resistance over skeletal aragonite particles is sometimes shared by oolitic envelopes (p. 298) and must be related either to a high organic content or to early calcitisation.

Organic perforations. Boring algae and fungi are responsible for the production of micrite envelopes as described above, but they also produce less dense perforation in which individual tubules can be distinguished. Fungal perforations possess diameters of up to 4 μm; larger borings, up to 15 μm diameter, are probably of algal origin. The perforations may be micrite filled or cemented by calcite, pyrite, limonite, etc. Much larger perforations ranging from a few hundred micrometres to several millimetres in diameter result from the boring action of certain bivalves, gastropods, sponges, and arthropods (fig. 25.1A).

Petrography of limestones

Limestones may be classed according to their grain constituents or according to their texture. Interpretation of textures is often difficult because of the extensive diagenetic modifications experienced by many limestones but the emphasis is here placed on texture because of its importance in terms of mechanics of deposition, diagenesis, and fluid reservoir properties. The distinction between cement, primary mud (usually seen in recrystallised form), and organically bound fabrics is fundamental to textural interpretation and the diagnostic features are discussed in the following sections.

In all limestone studies it is important to bear in mind the metastable nature of two important primary minerals, aragonite and magnesian calcite; their instability has a profound effect on the nature and timing of limestone diagenesis.

Grainstones

The most characteristic grain assemblages are those composed largely of ooliths or skeletal fragments. Fragmental components commonly show considerable rounding and the whole grain assemblage is often well sorted – or very well sorted in most oolitic, crinoidal, and pelletal varieties. Such assemblages are not specific to grainstones, and positive identification of a grainstone can be made only where voids or cement fabrics preclude the existence of a primary mud matrix.

Difficulty arises in attempting to distinguish cement fabrics from those of neomorphic spar (7). Neomorphism can usually be ruled out where spar is associated with, and yet sharply distinct from, patches of interstitial micrite and also where the texture of the spar is distinctly related to grain surfaces (fig. 25.1). Features which can be regarded as fully reliable indicators of cement fabrics are the presence of well-defined fringes of uniform thickness (fig. 25.1A) and the occurrence of enfacial angles in high proportion (fig. 25.1B and *see* p. 231).

Many limestones display two clearly defined phases of cementation separated by a phase of compaction. Very often the two generations of calcite show a difference in iron content, the later generation usually being the more iron rich. Less commonly the early cement is sideritic (fig. 25.1A). The derivation of carbonate cement and its introduction into a host sediment are problems which cannot normally be solved by petrographical studies, but it may be possible to observe local controls on cementation. For example nucleation may be influenced by the mineralogy and fabric of grain surfaces and on this basis it may be possible to assess the timing of cementation relative to the aragonite–calcite transformation.

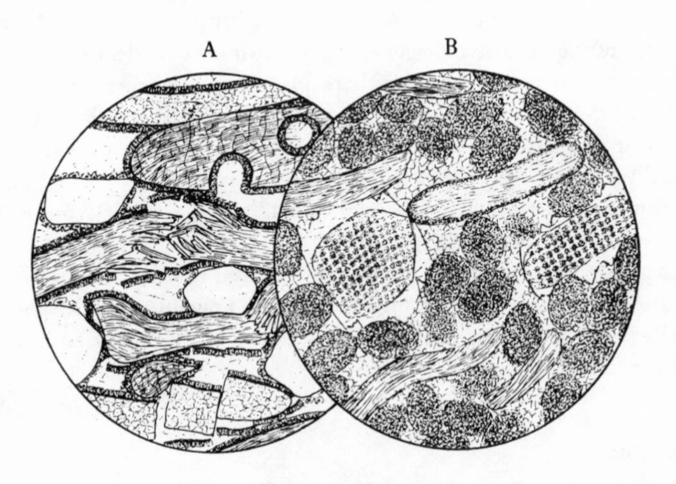

Fig. 25.1. × 20.

A. Shelly grainstone, showing an early siderite fringe cement and a later void-filling cement of calcite. Fracturing of grains and dislocation of associated fringes has occurred between phases of cementation. Aragonitic shells now composed of neomorphic calcite (stippled). Jurassic, Yorkshire.

B. Pelletal grainstone, showing syntaxial rim cementation on an echinoderm nucleus and mosaic cement elsewhere. Carboniferous, Yorkshire.

The control of grain surfaces on nucleation is expressed in two ways, by variation in crystal habit and by variation in bulk. Both aspects are illustrated by the widely observed contrast between cementation on echinoderm nuclei and that on associated polycrystalline grains. This contrast is particularly evident in the case of fringe cements where the relative ease of nucleation on echinoderm grains is shown in the development of bulky syntaxial overgrowths in contrast to relatively meagre fringes on other grains whose surfaces have been the sites of much more competitive nucleation (fig. 25.1B). Where echinoderm grains are associated with non-carbonate grains early cementation may be entirely restricted to syntaxial overgrowth (fig. 27.3A). The absence of contrast in nucleation as described above is generally related to the presence of micrite envelopes, which inhibit syntaxial overgrowth.

Less obvious control of nucleation is displayed by other skeletal constituents as for example where coarsely prismatic layers in bivalves receive more cement than lamellar layers. In some cases it is even possible to distinguish a contrast between the amount of cement received by calcitic and aragonitic components, the former providing better conditions for nucleation of calcite cement. Such preference would not be shown after the aragonite to calcite transformation.

The compactional fracturing of grains provides a useful basis for the relative timing of cement events. Where more than one generation of cement is present it is usual for early fringe cements to precede the main phase of compaction and for the fringes to become involved in the grain fracture with buckling and detachment from their host grains (fig. 25.1A). Subsequent wholesale cementation fills both the remaining primary pore space and the compactional fractures; its strength prevents further fracturing.

The timing of wholesale cementation is less easily determined in the absence of compactional fracture. It may be possible to obtain a timing relative to intergranular solution, and it is usual for the latter to be inhibited in cemented portions of limestones. This is not surprising as although superficial grain welding may take place near the surface, large-scale interpenetration does not normally develop until depths of thousands of metres have been attained.

As already mentioned it is possible that some limestones have been originally cemented by aragonite rather than calcite. It is unlikely that neomorphic calcite would preserve relict textures of finely crystalline aragonite cements but more coarsely crystalline radially fibrous aragonite cement may be represented by ***radiaxial calcite*** (8). This consists of groups of crudely radial crystals, each of which displays curved twin lamellae and cleavage (fig. 25.2B). Radiaxial calcite is distinguished from radial fibrous calcite by the movement of the extinction shadow across the crystals in the same sense as the rotation of the stage.

Grainstones also provide a useful insight into neomorphism within the grains themselves, particularly in relation to the processes of compaction and cementation. Neomorphism principally involves the aragonite to calcite transformation, but also involves the alteration of magnesian calcite to low-Mg calcite. In the former case the transformation can be dated relative to compaction by the nature of the fracture of aragonitic grains. For example shells which are now composed of neomorphic

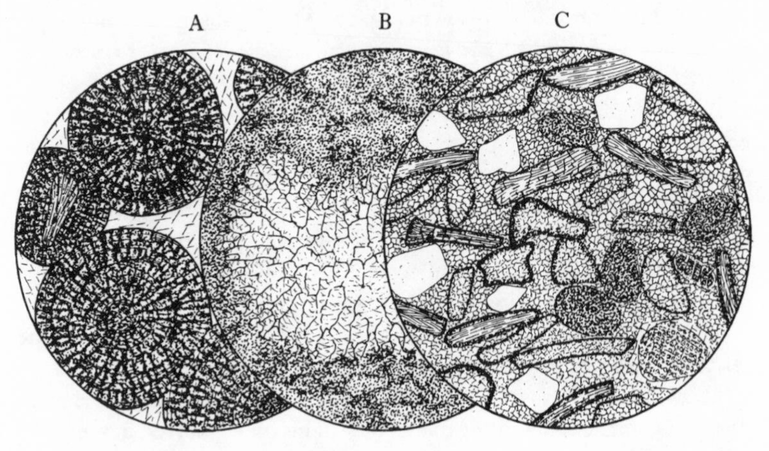

Fig. 25.2. × 20.

A. Colitic grainstone showing intergranular solution prior to cementation. Jurassic, Northamptonshire.
B. Radiaxial cement in a cavity in micritic limestone. Carboniferous, Yorkshire.
C. Shelly grainstone showing cement-filled casts of aragonitic shells whose form has been retained by the strength of micrite envelopes. Jurassic, Lincolnshire.

calcite may show fracture corresponding to weaknesses along primary laminae (fig. 25.1A) or transverse displacement along primary prism boundaries. The presence of spalled concentric laminae in calcitised ooliths similarly indicates that fracture preceded calcitisation.

Textures do not always reveal the relative timing of calcitisation and calcite cementation. The continuity of individual crystals across a grain–cement boundary may indicate simultaneous replacement and cementation but can also be interpreted as resulting from syntaxial nucleation of one generation of calcite upon another. Nevertheless where both neomorphic and cementing calcite are of ferroan composition a close genetic relationship is indicated.

It is important in considering textures of the grains themselves to distinguish between neomorphic calcite spar and cement spar deposited in moulds resulting from solution of components from within micrite envelopes. Apart from the standard criteria for distinguishing between textures of cementation and neomorphism (pp. 231–3) the neomorphic calcite of carbonate grains can usually be distinguished by the presence of organic inclusions, which may define a 'ghost' structure. Where ferroan calcite is present that of neomorphic origin is often less iron-rich than the associated cement.

Visual porosity in grainstones is usually of primary intergranular type, having survived elimination by cementation and intergranular solution. Primary intergranular porosity may also be retained in the cavities and pores of skeletal grains. Secondary porosity mostly results from selective solution of primary or authigenic constituents, such as the solution of opaline grains within a calcite matrix.

The most common non-carbonate cementing mineral is silica (chapter 27). Siderite (fig. 25.1A) also occurs as a rare cement and both siderite and dolomite may develop in late diagenesis as coarsely crystalline rhombs or aggregates cutting across the cement–grain fabric.

Mudstones, wackestones, and packstones

This group is defined by the presence of a primary carbonate matrix. The matrix will normally have been deposited as aragonite mud and subsequently converted to calcite micrite (fig. 25.3A) in which the primary microparticulate texture is replaced by a microcrystalline texture of lower porosity. Retexturing usually continues beyond the calcite micrite stage with the formation of fine to coarse neomorphic spar.

Neomorphic spar appears to develop in two ways. In one process recrystallisation begins at a number of centres within the matrix and then progresses outwards to form patches of uniformly sparry calcite lying within unaltered micrite matrix (fig. 25.4A). In the second form of recrystallisation the micrite appears to have dissociated partly or completely into 'clots' some 30 μm in diameter which are embedded in a matrix of fine spar (fig. 25.4A). This texture is known as ***structure grumeleuse***; its origin is uncertain and may be related to diagenetic segregation alone or possibly to some primary pelletal texture (9).

Recrystallisation is sometimes influenced by included grains either by the formation of neomorphic sparry fringes along grain margins or by the development of rims of syntaxial recrystallisation. The latter differ from cement rims in possessing inclusions and irregular and often diffuse outer margins.

It is difficult to determine the original fabric in coarsely recrystallised sparites as both grains and matrix have usually undergone recrystallisation. Nevertheless grain outlines can often be made out by the distribution of inclusions, which are usually fine organic particles concentrated in ooliths and in micrite envelopes. Mudstones and wackestones can thus be identified if the grain boundaries are clear enough. Packstones may be confused with grainstones if the neomorphic spar is mistaken for cement, but the absence of characteristic cement textures, relatively loose packing, and general cloudiness due to inclusions should allow correct identification.

Compactional effects in micritic limestones are usually limited to collapse of articulated bivalved shells and fracture of shells at structurally weak point contacts. Intergranular solution is usually inhibited by the matrix, especially after calcite neomorphism, but more widespread solution surfaces commonly develop in later diagenesis. Large-scale solution may produce well-defined clay seams where muddy limestone horizons have been selectively decalcified. Individual stylolitic seams (fig. 22.8B) also show relative concentration of insoluble constituents (terrigenous grains, organic matter, and authigenic pyrite). Although stylolites may cut across carbonate grains without any modification, coarsely crystalline grains such as echinoderm fragments offer more

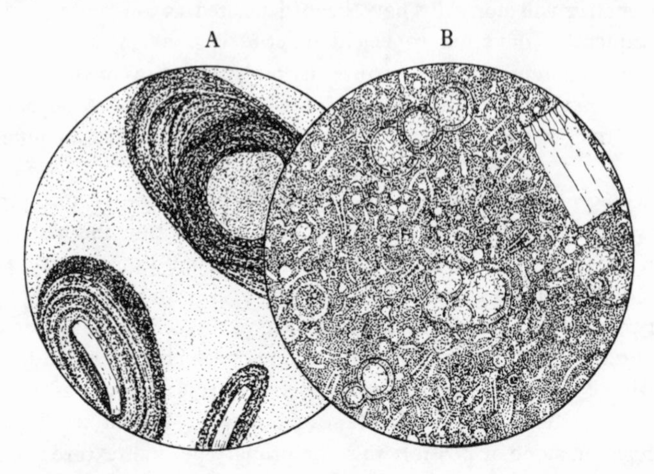

Fig. 25.3.

A. Pisolitic wackestone; pisoliths of algal origin set in a matrix of micrite. Jurassic, Lincolnshire. × 10.

B. Chalk, showing foraminiferans and prismatic fragments of *Inoceramus*. Cretaceous, Norfolk. × 20.

resistance to solution and cause local splitting of the stylolitic seam.

Chalk differs from most micrites in having been deposited as finely crystalline calcite and not as aragonite. In consequence it has often undergone little neomorphism and the primary microparticulate texture is preserved. The calcite particles are crystals derived from the breakdown of algal coccospheres. Larger particles include foraminiferans (fig. 25.3B), occasional silt and sand grains, glauconite grains, and scattered macrofossils. The macrofossils are typically calcitic, the aragonitic forms having been leached out prior to compaction.

Cementation of chalk on a large scale is probably caused by solution transfer. More restricted cementation has taken place in relation to hardgrounds (10) in which the cementing calcite is apparently of very early diagenetic origin.

Boundstones

These are produced by precipitation or trapping of calcium carbonate by sessile colonial organisms. Corals and algae are most commonly responsible for boundstone textures; the former may display the typical internal corallite structures and the latter often show lamination due to

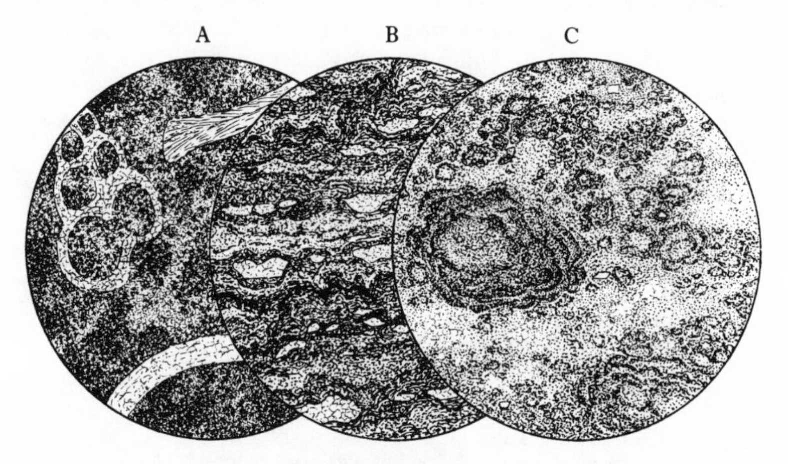

Fig. 25.4. × 20.

A. Structure grumeleuse.
B. Boundstone texture in a stromatolite.
C. Cornstone (caliche) showing ill-sorted and diffuse 'grains', some of which display a crude concentric layering. Old Red Sandstone, Dumfriesshire.

varying amounts of organic matter as in stromatolites (fig. 25.4B). Some structureless rocks are also believed to be of boundstone origin; the lack of structure may be related to recrystallisation (and also to dolomitisation) but often it may simply reflect a rapid influx of detrital sediment causing masking of any lamination of biological origin.

Many boundstones possess cavities which are too large to have originated as intergranular voids. In some cases these cavities reflect the open framework of the original organic structure but in others the cavities appear to be secondary, resulting from the decomposition of algal tissue. A possible example of the latter is the *Stromatactis* structure in which large irregularly shaped cavities occur sporadically through fine-grained limestones, with no apparent relationship to organic accretionary texture (11). Boundstone cavities are partially or wholly cemented by sparry calcite but in many cases cementation has been preceded by the accumulation of detrital mud, producing a geopetal texture (p. 235).

Caliche

In this essentially concretionary limestone the calcite may occur as a simple intergranular impregnation but often has developed by replacive or displacive crystallisation to form almost pure limestone bodies within siltstone or sandstone hosts. The calcite is mostly finely crystalline with local patches and veins of coarse spar. The calcite commonly displays a variety of textures, including a clotted texture similar to *structure grumeleuse* (p. 279), laminated texture, and accretionary textures which may be poorly defined (fig. 25.4c) or which may display sharply defined pisolitic structure.

References and notes

(1) The literature on limestones is very extensive, but several recent works provide useful summaries of current thought on their classification, petrology, and interpretation: Bathurst, R. G. C., 1971. *Carbonate sediments and their diagenesis*, Developments in sedimentology 12. London: Elsevier. Bricker, O. P., ed., 1971. *Carbonate cements*. Baltimore: Johns Hopkins Press. Friedman, G. M., ed., 1969. *Depositional environments in carbonate rocks*, Spec. Publs Soc. econ. Paleont. Miner., Tulsa 14.
Illustrations of a variety of oolitic, algal, concretionary, and brecciated textures are to be found in Cayeux, L., 1935. *Les roches sedimentaires de France: roches carbonatées*. Paris: Masson. This has recently been translated and revised by Carozzi, A. V., 1970. *Carbonate rocks*. New York: Hafner.

(2) Folk, R. L., 1959. Practical petrographic classification of limestones. *Bull. Am. Ass. Petrol. Geol.* **43**, 1–38.

(3) A summary of staining techniques is given by Friedman, G. M., 1971. Staining, *in* Carver, R. E., ed., *Procedures in sedimentary petrology*. New York: Wiley, pp. 511–30.

(4) Detailed thin section descriptions of skeletal grains are given by the following: Bathurst, R. G. C., 1971. *Carbonate sediments and their diagenesis*, pp. 1–76. Horowitz, A. S. & Potter, P. E., 1971. *Introductory petrography of fossils*. New York: Springer-Verlag. Majewske, O. P., 1969. *Recognition of invertebrate fossil fragments in rocks and thin sections*. Leiden: Brill.

(5) Hudson, J. D., 1962. Pseudo-pleochroic calcite in recrystallised shell-limestones. *Geol. Mag.* **99**, 492–500.

(6) Bathurst, R. G. C., 1966. Boring algae, micrite envelopes and lithification of molluscan biosparites. *Geol. J.* **5**, 15–32.
(7) Bathurst, R. G. C., 1971. *Carbonate sediments and their diagenesis*, pp. 417–19.
(8) Bathurst, R. G. C., 1959. The cavernous structure of some Mississippian *Stromatactis* reefs in Lancashire, England. *J. Geol.* **67**, 506–21.
(9) Cayeux, L., transl. Carozzi, A. V., 1970. *Carbonate rocks*, pp. 253–4. Beales, F. W., 1965. Diagenesis in pelletted limestones, *In* Pray, L. C. & Murray, R. C., eds., *Dolomitisation and limestone diagenesis: a symposium*. Spec. Publs Soc. econ. Paleont. Miner., Tulsa 13, 49–70.
(10) Bromley, R. G., 1967. Some observations on burrows of thalassinidean Crustacea in Chalk hardgrounds. *Q. Jl geol. Soc. Lond.* **123**, 157–82.
(11) Lowenstam, H. A., 1950. Niagaran reefs of the Great Lakes area. *J. Geol.* **58**, 430–87.

26

Dolomitic rocks

Introduction

The mineral dolomite, $Ca(Mg,Fe)(CO_3)_2$, is most closely associated with calcite with which it forms a wide range of lithologies intermediate between pure limestone and pure dolomite rock. The rock term ***dolomite*** (or ***dolostone***) is usually applied to rocks containing more than 50 per cent of carbonate minerals and in which dolomite exceeds calcite. Rocks with a lesser dolomite content, ranging down to 10 per cent of the carbonate fraction, are termed ***dolomitic limestones***.

Origin

Petrographical and field evidence indicate that many dolomites have resulted from replacement of limestones, but many of the purer dolomites lack such evidence and have therefore been regarded as of possible primary origin (1). The possibility of primary dolomite precipitation has been investigated both in the laboratory and in present-day environments. Experimental work indicates that very slow crystallisation is required for the direct precipitation of dolomite and that under surface conditions precipitation of a less ordered calcium-rich ***protodolomite*** is to be expected (2) with subsequent diagenetic neomorphism to dolomite. Studies on recent sediments support this mechanism as protodolomite has been recorded in environments which favour rapid precipitation of magnesium-rich carbonates.

The primary origin of these naturally occurring protodolomites has, however, yet to be demonstrated and in most cases it can be shown that dolomitic sediments result from replacement of primary calcium carbonate. This may take place soon after deposition as a result of Mg^{2+} enrichment of interstitial waters through the evaporative precipitation of Ca^{2+} in the form of gypsum. The process may in some cases be aided by the metastable nature of the calcium carbonates (aragonite and magnesian calcite) and perhaps also by organic enrichment in magnesium as found in mats of blue-green algae.

Dolomitisation at somewhat greater depths is believed to take place by the seepage-reflux mechanism (3). This involves the downward and seaward migration of brines that have become concentrated by evaporation in backshore lagoons. This process may have caused the dolomitisation of relatively large carbonate units such as reef masses.

Dolomite has also been reported from deep-sea sediments where it usually occurs as well formed crystals. Whether this dolomite is primary or diagenetic is not known, but its well-crystallised nature probably reflects very slow crystallisation.

Mineralogy

Dolomitic rocks display a variety of mineralogical characters depending on lithological association, and the only specific component is dolomite itself. Ionic substitution is more pronounced than in calcite, and the substitution of Fe^{2+} for Mg^{2+} is of particular importance in the sedimentary dolomites. The range in $FeCO_3$ content in natural dolomites is 0–30 per cent, but the higher values are associated with dolomites of hydrothermal origin.

The term *ankerite* is used for very iron-rich dolomites but the definitive Mg:Fe ratio is variously placed at 2:1 and 1:1, neither of which limits has an obvious petrographical expression. In thin section studies it is simpler to refer to all iron-rich dolomite minerals as *ferroan dolomites* (4). They may be distinguished from the non-ferroan dolomites by staining; with standard techniques (5) the lower limit of detection is about 1 or 2 per cent. In general the intensity of stain increases with iron content, but as other factors may influence solubility (and hence stain formation) colour intensity cannot be relied on as a certain measure of relative iron content.

Petrography

Dolomite is largely if not wholly of diagenetic origin and unlike calcite does not form distinctive primary grain types. It occurs as subhedral and euhedral aggregates or as isolated well-defined rhombs. The rhombs often display inclusions arranged in zones parallel to the crystal faces. The inclusions, of insoluble organic matter and clay particles, are residual from the host rock. Commonly a central cloudy zone is present, surrounded by a clear rim; this rim may represent cementation in a marginal cavity formed by solution of adjacent calcium carbonate or it may represent slower and more complete replacement. More rarely two or more zones of cloudy dolomite are present, perhaps representing variation in rate of crystallisation. Rhombohedral zones of finely divided iron hydroxide probably represent exsolution of ferrous iron during weathering of ferroan dolomite.

Dolomitic limestones

Partial dolomitisation of limestone is selective (6), with the less resistant components usually being those of primary aragonitic or magnesian calcitic composition. Ooliths, algae, gastropods, etc. may be partly or wholly replaced by dolomite while the matrix remains unaffected. The dolomite occurs as isolated rhombohedra within the grains or as crystal aggregates bounded by the original grain margin (fig. 26.1A).

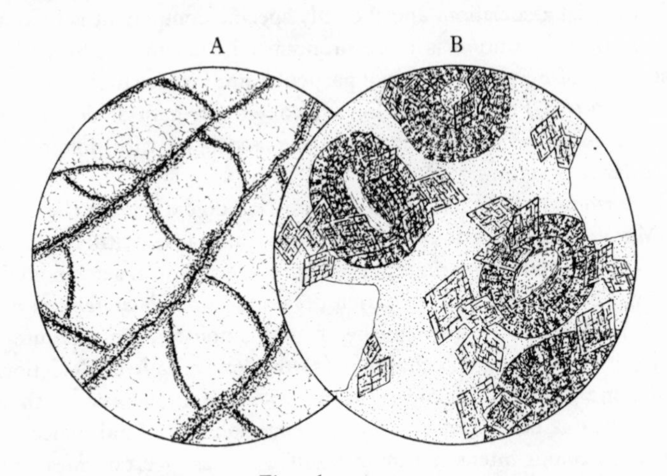

Fig. 26.1. × 20.

A. Dolomitised rugose coral; dolomite (heavy stippling) has replaced the septa and calcite cementation has filled the cavities between septa and dissepiments.
B. Dolomitic limestone; dolomite rhombs have replaced both ooliths and matrix without selectivity. Inclusions in the dolomite reveal the original outlines of the ooliths.

Selective replacement does, however, depend on the previous diagenetic history of the limestone and if aragonite and magnesian calcite have already been converted to low-Mg calcite, then such constituents will be as resistant as the original low-Mg calcite components; the distribution of dolomite may thus be independent of primary texture (fig. 26.1B). The nature and degree of selective replacement may thus reveal the timing of dolomitisation relative to calcium carbonate diagenesis. In limestones composed wholly of low-Mg calcite dolomite may still be selective, favouring more finely crystalline material. By observing whether the variation in crystallinity is primary or diagenetic it should be possible to date the dolomitisation relative to the aragonite–calcite transformation.

Dolomite tends to form idiomorphic crystals and typically does not retain the fabric of replaced limestones or the internal structure of their grain constituents. Opaque inclusions sometimes reveal the pre-existing texture but their zonal distribution usually disrupts original textural patterns beyond recognition. Exceptions have, however, been described in which dolomite retains delicate shell structure (7).

Partial dolomitisation of limestone has little effect on porosity. Although the molecular volume of dolomite is 12 to 13 per cent less than that of calcite, dolomite rhombohedra occurring within a limestone matrix normally lack associated porosity. Evidence of loss of such porosity by cementation is also lacking. Such replacement is therefore of a volume-for-volume nature, and must involve introduction of further carbonate as well as magnesium.

Many dolomitic limestones exhibit mottling, in which patches of relatively pure dolomite develop within otherwise unaltered and apparently homogeneous limestone. Such mottling may be purely diagenetic, resulting from the development of scattered centres of dolomite crystallisation but it may also reflect slight differences in primary composition or porosity caused by burrowing. Late diagenetic dolomitisation sometimes takes the form of veining and may lead to the development of a 'pseudo-breccia' in which angular relics of limestone are set in an anastomosing network of dolomite veins.

Dolomites

Some dolomites are merely extensions of dolomitic limestones, retaining some evidence of a pre-existing limestone fabric. These may consist of dolomite enclosing calcareous grains or residual patches of calcareous matrix; solution of the calcareous components may lead to the development of a high mouldic porosity.

Pure dolomites often display a markedly equicrystalline texture in which a few crystals at least display partial development of crystal outline (fig. 26.2A). Where well-defined rhombs lie within an otherwise anhedral mosaic two generations of dolomite are represented.

Some dolomites possess a highly porous *sucrosic texture* consisting of a loosely interlocking aggregate of rhombs with well-defined intercrystalline porosity (fig. 26.2B). The porosity largely results from volume loss caused by the replacement of calcite by the denser dolomite (8). This porosity may exceed 15 per cent, and may be augmented by mouldic porosity. The development of moulds follows dolomitisation and later cementation by dolomite is rare and limited only to a single layer of crystal overgrowths nucleated on matrix crystals forming the mould margin.

Heterogeneity of the host limestone may be reflected in the dolomite by variation in crystal size, porosity, or proportion of inclusions. Nevertheless, the tendency for dolomite to form equal-sized crystals is very pronounced, and inclusions are often rearranged crystallographically.

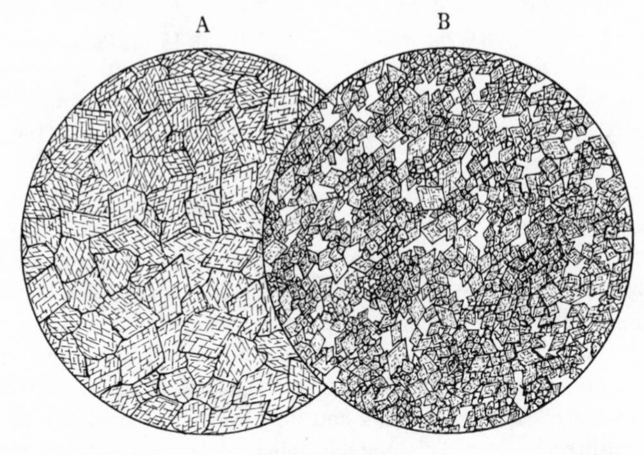

Fig. 26.2 × 20.

A. Equicrystalline dolomite mosaic in which some crystals show partial development of crystal faces.
B. Sucrosic dolomite; crossed polars (pores appear dark).

Rocks which display alternation of dolomite and calcite layers are generally interpreted as representing alternating phases of very early diagenesis. It can sometimes be demonstrated that the dolomite has formed by selective replacement of magnesium-rich algal mats which were interbedded with lime mud. These algal layers may show internal structure and can sometimes be distinguished by a lack of detritus and by a relative abundance of finely divided carbonaceous inclusions.

Dolomites associated with evaporitic sequences often lack primary textures and although this does not preclude an origin by limestone replacement, it is possible that they represent primary evaporative precipitation of dolomite or protodolomite. Many Precambrian dolomites similarly lack limestone textures and are thought to be of primary origin, reflecting somewhat different chemical environments from those of the Phanerozoic.

Dedolomites

Dolomite is normally a resistant mineral during diagenesis, but where it is affected by surface water circulation it may be leached out or replaced by calcite (9). Such processes constitute ***dedolomitisation***, and should be considered when examining samples collected from surface outcrops. Dedolomitisation is often attributed to reaction of dolomite with calcium sulphate solutions derived from evaporite sequences or from the oxidation of pyrite; in ferroan dolomites, however, the presence of ferrous iron alone may account for the instability of the dolomite.

Dedolomitisation is most clearly displayed by the alteration of dolomite rhombs. Calcite may wholly or partly replace the rhombs either as single crystals or as aggregates and in ferroan dolomites the calcitisation may be accompanied by precipitation of iron hydroxide. Sometimes the dolomite is not calcitised but is leached out to form rhombic pores which may later become calcite-cemented. The development of pores indicates chemical conditions unfavourable to both dolomite stability and calcite precipitation.

Variation in texture of replacive calcite may apparently be controlled by the distribution and optical orientation of relict calcite inclusions within the dolomite rhombs. These inclusions act as nuclei for the replacive calcite so that relict inclusions derived from a single crystal may cause regeneration of a single crystal by syntaxial nucleation. In extreme cases the broad pre-dolomitisation fabric of the original limestone may be regenerated during calcitisation (10).

Detrital dolomite

Because dolomite does not form primary precipitated grains, detrital dolomitic constituents (11) are rare. They usually result from subaerial erosion of dolomitic rocks and, depending on the fabric of the host rock, may be polycrystalline or may consist of abraded or cleaved single crystals. Diagenesis can alter the texture of detrital dolomites by intergranular solution or by cementation and in either case the detrital character of the rock may be obscured.

References and notes

(1) The following references provide reading on general aspects of the petrography and origin of dolomites: Bathurst, R. G. C., 1971. *Carbonate sediments and their diagenesis*, Developments in sedimentology 12, 517–43. Cayeux, L., transl. Carozzi, A. V., 1970 *Carbonate rocks*, pp. 300–415. Pray, L. C. & Murray, R. C., eds., 1965. *Dolomitization and limestone diagenesis: a symposium*, Spec. Publs Soc. econ. Paleont. Miner., Tulsa 13.

(2) Goldsmith, J. R. & Graf, D. L., 1958. Structural and compositional variations in some natural dolomites. *J. Geol.* **66**, 678–93.

(3) Adams, J. E. & Rhodes, M. L., 1960. Dolomitisation by seepage refluxion. *Bull. Am. Ass. Petrol. Geol.* **44**, 1912–20.

(4) Goldsmith, J. R., Graf, D. L., Witters, J. & Northrop, D. A., 1962. Studies in the system $CaCO_3$–$MgCO_3$–$FeCO_3$. *J. Geol.* **70**, 659–88.

(5) Friedman, G. M., 1971. Staining, *in* Carver, R. E., ed., *Procedures in sedimentary petrology*, pp. 511–30.

(6) Selective dolomitisation of Mg-rich algal grains is illustrated in Schlanger, S. O., 1957. Dolomite growth in coralline algae. *J. sedim. Petrol.* **27**, 181–6.

(7) Murray, R. C., 1964. Preservation of primary structures and fabrics in dolomite, *in* Imbrie, J. & Newall, N., eds., *Approaches to palaeoecology*. New York: Wiley, pp. 388–403.

(8) Murray, R. C., 1960. Origin of porosity in carbonate rocks. *J. sedim. Petrol.* **30**, 59–84. Weyl, P. K., 1960. Porosity through dolomitisation: conversion-of-mass requirements. *J. sedim. Petrol.* **30**, 85–90.

(9) Evamy, B. D., 1967. Dedolomitisation and the development of rhombohedral pores in limestones. *J. sedim. Petrol.* **37**, 1204–15.

(10) Shearman, D. J., Khouri, J. & Taha, S., 1961. On the replacement of dolomite by calcite in some Mesozoic limestones from the French Jura. *Proc. Geol. Ass.* **72**, 1–12.

(11) Sabins, F. F. Jr., 1962. Grains of detrital, secondary and primary dolomite from Cretaceous strata of the western interior. *Bull. geol. Soc. Am.* **73**, 1183–96.

27
Siliceous rocks

Introduction

Non-clastic siliceous rocks are typified by *chert*, a tough and usually dense rock composed of amorphous or finely crystalline precipitated silica. Most cherts are vitreous, and break with a splintery or conchoidal fracture; flint is a particularly finely crystalline and vitreous variety found as concretions in chalk. *Porcellanous chert* or *porcellanite* is a porous rock of low density and with dull fracture; it often contains fine clay or carbonate inclusions. Cherts composed largely of the remains of siliceous organisms are given the names *radiolarite*, *diatomite*, and *spiculite* (1). Where organic remains are less abundant the terms *radiolarian*, *diatomaceous*, and *spicular chert* are used. Organic and porcellanous cherts are usually of Tertiary or Mesozoic age as in older deposits burial and increased temperature have usually caused recrystallisation to vitreous chert.

In the field two types of chert may be distinguished: the *bedded cherts* and the *nodular cherts*. The bedded cherts are laterally extensive; contact with adjacent lithologies is usually sharp, representing a change in the nature of the primary deposit. Nodular cherts are of limited lateral extent; they may be irregularly disposed in relation to bedding and grade (usually rapidly) into the host rock. Other chert types include terrestrial spring deposits and pockets of chert occurring within pillow lavas.

Origin

Some cherts, particularly nodular varieties, clearly result from silicification of limestones or detrital silicate hosts. The origin of many bedded cherts is less certain; petrographical characters and field associations indicate a primary control on chert formation, although the precipitation of silica itself is usually diagenetic.

Natural waters are undersaturated with respect to silica (2), and sea-water more so than river-water. The drop in silica concentration as rivers enter the sea is largely due to dilution but also to organic fixation. The organically fixed silica is liable to resolution when the host organisms die (3) so that only the more robust bodies tend to be incorporated into the sediment. Nevertheless biogenic precipitation is the only mechanism by which non-detrital silica might accumulate under normal conditions. Inorganic precipitation may result from submarine volcanic and hydrothermal activity but this has not been substantiated by present-day observations, although it is theoretically possible because of the very high silica concentrations reached in some hydrothermal waters. Inorganic precipitation may have been of greater importance in Precambrian times, when a silica-fixing biota may not have existed.

Measurements of silica concentrations in pore waters have mostly revealed enrichment relative to sea-water, but not to the point of saturation. Studies of younger cherts have, however, provided insight into some aspects of silica diagenesis, and in particular the occurrence of an intermediate cristobalitic stage in the conversion of opal to quartz (4).

Bedded cherts

These occur in a variety of both deep and shallow-water facies but are virtually restricted to marine sequences (5). The basic essentials for the formation of bedded cherts appear to be either sudden and overwhelming precipitation of silica or slow continuous silica deposition in areas starved of other sedimentary components.

Most bedded cherts are associated with shales, mudstones, greywackes, and sometimes interbedded lavas. Many of these sequences represent deep-water sedimentation (6) but others are clearly of shallow-water origin (7). Much of the silica appears to be primary and chemical considerations favour biochemical rather than physicochemical primary accumulation. The presence of radiolaria, or diatoms, sometimes in abundance, may reflect an original biogenic deposit in which the silica of the smaller, more delicate, and broken bodies has been redistributed to form the chert matrix (8). Evidence for inorganic precipitation is the association of many bedded cherts with lavas, but this association is rarely intimate and is generally regarded as being incidental or at most reflecting an indirect relationship. For example volcanic or hydrothermal activity might increase the concentration of silica in bottom waters and thereby inhibit solution of organic silica; in this way chert formation is promoted but without direct inorganic precipitation.

In some cherts the silica may be wholly diagenetic in origin, as in cherts formed by the diagenetic alteration of tuffs. It is also possible that some cherts have accumulated by redistribution of organic bodies by turbidity currents (9).

Bedded cherts of near-shore facies include the spiculitic cherts and those derived from the replacement of non-siliceous sediments (particularly limestones). The spiculitic cherts are essentially primary, but being for the most part grainstones, must have undergone considerable diagenetic enrichment in silica. Bedded chert resulting from the replacement of limestone have been recognised on the following criteria: gradational boundaries into limestone (10), presence of disseminated carbonate inclusions, and presence of structures characteristic of limestones (11).

Nodular cherts

Nodular cherts occur principally in limestones, but also in shales and sandstones. In many cases the origin of the nodules by replacement is clearly indicated by field relations and by 'ghosts' of primary texture. Suggested sources of the secondary silica include: solution of biogenic silica in adajcent deposits, silica liberation during clay and other silicate diagenesis, and large-scale movement of pore solutions from one formation to another.

Petrography of chert constituents

X-ray studies on cherts reveal the presence of amorphous silica (usually biogenic), cristobalite, tridymite (rare), and quartz. Under the microscope these distinctions can rarely be made and a number of 'minerals' have been established for purposes of thin section description. These include opal, chalcedony (chalcedonic quartz), microquartz (microcrystalline quartz), normal quartz, and lutecite (12).

Opal. This term is applied to isotropic silica; this may be truly amorphous but is commonly cryptocrystalline (as revealed by X-rays). Opal is typically brown due to the dispersion of light by abundant fluid-filled inclusions; these produce a milky appearance in reflected light.

Chalcedony. This displays a fibrous texture under crossed polars (fig. 27.2C). The 'fibres' are mostly length-fast and thus do not represent simple acicular quartz crystals (13). Chalcedony is often brown due to fluid inclusions, and variation in their abundance may be expressed in colour banding. Chalcedony associated with evaporites is length-slow (14), although in some cases at least there may have been confusion with lutecite (*see* below).

Microquartz. This occurs as finely crystalline quartz mosaics in which individual crystals show undulose extinction. Microquartz grades into both normal quartz and chalcedony.

Normal quartz. This shows all the features typical of low-temperature quartz, but grades into chalcedony, microquartz, or lutecite. It is essentially identified on its uniform extinction, but undulose extinction is sometimes developed in crystals too large to be included in the other silica types.

Lutecite. This variety is not always recognised in petrographical studies (15). Like chalcedony it displays a fibrous appearance under crossed polars but the fibres are length-slow. The texture often appears feathery rather than truly radial (fig. 27.2C).

The basic fabric of the various silica types is modified according to their occurrence. Chalcedony may form fringes, in which case the fibres are roughly perpendicular to the host surface or it may occur as spherulites. Microquartz generally forms random aggregates except where it grades into an oriented normal quartz mosaic. Normal quartz cements often show elongation of early crystals normal to the surface of host grains, whether or not there is an intervening fringe of chalcedony. Replacive quartz aggregates may display crystal elongation and crystallographic orientation in relation to the structure of the host grain (e.g. prismatic structure of bivalves). When lutecite occurs as overgrowths on normal quartz its radiating habit may be suppressed.

Petrography of cherts

Bedded cherts

Radiolarian cherts are representative of cherts of pelagic type. The number of radiolaria varies from scattered individuals to dense accumulations (radiolarite) and other distinctive particles are typically absent. In finely crystalline varieties the wall structure of the radiolaria may be visible, and spines may be seen projecting from the surface. In exceptionally unaltered cherts the matrix may be composed largely of radiolarian debris but recrystallisation normally causes progressive recrystallisation and destruction of the organic textures (fig. 27.1A), the end product being diffuse and indeterminate areas of clear silica (fig. 27.1B). The cement fill of the radiolaria usually consists of one or more chalcedony aggregates centred on the inner surface of the test wall but microquartz or even single crystals of normal quartz are sometimes present.

The matrix of pelagic cherts usually consists of microquartz with minor amounts of finely crystalline chalcedony but exceptionally unaltered cherts may be cryptocrystalline and almost isotropic. Highly recrystallised cherts consist of a normal quartz mosaic. Thermal metamorphism may cause 'spotting' by the development of chalcedony spherulites. The overall appearance of the matrix in ordinary light depends largely on the pigmenting material – haematite (red), illite or chlorite (green), or carbonaceous matter and pyrite (grey or black).

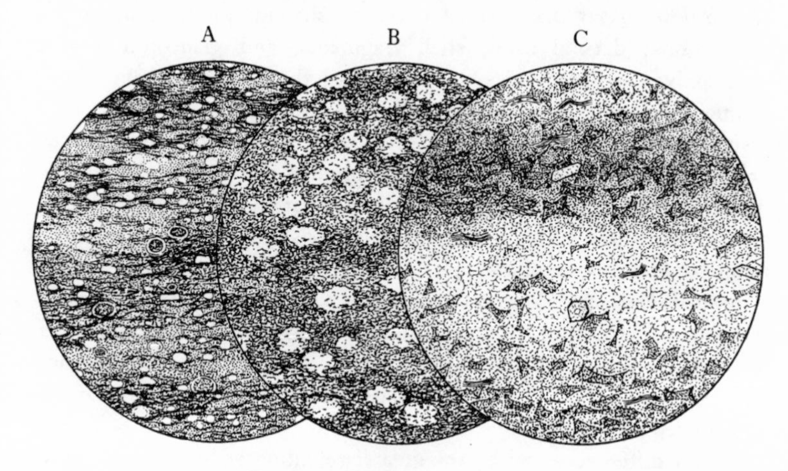

Fig. 27.1. Bedded cherts; × 10.

A. Radiolarian chert showing moderate to well preserved radiolaria. The chert contains abundant horizontal solution films, some of which cut across radiolaria.

B. Radiolarian chert recrystallised and deformed such that the radiolaria are represented by diffuse elliptical bodies of clear microquartz. Carboniferous, Devonshire.

C. Tuffaceous chert, showing 'ghosts' of glass shards and scattered euhedral apatite crystals of igneous origin. Ordovician, Dumfriesshire.

Authigenic accessory minerals are not common in bedded cherts of pelagic origin although replacement by sharply defined rhombs of dolomite has sometimes taken place, possibly before silica recrystallisation.

Tuffaceous cherts may be distinguished by the presence of shards, which are often brownish and relatively finely crystalline. As recrystallisation proceeds the shards become obliterated and the originally tuffaceous nature is made apparent only by accessory volcanic crystals which typically include angular fragments of quartz and feldspar, flakes of biotite, and idiomorphic prisms of zircon and apatite (fig. 27.1C).

Deformation is common in cherts which have undergone mild regional metamorphism, a distinctive feature being the distortion of radiolarian outlines (fig. 27.1B). Solution surfaces may also develop; these appear as laminae with high concentrations of clay and pigmenting materials which may be seen to cut radiolarian or other primary structures (fig. 27.1A). Cherts may also deform by brecciation which is related either to true open fracturing or to recrystallisation. The latter process is often accompanied by loss of pigmentation.

Spiculitic cherts are characteristic of shallow-water, high-energy associations; detrital sand, shell fragments, and glauconite grains commonly accompany the spicules. Because the host rock is chert, it is frequently assumed that the spicules were originally siliceous. This assumption has been questioned in relation to cherts of replacement origin (16) and wherever possible the original composition should be determined on the basis of petrographical relationships. Unfortunately spicules of certain siliceous composition may show a wide range in the nature of their preservation (17).

Bedded cherts of replacement origin vary petrographically according to the nature of the host rock. Inclusions of carbonate usually indicate a limestone host, and sometimes the presence of ooliths and other distinctive grain types points clearly to replacement of limestone. Two generations of replacive silica are sometimes present and can be distinguished by colour differences and by the greater retention of limestone texture in the earlier generation of silica (fig. 27.2A).

In Precambrian sequences chert is present in great abundance in iron-formation. Precambrian beds also include stromatolitic cherts in which the stromatolites are associated with ooliths and pisoliths and are mostly of algal origin (18). Areas of turbid mosaic quartz represent an original mud matrix and in the absence of carbonate inclusions it may be supposed that the mud was originally siliceous. Areas of clear microquartz, showing cement textures, represent interstitial cement. Grains are preserved in microquartz with dusty inclusions defining primary textures.

Nodular cherts

These differ from most bedded cherts in that they often possess distinctive grain types related to a replaced host rock. These grains provide information as to the timing of silicification relative to other diagenetic events. The most distinctive primary components encountered are ooliths, calcareous shell fragments, and siliceous sponge spicules.

Ooliths are typically replaced by microquartz, although in some cases normal quartz or (more rarely) chalcedony are also present. Silicification is often selective so that isolated silicified ooliths may occur within a calcareous matrix or *vice versa*. The oolitic structure is usually destroyed by silicification and is revealed by variation in crystal size or by inclusions. Fine calcite inclusions are commonly present and are usually

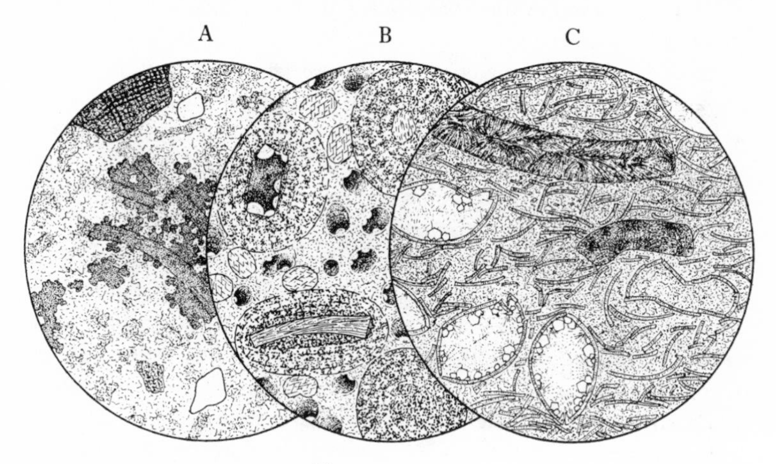

Fig. 27.2. × 20.

A. Bedded chert of replacement origin showing two phases of silicification: (i) local replacement by brown globular chert, preserving primary textures; (ii) wholesale replacement by microquartz with destruction of textures. Carboniferous, Yorkshire.

B. Nodular chert; silicified ooliths and ovoid *Rhaxella* spicules set in a matrix of silicified carbonate mud. The spicules show two preservations, one in silica cement, the other in calcite cement. Jurassic, Yorkshire.

C. Chert with abundant ostracodes, articulated and disarticulated. Ostracodes are unsilicified but aragonitic bivalve shells are replaced by lutecite. Shell cavities are cemented by early incomplete calcite fringes and later cavity-filling chalcedony. Silica fibres diagrammatic. Jurassic, Dorset.

arranged in a crude radial pattern; less commonly the inclusions show a radial orientation of their c-axes. Both features indicate that silicification took place during or after the development of calcitisation textures within the oolith. In rare cases earlier silicification (by brown opal or its recrystallisation products) preserves the delicate concentric layering of the aragonitic envelope.

Aragonitic shells are liable to extensive alteration as in non-cherty limestones. In grainstones they are often completely dissolved, their form being retained by micrite envelopes. The silica fill of the moulds consists of chalcedony rims or more extensive areas of quartz or microquartz; similar features are seen in aragonitic shells which have acted as nuclei to ooliths (fig. 27.2B).

Alternatively aragonitic shells may undergo replacement by silica, especially in muddy limestone hosts. In bivalve shells the prismatic layer is the more resistant, sometimes remaining after the lamellar layer is completely silicified. Pseudopleochroism in relict neomorphic calcite and sometimes even in the chert itself indicates an originally aragonitic composition.

Calcite shells are usually relatively resistant to silicification but susceptibility varies to some extent with the type of shell; for example the high-Mg calcite of echinoderms is less resistant than the low-Mg calcite of bivalves. Replacement is normally by normal quartz, although chalcedony or lutecite is common in some formations. Silica invades lamellar structure from the outer margin inwards and is not much influenced by shell structure; in prismatic layers the silicification tends to proceed along prism boundaries which often influence the growth of the quartz crystals. Primary internal shell cavities are filled by quartz or chalcedony cement, sometimes accompanied by early calcite fringe cement (fig. 27.2C).

In grainstones echinoderm fragments are commonly resistant to silicification because of the precipitation of calcite both as syntaxial rims and within the internal pore system. Sometimes, however, the echinoderm fragment and, to a lesser extent, its overgrowth have been replaced (fig. 27.3A). In muddy sediments silicification is more common because retention of the original porosity (fig. 27.2A) renders the grains more permeable to the silicifying fluids.

Spicules may be of primary opaline or calcitic composition (p. 296). In spiculitic nodules of sandy facies a dense chert core is often surrounded by a highly porous cortex. In the core the spicules are preserved in microquartz, with no trace of original structure; in the cortex they are preserved as cavities, which are responsible for the high porosity (fig. 27.3A). Solution of the cortex spicules clearly occurred after the main phase of chalcedony cementation; where associated calcite constituents are not dissolved out it can be assumed that the spicules were primarily opaline and that their solution reflects the relative instability of opal compared with crystalline chalcedony. The microquartz of the spicules in the core must therefore result from recrystallisation.

In limestones spiculitic nodules may display a similarly complex diagenetic sequence except that any solution moulds in the cortex have been filled by late-phase calcite cement (fig. 27.2B). This leaching of spicules must again represent instability of the biogenic opal during the later diagenesis of the surrounding limestone.

The presence or absence of a primary mud matrix in the host rock exerts an important control on the course of chertification. Carbonate muds are usually replaced by microquartz or normal quartz although isolated chalcedony spherulites sometimes occur. The individual quartz crystals may contain fine carbonate inclusions at their centre. In flints carbonate inclusions are typically absent except for scattered organic remains; the groundmass sometimes consists of a brown cryptocrystalline base in which are developed minute spheres of finely crystalline microquartz. These spheres often coalesce to form pale or colourless chert (fig. 27.3B).

Chert cements can usually be distinguished from matrix replacements by the presence of chalcedony rims and inclusion-free quartz or microquartz mosaics.

Relative timing of silica and carbonate diagenesis. Cherts developed in grainstone hosts most clearly reveal the interaction between silica and

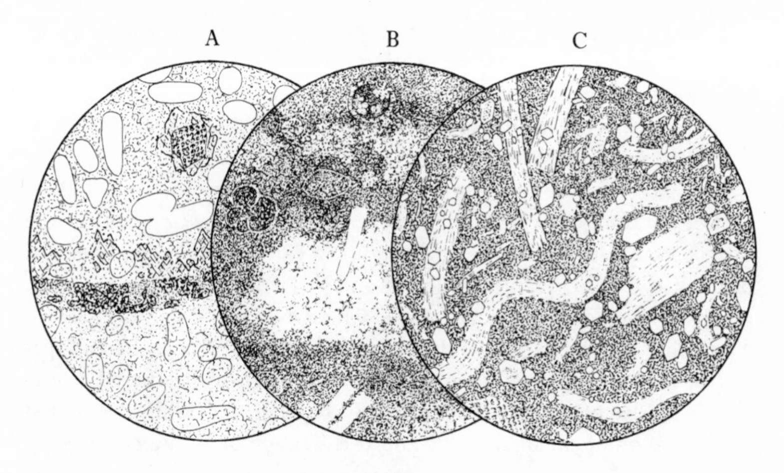

Fig. 27.3. × 20.

A. Nodular spiculitic chert; a dense core (lower part) contains spicules preserved in microquartz and highly corroded echinoderms whereas the cortex (upper part) contains empty spicule moulds. Cretaceous, Dorset.
B. Flint, showing variation in colour due to differing degrees of recrystallisation. Cretaceous, Norfolk.
C. Idiomorphic quartz developed in shelly packstone. Carboniferous, Yorkshire.

carbonate diagenesis. The earliest diagenesis in the chert nodule usually parallels that of the host limestone. Where fringe cements and syntaxial overgrowths are present in the host rock, they are present in the chert nodule (fig. 27.3A); where early cements are absent, other early diagenetic features may be detected such as the solution of aragonitic shells and the retention of moulds in micrite envelopes. Furthermore both nodule and host display early compactional fracturing of weak grains and associated cement fringes.

It is not until after fringe cementation and fracture that the diagenetic histories of the nodule and of the host rock diverge. In the nodule silica is precipitated as a cement in both primary and diagenetic voids while the host rock remains unaffected. Sometimes silica cementation takes place in two phases usually represented by an early chalcedony fringe and later wholesale cementation by radial chalcedony or mosaic quartz. Only very rarely does carbonate precipitation intervene, and then only on a small scale.

Wholesale calcite cementation of the host rock takes place after silica cementation is complete. This is indicated by textures in the cortex of some nodules; moulds formed by solution of opaline spicules have been retained in chalcedony cement and subsequently cemented by calcite (fig. 27.2B).

The relative timing of replacement of carbonate grains is less easy to determine. The common preservation of neomorphic radial texture in ooliths (p. 270) implies that silicification followed the aragonite–calcite transformation and it is possible that the late stage leaching of opal from nodule cortexes provides silica for these late replacements.

Not all grainstone cherts follow the above pattern. Some show replacement of grains without silica cementation and *vice versa*. In the former case silicification is probably very early, with preservation of primary aragonitic textures in ooliths and shell fragments. Also the main phases of silicification may be supplemented by minor changes of a later date. Silica may be mobilised during weathering and redeposited as opal in cracks and vugs left by solution of calcite. In unweathered cherts other late diagenetic modifications may be seen, including the partial recalcitisation of silicified grains and the invasion of calcite cements by silica (19).

Dolomite shows the same variable time relationships to silicification as calcite, but the petrographical relationships are different. Evidence from the cherts themselves and from a comparison with surrounding limestones has revealed that more than one phase of dolomitisation may occur, sometimes alternating with phases of silicification (20).

Idiomorphic quartz crystals

Idiomorphic authigenic quartz is most commonly developed as overgrowths on detrital grains (p. 255) but wholly authigenic crystals occur in limestones and to a lesser extent in other chemical rocks. The crystals typically occur as scattered individuals or as small crystal aggregates. The quartz may show selective replacement but is more often randomly distributed (fig. 27.3C) in relation to the host fabric. The development of idiomorphic quartz and true chertification appear to be antipathetic, perhaps because the former is a much later and slower process of silica redistribution.

Selective silicification of fossils

Little is known of the controls on selective silicification but it is clear that no single mechanism is involved. Preservation varies from the retention of delicate internal and surface structure to the production of shapeless masses of silica. Normal quartz is the most common mineral variety.

A comprehensive study (21) has shown that a consistently developed order of silicification can be established for a particular lithological association. In this case the order of increasing resistance to silicification is given as: 1. Bryozoans, tetracorals, tabulate corals, punctate brachiopods; 2. Impunctate brachiopods; 3. Molluscs; 4. Echinoderms; 5. Foraminiferans; 6. Calcareous sponges and dasycladacean algae. Beyond this point silicification ceases to be selective. The relative stability of aragonitic forms over some calcitic forms indicates that selective silicification follows the aragonite-to-calcite transformation.

References and notes

(1) The use of the term 'chert' is sometimes restricted to vitreous varieties only, thereby excluding porcellanous and organic varieties.

(2) Krauskopf, K. B., 1959. The geochemistry of silica in sedimentary environments, *in* Ireland, H. A., ed., *Silica in sediments*. Spec. Publs Soc. econ. Paleont. Miner., Tulsa 7, 4–19.

(3) Berger, W. H., 1968. Radiolarian skeletons: solution at depths. *Science* **159**, 1237–39.

(4) Sherwood, W. W., Bennett, F. B., & Weaver, F. M., 1972. Chemically precipitated sedimentary cristobalite and the origin of chert. *Eclog. geol. Helv.* **65**, 157–63.

(5) Bedded cherts also occur in association with certain alkaline lake deposits through the leaching of sodium silicate precursors: Surdam,

R. C., Eugster, H. P., & Mariner, R. H., 1972. Magadi-type chert in Jurassic and Eocene to Pleistocene rocks, Wyoming. *Bull. geol. Soc. Am.* **83**, 2261–6.

(6) Grunau, H. R., 1965. Radiolarian rocks in space and time. *Eclog. geol. Helv.* **58**, 157–208.

(7) Folk, R. L., 1973. Evidence for peritidal deposition of Devonian Caballos Novaculite, Marathon Basin, Texas. *Bull. Am. Ass. Petrol. Geol.* **57**, 702–25.

(8) Fagan, J. J., 1962. Carboniferous cherts, turbidites, and volcanic rocks in Northern Independence Range, Nevada. *Bull. geol. Soc. Am.* **73**, 605.

(9) Swarbrick, E. E., 1966. Turbidite cherts from northeast Devon. *Sedim. Geol.* **1**, 145–57.

(10) Hey, R. W., 1955. Cherts and limestones from the Crow Series near Richmond, Yorkshire. *Proc. Yorks. geol. Soc.* **30**, 291.

(11) Folk, 1973, p. 710.

(12) The crystalline silica varieties are illustrated in Wilson, R. C. L., 1966. Silica diagenesis in Upper Jurassic limestones of southern England. *J. sedim. Petrol.* **36**, Fig. 2, p. 1038.

(13) Folk, R. L. & Weaver, C. E., 1952. A study of the texture and composition of chert. *Am. J. Sci.* **250**, 498–510.

(14) Pitman, J. S. & Folk, R. L., 1971. Length-slow chalcedony after sulphate minerals in sedimentary rocks. *Nature, phys. Sci.* **230**, 64–5.

(15) A description of lutecite is given by Wilson, R. C. L., 1966, p. 1401.

(16) Hey, R. W., 1955, pp. 293–5.

(17) Wilson, V., 1933. The Corallian rocks of the Howardian Hills (Yorkshire). *Q. Jl geol. Soc. Lond.* **89**, 485–86.

(18) Some stromatolitic cherts have been compared with the deposits of siliceous hot springs: Walter, M. R., 1972. A hot spring analog for the depositional environment of Precambrian iron formations of the Lake Superior region. *Econ. Geol.* **67**, 965–80.

(19) Walker, T. R., 1962. Reversible nature of chert-carbonate replacement in sedimentary rocks. *Bull. geol. Soc. Am.* **73**, 237–41.

(20) Dietrich, R. V., Hobbs, C. R. B., & Lowry, W. D., 1963. Dolomitisation interrupted by silicification. *J. sedim. Petrol.* **33**, 646–63. Swett, K., 1965. Dolomitisation, silicification, and calcitisation patterns in Cambro-Ordovician oolites from northwest Scotland. *J. sedim. Petrol.* **35**, 928–38.

(21) Newell, N. D. *et al.*, 1953. *The Permian reef complex of the Guadelupe Mountains region, Texas and New Mexico.* San Francisco: Freeman, pp. 171–4.

28

Phosphatic rocks

Introduction

Phosphatic constituents include those formed by organisms (bones, teeth, scales, shells) and those formed by essentially inorganic processes (mudstone, pellets, ooliths, nodules, phosphoclasts, concretions). These components contribute to a varied assortment of rock types, including bedded and concretionary deposits.

The term *phosphorite* is applied to rocks composed essentially of calcium phosphate minerals; it is not defined quantitatively. The terms *rock phosphate* and *phosphate rock* have a similar meaning. Rocks in which phosphate is a distinct, but minor, component are described as phosphatic. No formal classification is in general use and the phosphorites are usually named according to the dominant constituent type – e.g. pelletal phosphorite.

Origin

Facies associations and included fossils show most phosphorites to be marine. The distribution of present-day and fossil phosphorites reveals their restriction to warm climatic zones; they are located on the continental shelf in two main environmental settings (1): one where cold equatorially-directed currents upwell and the other where cold water is brought to the surface by warm polar-directed currents (2).

The concentration of phosphate is partly the result of a low influx of terrigenous detritus, which in turn is related to low rates of run-off and to the accumulation of phosphorites on topographical highs. The mechanism of chemical concentration of phosphate must require somewhat specialised conditions as the geochemistry of calcium phosphate is similar to that of the far more abundant calcium carbonate. The fractionation of phosphate and carbonate is probably achieved by organic agencies (3). Bottom waters may become enriched in phosphate by the accumulation of phosphate-rich organic matter to the point where the phosphate radical substitutes for the carbonate radical in surface calcareous deposits and perhaps even to the point where direct precipitation can take place. Poor circulation seems to favour the attainment of high phosphate concentrations, but the general absence of soft organic matter points to oxidising conditions at the sediment surface. A low pH may also be necessary to promote the relative solution of carbonate (4). The concentration of phosphate is apparently enhanced by conditions which favour high organic productivity – particularly in areas of upwelling but also in estuarine environments (5).

Petrographical features indicate that direct precipitation of phosphate is of minor importance and that the bulk of phosphate is precipitated interstitially and by replacement of fine-grained silicate and carbonate sediment (6, 7). When phosphatisation of available carbonate is complete direct precipitation may take place to form ooliths, nodules, and thin accretionary layers (8).

Phosphorites of different origins show some petrographical features in common but differ markedly in their facies associations. Many phosphorites of 'cold-current' type are associated, sometimes intimately, with chert and black shales or mudstones; phosphorites of 'warm-current' type usually occur in association with sandy or calcareous facies. Reworking affects shallow-water deposits of either type and may lead to the concentration of lean phosphorites to form relatively rich ores.

Mineralogy

Sedimentary phosphates are mostly composed of varieties of apatite. These include fluorapatite, $Ca_5(PO_4)_3F$, and hydroxylapatite, $Ca_5(PO_4)_3OH$, which in sediments are modified by substitution of carbonate for phosphate to form *francolite* (carbonate fluorapatite) and *dahllite* (carbonate hydroxylapatite). Cryptocrystalline phosphate, which gives the X-ray patterns of apatite, is termed *collophane*.

Petrography of phosphorite constituents

Despite the variety of facies association displayed by phosphorites the number of primary component types, as described below, is small.

Phosphate pellets. Pellets or *ovules* are structureless grains composed of collophane or microcrystalline apatite (fig. 28.1). They vary in shape from spherical through ovoidal and discoidal to tabular and irregular and may possess rough or smoothly rounded surfaces. Their colour ranges from light brown to black, and surface colour may differ from that of the bulk of the grain. Dark colours are caused by inclusions of pyrite and organic matter; fine detrital silicate particles and microfossils may also be included. Diffuse concentric bands of organic staining are sometimes displayed; these do not conform to changes in crystallinity and are therefore not oolitic and probably form during weathering.

Some present-day shelf pellets are restricted to sand-grade sediments, those of silt size being rare; similar size restriction is displayed in fossil phosphorites. The recent pellets have been ascribed to replacement of sand-sized mud grains, possibly faecal pellets (9), or calcitic micrite pellets (10). Some small pellet-like bodies associated with reworked phosphorites may be of clastic origin. Dark pellets are indicative of formation under reducing conditions but in weathered rock these may become bleached.

Phosphate ooliths. These grains show a well-defined concentric structure, often emphasised by variation in crystallinity. The typical oolitic envelope consists of collophane alternating with francolite; the latter forms radially oriented length-fast crystallites. The francolite laminae are often impersistent and the lamination in general may be rather irregular, with asymmetrical and non-spherical envelopes being common. Nuclei commonly consist of pellets but may also consist of microfossils, fossil fragments, glauconite grains, and detrital silicate grains. Superficial ooliths are abundant in some formations; these consist of a very thin anisotropic francolite layer enclosing a pellet nucleus.

The ooliths are clearly accretionary in origin. The asymmetry and non-spherical shape of some envelopes indicates that continuous agitation is not essential to accretion. The alternation of collophane and francolite laminae may not be original as the latter may have developed by recrystallisation along preferred laminae within the collophane envelope (11).

Phosphate nodules. True nodules (i.e. large bodies with accretionary structure) consist of a laminated envelope surrounding a large nucleus; they commonly show one-sided accretion. Such nodules have been described from sea-floor phosphorites (12) and, rarely, from fossil deposits. Nuclei include structureless collophane mudstone, and pelletal, oolitic, and brecciated varieties of phosphorite. Nodules with relatively thick envelopes are essentially very large pisoliths and are often associated with oolitic deposits.

Phosphoclasts. These result either from fragmentation of pre-existing phosphorite or by replacement of non-phosphatic intraclasts. Many 'nodules' described in the literature are not accretionary and would therefore be better described as phosphoclasts.

Bone fragments (13). These are distinguished primarily by the coarse perforations of the Haversian canals (fig. 28.2C). The bone matter is pale to dark brown and often patchy in colour. It is slightly anisotropic and consists of delicate lamellae arranged concentrically around the canals and, in the dense outer part, parallel to the surface. The canals may be filled with collophane mud but often contain detrital mud, calcite mud, or cement.

Teeth. These consist of thick internal dentine and a sharply defined outer layer of enamel. The dentine is mostly dense but becomes vesicular near the base of the tooth; it is brown and sometimes growth layered. Dentine is weakly birefringent and is composed of delicate length-fast lamellae which are complexly arranged except where they run parallel to the outer surface. The enamel layer is colourless, more highly birefringent (to first order white) and consists of prismatic crystals arranged normal to the surface (so that the layer as a whole is length-slow).

Scales. These mostly possess a bipartite structure similar to that of teeth (fig. 28.2B). The inner layer is pale brown and more or less structureless; it is composed of very low birefringent straight or wavy laminae lying parallel to the flat surfaces. As in teeth, the inner layer is length-fast and the outer enamel layer length-slow. The enamel layer is missing in some types of scale and even when originally present may become detached during transport.

Phosphatic brachiopods. These consist of alternating dark phosphate-rich layers and pale organic-rich layers (14). The lamination is usually visible under ordinary light and is either parallel (fig. 28.2C) or oblique to the surface. Fine punctae normal to the lamination may be visible at high magnification. The shells show uniform birefringence (grey to white) and extinction parallel to the lamination. Most sections are length-slow, but sometimes length-slow and length-fast layers alternate or, more rarely, the entire shell is length-fast. Scales can usually be distinguished from brachiopods by their bipartite structure, lack of colour banding, and optical orientation.

Phosphatised calcareous shells. These are preserved either in clear collophane retaining faint traces of original structure (15) or in structureless dark collophane. Echinoderm fragments often show gradational phosphatisation ranging from phosphate impregnation of pores to total replacement with inclusions delineating the original structure. In grainstones the primary calcite may be phosphatised and the pores filled by later calcite cement. Some rocks show solution of included shells, whose moulds may be lined with a thin layer of collophone or apatite cement.

Phosphate mud. This forms the groundmass of many phosphorites and also forms thin beds and concretions. It consists of brown collophane, often with incipient crystallisation. Inclusions sometimes indicate that the phosphate is a replacement of a carbonate or detrital silicate mud.

Phosphate cement. This is relatively rare in the marine phosphorites and where present it is as very thin rims. Most commonly the phosphorites have original mud matrices or consist of open porous grain aggregates.

Petrography of the marine phosphorites

The phosphorites are described here in terms of two major associations: the granular phosphorite association and the pebbly phosphorite association. Although there is some overlap these two categories largely represent natural divisions.

Petrography of the granular phosphorite association

The rocks of this type are composed essentially of sand-grade phosphate grains although pisoliths, nodules, organic grains, or phosphoclasts occur in varying proportions and locally attain dominance. Pellets and ooliths normally show a high degree of sorting both in grainstone and muddy facies. The good sorting of many oolith assemblages is a reflection of uniformity in size of pellet nuclei; ooliths with relatively thick envelopes may show a much greater range in grain size (16).

The groundmass of the granular phosphorites often provides a better indication of depositional environment than the grains themselves. Detrital silt and clay, fine sand, micritic calcite and dolomite, chert, and collophane all occur as major matrix constituents; calcite and, rarely, phosphate occur as cements. The granular phosphorites often display a complex sequence of replacement (17).

Granular phosphorites with a detrital matrix. These usually consist of phosphate grains set in a matrix of detrital mud which sometimes displays patchy replacement by collophane. The phosphate grains generally show little alteration.

Granular phosphorites with a collophane matrix. These consist of phosphate grains set in structureless collophane mud which often contains abundant inclusions of clay, silt, and fine sand (fig. 28.1A), indicating replacement of a detrital mud host. Collophane may also replace calcite mud, where a complete range in degree of replacement can be found (18).

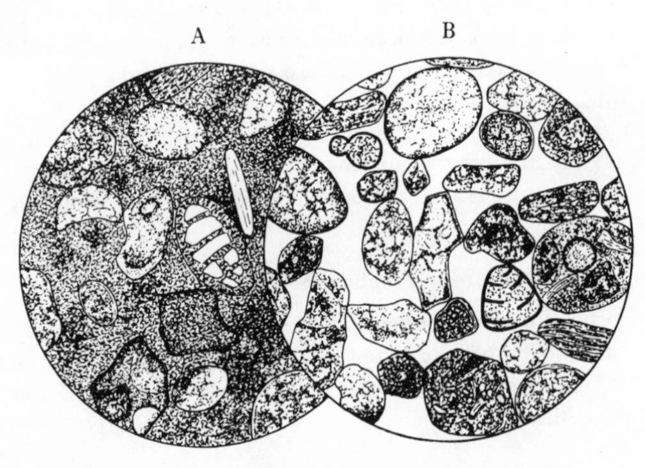

Fig. 28.1. × 20.

A. Pellet phosphorite with packstone texture; collophane pellets are set in a collophane matrix. Thick-walled foraminiferan at right of centre.
B. Pellet phosphorite with grainstone texture; grains include pellets and impregnated foraminiferans.

Granular phosphorite with a carbonate matrix. These occur where phosphorites are interbedded with or grade into micritic limestones (19), and all gradations exist between the two extremes. The phosphate may show replacement by microcrystalline calcite. Dolomitic phosphorites show similar relationships.

Granular phosphorites with a cherty matrix. In these rocks phosphate grains are set in a matrix of microquartz (20). The chert may contain fine detrital inclusions and may be intimately associated with collophane. Grain packing clearly indicates that the chert represents an original matrix, possibly a detrital silicate mud. Silicification of phosphatic constituents occurs on a minor scale.

Granular phosphorites with an original grainstone texture. These typically lack significant interstitial cement (fig. 28.1B), and the phosphate grains usually show compactional deformation. Extensive cementation usually involves calcite, and tends to occur in marginal phosphorite facies.

Phosphate mudstone. This typically consists of a collophane matrix containing dispersed detrital grains, pellets, organic grains, or phosphoclasts, and appears to represent replacement of detrital mud. Some thin beds of relatively pure phosphate mud may be primary. The collophane may show incipient crystallisation to apatite as randomly oriented laths.

Petrography of pebbly phosphorites

The pebble constituents include reworked concretions, phosphatised intraclasts, and skeletal grains. The distinction between concretions and phosphoclasts is often not possible in thin section. The presence of a crude concentric zonation probably indicates a concretionary pebble but superficial zonation may result from diagenetic modification of the

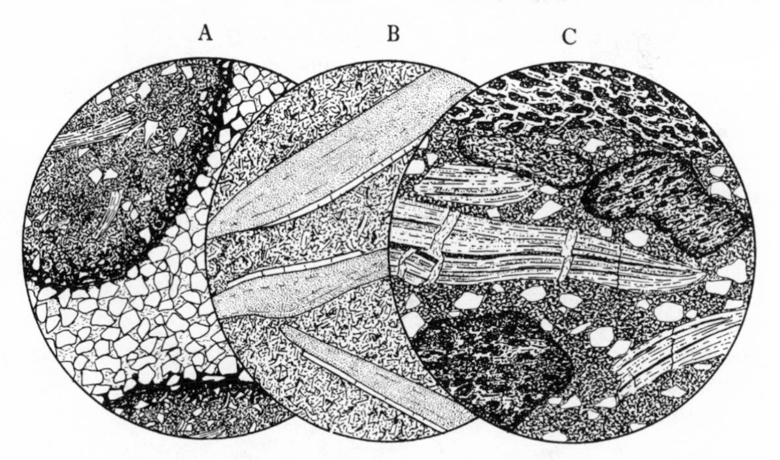

Fig. 28.2 × 20.

A. Pebbly phosphorite; phosphate pebble set in matrix of very fine sand. The surface of the pebble has been the site of pyrite precipitation during diagenesis. Cretaceous, Cambridgeshire.

B. Fish scales set in a matrix of phosphatic chalk (coprolitic). The outer enamel layer of the scales is locally broken.

C. Pebbly phosphorite, rounded bone fragments and unrounded phosphatic brachiopod shell set in a silty matrix. Silurian, Shropshire.

pebbles themselves (fig. 28.2A). The pebbles consist of collophane and, depending on the nature of the host rock, contain inclusions of detrital silicates, shell fragments, glauconite grains, and other precipitated components. Organic components can usually be identified although abrasion may destroy some of the characteristic structural features. Pebble beds composed almost entirely of abraded bones and teeth represent prolonged non-deposition and reworking; they may also contain less abraded remains of organisms living at the time of final accumulation (figs. 28.2B, C).

The matrix generally consists of detrital or calcareous grains, often accompanied by glauconite. Cementation by calcite is common. Replacement of phosphate pebbles by calcite may occur, especially where a calcite matrix has become coarsely recrystallised.

Thick and extensive pebbly phosphorites originate by reworking of pre-existing phosphorites and so may contain a wide variety of grain types: nodules, concretions, pellets, ooliths, and organic grains, together with phosphoclasts and the phosphatic remains of a contemporaneous fauna.

Guano and related deposits

Guano is a phosphate rock of variable lithology produced by the alteration of accumulations of excrement, mostly of birds. Such accumulations are significant mainly on oceanic islands in areas where both climate and high organic productivity are favourable to their formation and preservation. The guano itself possesses complex lithology, including massive and colloform structural varieties. During guano diagenesis, dissolved phosphates are carried down into underlying formations, which are usually igneous rocks or limestones. The latter in particular are susceptible to replacement by phosphate (21). The resulting deposits show a variety of textures, some inherited from the limestone and others by direct precipitation of phosphate. In particular cementation by fibrous apatite and collophane is a common feature, in contrast to normal marine phosphorites.

References and notes

(1) Sheldon, R. P., 1964. Paleolatitudinal and paleogeographic distribution of phosphorite, *Prof. Pap. U.S. geol. Surv.* **501**-C, 106–13.

(2) McKelvey, V. E., 1967. Phosphate deposits. *Bull. U.S. geol. Surv.* **1252**-D.

(3) Gulbrandsen, R. A., 1969. Physical and chemical factors in the formation of marine apatite. *Econ. Geol.* **64**, 365–82.

(4) Berge, J. W., 1972. Physical and chemical factors in the formation of marine apatite. *Econ. Geol.* **67**, 824–7.
(5) Pevear, D. R., 1966. The estuarine formation of the United States Atlantic coastal plain phosphorite. *Econ. Geol.* **61**, 251–6.
(6) D'Anglejan, B. F., 1967. Origin of marine phosphorites off Baja California, Mexico. *Mar. Geol.* **5**, 15–44.
(7) Tooms, J. S. & Summerhayes, C. P., 1969. Geochemistry of marine phosphate and manganese deposits. *A. Rev. Oceanogr. Mar. Biol.* **7**, 73–4.
(8) Bromley, R. G., 1967. Marine phosphorites as depth indicators. *Mar. Geol.* **5**, 503–9.
(9) D'Anglejan, 1967, p. 42.
(10) D'Anglejan, B. F., 1968. Phosphate diagenesis of carbonate sediments as a mode of *in situ* formation of marine phosphorites: observations on a core from the eastern Pacific. *Can. J. Earth Sci.* **5**, 81–7.
(11) Lowell, W. R., 1952. Phosphatic rocks in the Deer Creek–Wells Canyon area, Idaho. *Bull. U.S. geol. Surv.* **982**-A, 20.
(12) Dietz, R. S., Emery, K. O. & Shepard, F. B., 1942. Phosphorite deposits on the sea floor off southern California. *Bull. geol. Soc. Am.* **53**, figs 1, 2.
(13) Rogers, A. F., 1924. Mineralogy and petrography of fossil bone. *Bull. geol. Soc. Am.* **35**, 535–56.
(14) Graham, D. K., 1970. Scottish Carboniferous Lingulacea. *Bull. geol. Surv. Gt. Br.* **31**, 165–7 and pl. 20.
(15) Mansfield G. R., 1927. Geography, geology, and mineral resources of part of southeastern Idaho. *Prof. Pap. U.S. geol. Surv.* **152**, pls, 67, 68. (The replaced shells are accompanied by originally phosphatic shells which display colour banding.)
(16) Mansfield, 1927, pl. 63.
(17) Cook, P. J., 1970. Repeated diagenetic calcitisation, phosphatisation, and silicification in the Phosphoria Formation. *Bull. geol. Soc. Am.* **81**, 2107–16.
(18) Bromley, 1967, p. 507.
(19) Russell, R. T. & Trueman, N. A., 1971. The geology of the Duchess phosphate deposits, northwestern Queensland, Australia. *Econ. Geol.* **66**, 1207.
(20) Russell & Trueman, 1971, figs. 20, 22, 23.
(21) Braithwaite, C. J. R., 1968. Diagenesis of phosphate carbonate rocks on Remire, Amirantes, Indian Ocean. *J. sedim. Petrol.* **38**, 1194–1212.

29

Ferruginous rocks

Introduction

The iron-rich rocks are classified on gross lithological character into *ironstone* and *iron-formation* (1).

Ironstone includes bedded and concretionary rocks of mudstone or oolitic texture; it typically lacks the disseminated or interbedded chert that is definitive of iron formation. Iron occurs in silicate, carbonate, oxide, hydroxide, and sulphide phases of which the first four commonly act as major components. Individual beds are usually not more than a few metres thick.

Iron-formation displays a similar range in mineral facies as ironstone, but with the sulphide sometimes acting as a major component. A distinctive feature is the intimate association of chert with the iron minerals. This chert usually constitutes 50–60 per cent of the rock. The chert is either interlaminated with the iron minerals or acts as a matrix to granular iron components. Iron-formation typically forms units of many tens of metres in thickness which may extend over tens of thousands of square kilometres. Despite the high silica content, the iron formations are important ores on account of their massive development. Iron-formation is virtually restricted to Precambrian strata.

Glauconitic rocks are treated separately as they possess distinctive mineralogy and facies associations. Glauconite occurs as granules, which may accumulate to form *greensand*, and as a finely divided matrix constituent.

Origin

It may be assumed that the Phanerozoic ironstones formed under a geochemical regime similar to that of the present time. In normal oxygenated surface environments ferric hydroxide is the stable mineral phase, and iron is therefore transported in a particulate state. The highest concentrations of iron in natural suspensions are associated with fine clay particles, on which surface films of iron hydroxide may account for several per cent by weight of each particle. Chamosite, an important mineral in the Phanerozoic ironstones, may have originated by the reconstitution of such fine clay and associated hydroxide pellicles (2). Such transformations require very little detrital sedimentation, probably on offshore shoal areas in warm seas with sluggish circulation (3). These conditions would favour the settling out of the finest clays alone, but where fractionation is less extreme the hydroxide coatings may survive into early diagenesis, when they are dissolved and the iron reprecipitated as pyrite or siderite.

Oolitic structure is common in bedded ironstones but because of the absence of modern analogues its origin is obscure. Evidence for high energy conditions is often lacking, and the ooliths may have formed under relatively quiet conditions by surface reworking (4) or diagenetic modification of iron-rich muds or gels.

The contrast between iron-formation and ironstone suggests that the geochemistry of iron during the Precambrian differed from that of Phanerozoic times. The absence of recognisable detrital matter, the low aluminium content, and the constant association with chert indicate direct precipitation from solution. The existence of iron in surface solution may reflect the absence or paucity of oxygen in the atmosphere (5), together with a high proportion of carbon dioxide. The thickness, areal extent, and internal continuity of iron formations point also to great environmental uniformity and stability. Iron formations have been ascribed to sedimentation in lacustrine (6), restricted marine, and evaporating basin (7) environments; others are associated with sediments of open marine facies.

Glauconite forms at the present day in warm shallow seas, where it apparently originates by replacement of a variety of host materials, including faecal pellets, mud infills of shells, and fine mud in general (8). It may form in sandy or muddy facies, but many glauconite sands represent reworking of earlier glauconitic muds. Glauconite is virtually restricted to marine formations.

Mineralogy

To some extent iron mineralogy varies with age; thus among the silicates chamosite is dominant in ironstones and greenalite, minnesotaite, and stilpnomelane in iron formations. Also glauconite is largely restricted to Phanerozoic beds. Of the iron-formation silicates greenalite is generally considered to be primary, stilpnomelane diagenetic, and minnesotaite metamorphic (9).

The greater abundance of haematite relative to limonite in iron-formation may be a reflection of long-term dehydration, but primary precipitation of haematite in the Precambrian cannot be ruled out. The abundant magnetite of some iron formations has been interpreted as primary, but is now generally thought to be secondary (10).

Sedimentary iron minerals include both ferrous and ferric iron. Limonite and haematite are ferric, and the ferric ion is dominant in glauconite. Ferrous iron occurs alone in siderite and pyrite, and is dominant in chamosite. In magnetite and greenalite ferrous and ferric iron are both major constituents. Provided the mineral paragenesis is understood, a mineral assemblage may thus yield useful information on the degree of oxidation-reduction in the depositional and diagenetic environments (11). At the same time the importance of ion availability in determining mineralogy should not be underestimated (12).

Petrography of ironstone constituents

Chamosite ooliths. These consist of a delicately laminated envelope often enclosing a relatively large nucleus; in some cases a nucleus appears to be absent. The ooliths vary in colour from green to olive brown, depending on the amount of organic matter. Nuclei included rounded grains of chamosite mudstone (possibly faecal pellets), shell fragments, broken ooliths, terrigenous sand grains, and large tabular crystals of chamosite (see below).

Two types of oolitic lamina are present. 'Oriented' laminae show anisotropism due to the tangential orientation of the chamosite flakes; these laminae are relatively continuous and are normally pale and clear. 'Unoriented' laminae are isotropic due to the random orientation of the crystals and appear dark and turbid; they are relatively discontinuous and may form isolated lenses. Individual laminae are only a few micrometres thick, but alternating bands are visible at low magnification, corresponding to variations in the proportion of dark and pale lamina types. The overall layering is usually smooth and regular.

In some ironstones the ooliths are more or less spherical (fig. 29.2A), but in many cases they possess oblate ellipsoidal forms (fig. 29.2B). The latter appear elliptical in section apart from occasional large equatorial sections. This shape appears to be primary (13) and not due to compaction, which causes more intense deformation (fig. 29.2B).

Limonite and haematite ooliths. Haematite ooliths are usually opaque but limonite ooliths are commonly translucent. Some ferric iron ooliths are highly spherical; the envelope displays a very regular concentric layering and ooliths are small or apparently absent. These may represent primary accretion of iron in the oxidised state. In other cases the ooliths are ellipsoidal and display minor irregularities within the envelope, which often encloses a relatively large nucleus; these features suggest that the ooliths originated as chamosite ooliths and indeed some ironstones show a gradation from pure chamosite ooliths to those in which the envelope is completely oxidised. The alteration probably takes place on the sea floor, and is not related to recent weathering.

Pisoliths. In oolitic ironstones these commonly consist of two or more ooliths united by a single thin envelope. Where they are a major component they are essentially overgrown ooliths in which the outer part of the envelope often displays irregular and asymmetrical structure (14).

Tabular chamosite crystals. Crystals reaching about 0.25 mm in length occur as minor components of some chamositic mudstones and as nuclei to chamosite ooliths (fig. 29.2). They possess distinct basal cleavage and are pleochroic. Elongation is normally in the direction of the cleavage but sometimes at right angles to it. The crystals probably originate by authigenesis in chamosite mud; the slight abrasion of crystals occurring as oolith nuclei indicates exhumation and mechanical reworking.

Chamosite mud. This normally appears green to brown, isotropic, and structureless. Less commonly, bedding-plane orientation of the crystals is revealed by aggregate polarisation. Chamosite mud is most common as a matrix in oolitic ironstones (fig. 29.2A).

Siderite 'mud'. This actually consists of cryptocrystalline or microcrystalline aggregates (fig. 29.1A, B). The crystals tend to be lozenge-shaped and often contain dark turbid cores.

Siderite spar. Sparry siderite forms pale to medium brown aggregates displaying equigranular or more rarely spherulitic texture. Cements are clear and inclusion-free in contrast to replacive spar.

Spherulitic siderite. This is characteristic of certain nodules and concretionary beds. Siderite spherulites generally consist of an outer brownish radially fibrous zone enclosing a mass of pale structureless siderite, which is often pyritic (fig. 29.1C). The fibrous zone contains finely divided inclusions and scattered silt and sand grains; the latter may also be concentrated between spherulites. Closely packed spherulites show polygonal outlines.

Iron silicate cement. This is rare in ironstones; it usually consists of fibrous fringes of chlorite rather than chamosite.

Intraclasts. Lithologies encountered include chamositic, limonitic, and haematitic mudstones and oolitic packstones. Phosphatic pebbles are common and some possess oolitic ironstone texture. Intraclasts often reveal a complex history of deposition, diagenesis, and reworking in the environment of ironstone formation.

Organic remains. Calcareous shells may be replaced by structureless isotropic chamosite (fig. 29.2A), by rhombs or aggregates of siderite, or by cryptocrystalline limonite or haematite. In a siderite matrix they are often preserved as solution moulds cemented by siderite, kaolinite, or calcite.

Authigenic kaolinite. In rocks with a siderite matrix kaolinite often occurs as a clear colourless cement in voids of diagenetic origin. It also occurs as a replacement of chamositic constituents where it is turbid and may contains inclusions reflecting the structure of the host grain (fig. 29.1A).

Petrography of ironstones

The ironstones naturally fall into two groups, the mudstones and the oolitic ironstones, although certain types of mudstone are associated with and grade into oolitic ironstones. The mudstones are divided according to the major mineral constituent. The oolitic ironstones are here classed according to the presence or absence of an original matrix, but for detailed work a more comprehensive classification has been devised (15). A number of detailed accounts have been given of oolitic ironstones (16).

Chamosite mudstone. This relatively rare rock type consists of chamosite mud with varying amounts of terrigenous mud and silt. Scattered chamosite ooliths and tabular chamosite crystals may be present.

Replacement by siderite is common and chamosite mudstones grade into siderite mudstones by this process.

Siderite mudstone. This occurs both in true bedded form (replacing iron silicate) and as nodules or concretionary beds. Replacement of chamosite mud is indicated by the inclusion of chamosite ooliths, tabular chamosite crystals, and sometimes residual patches of chamosite mud (fig. 29.1A). The ooliths often show marginal replacement by siderite and internal replacement by kaolinite. Included shells may be unaltered (fig. 29.1B), dissolved to a thin residual film (fig. 29.1A), or dissolved to form voids which have acted as sites for later cementation (fig. 29.1A).

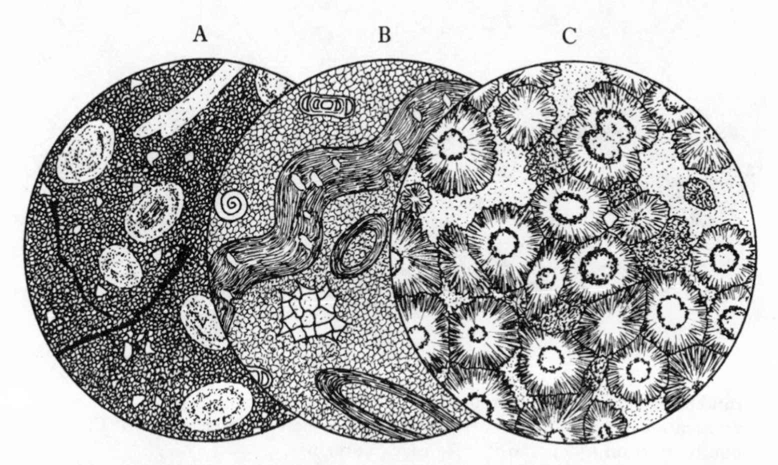

Fig. 29.1. Non-oolitic ironstones; × 20.

A. Siderite mudstone with scattered kaolinitised (originally chamositic) ooliths; aragonitic shells represented by pyritic solution films, calcitic shells by kaolinite-filled cavities. Jurassic, Yorkshire.

B. Siderite mudstone with abundant unaltered skeletal grains, including brachiopod shell, brachiopod spines, foraminiferans, and bryozoan fragment. Carboniferous, Northumberland.

C. Sphaerosiderite; siderite spherulites set in matrix of kaolinitic clay with scattered quartz grains. Jurassic, Yorkshire.

Concretionary siderite mudstones are commonly more silty and often more finely crystalline than those resulting from the replacement of chamosite mudstone.

Oolitic packstones and wackestones. The matrix may consist of unaltered mud composed of chamosite (fig. 29.2A), limonite, haematite, or less commonly terrigenous mud. Siderite commonly replaces chamosite mud but iron oxides are less commonly affected. Where replacement of chamosite mud is complete (fig. 29.2B) the primary matrix is reflected in the presence of disseminated pyrite, detrital inclusions, the strong body colour of the siderite, and oolith packing (ooliths about 60 to 65 per cent compared with 70 to 75 per cent in grainstones).

Chamosite ooliths set in a sideritised matrix often show marginal replacement in the form of a finely crystalline siderite 'rind' (fig. 29.2B);

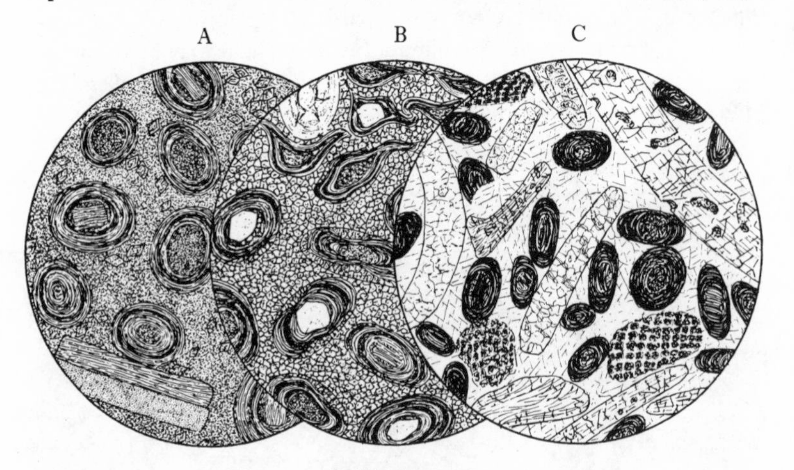

Fig. 29.2. Oolitic ironstones; × 20.

A. Chamositic chamosite oolite; chamosite ooliths set in a matrix of chamosite mud which shows partial replacement by small siderite rhombs. A bipartite aragonite/calcite shell (at bottom) displays selective replacement of the aragonitic portion by chamosite. Jurassic, Yorkshire.

B. Sideritic chamosite oolite; chamositic ooliths, showing marginal replacement by siderite, set in a matrix of sideritised chamosite mud. Compactional slip has caused distortion of the ooliths in the upper part but the phosphatised oolith (top left) is unaffected. Jurassic, Yorkshire.

C. Calcitic limonite oolite. Limonite ooliths and limonite impregnated echinoderm grains set in sparry calcite cement. Shell fragments show algal boring accentuated by limonitic impregnation. Jurassic, Lincolnshire.

internal replacement may also occur. Oxide ooliths are less commonly affected. Phosphate, unlike siderite, is commonly selective in its replacement; chamosite ooliths are again most susceptible and show total replacement with the retention of delicate envelope structure in pale brown isotropic collophane. Phosphatised ooliths do not possess siderite rinds, indicating that phosphatisation preceded sideritisation.

Compaction may cause plastic deformation of ooliths (fig. 29.2B), producing irregularly sigmoidal outlines with hooked ends – *spastoliths* (17).

Shell fragments may show replacement by chamosite, phosphate, siderite, pyrite, and amorphous iron oxides. Where solution has occurred the moulds are cemented by siderite, kaolinite, or calcite.

Oolitic grainstones. These are characterised by their close packing and by their clear crystalline cements of siderite, calcite (fig. 29.2C), or less commonly silica. The cemented grains, including ooliths, shell fragments, and terrigenous grains, generally show partial replacement by coarsely crystalline carbonate. Compaction causes mutual deformation of chamosite ooliths; oxide ooliths are more resistant but may show dislocation of the oolitic laminae.

The composition of the grain assemblage differs widely from bed to bed but is often varied due to reworking. Broken ooliths are common and associated skeletal material is often broken, abraded and perforated by boring organisms.

Sphaerosiderite. Siderite spherulites occur isolated in a mudstone matrix or as more closely packed aggregates forming well-defined concretions (fig. 29.1C). The latter may possess patches of residual matrix in which the clay flakes are reoriented by the displacive action of the siderite crystals. Spherulites may also accumulate mechanically as a result of reworking; such deposits can be distinguished by incipient oxidation, grain contacts, and often by admixture with terrigenous sand. Sphaerosiderite occurs in non-marine beds and marine beds affected by fresh water diagenesis.

Petrography of iron-formation constituents

In laminated iron-formation the minerals occur as discrete crystals or aggregates arranged in laminae; only in the granular facies do specific grain constituents occur. The latter include granules, ooliths, nodules, pisoliths, intraclasts, and pellets. The last two terms are of recent introduction (18) and overlap to some extent with the term 'granule'.

Granules. These are internally structureless spherical or irregular grains. They mostly fall in the size range 0.5 to 1.5 mm and consist of chert mixed with greenalite and haematite. Both iron minerals may be finely divided but greenalite may also occur as small spherulites. Microcrystalline siderite and stilpnomelane are commonly associated with the greenalite and in more altered formations coarsely recrystallised siderite and magnetite occur.

Intraclasts. Some formations contain fragments reaching several centimetres in length in which laminar, granular, and pelletal structure of pre-existing deposits may be recognisable. Smaller structureless fragments are less distinctive and grade into granules. Structureless intraclasts commonly show irregular contraction cracks.

'Pellets'. Small grains of chert containing disseminated greenalite or haematite and showing diffuse outer margins are commonly referred to as pellets. They range from 0.1 to 0.3 mm in diameter and are commonly flattened in the plane of the bedding. They have been interpreted as resulting from flocculation rather than fragmentation or accretion.

Ooliths and pisoliths. These are spherical or ellipsoidal and mostly exceed 0.5 mm diameter. The oolitic structure is defined by concentric layers rich in inclusions, usually of haematite.

Nodules. Large bodies of concretionary or accretionary origin are termed nodules. In the accretionary types a thin crudely layered envelope encloses a relatively large intraclast nucleus.

Matrix of granular deposits. The most common groundmass material is chert. This may display clear cement textures, as where chalcedony fringes are developed. Primary matrix is represented by microquartz containing finely divided inclusions.

Petrography of granular iron-formation (19)

Granular iron-formation may be divided into the silicate–carbonate–chert varieties and the haematite–chert varieties. The *silicate–carbonate types* consist of chert–silicate–carbonate granules set in either a silicate–carbonate or a silicate–carbonate–chert matrix. The actual mineral composition depends on the degree of alteration. The *oxide* types consist of haematite–chert granules, intraclasts, and ooliths set somewhat sparsely in a microquartz matrix (fig. 29.3B) or more densely in a silica cement. Irregular patches of siderite, magnetite, and minnesotaite (resulting from low-grade metamorphism) are often present replacing and veining earlier fabrics.

Petrography of laminated iron-formation (20)

The lamination in these rocks is caused by the alteration of iron-rich and iron-poor layers. The presence of chert in the former is often masked by the iron minerals. In thin section the boundaries show varying degrees of gradation. Individual units may be dominated by a single iron mineral, but several minerals often occur in close association. Many such mineral associations have been recorded and cannot all be covered here, but in all cases the mineral relationships are those typical of crystalline aggregates and can be interpreted with some knowledge of the characters of the various minerals involved.

Chert. This usually constitutes 50 to 65 per cent of the rock. In iron-poor laminae it consists of interlocking crystals with disseminated ferruginous inclusions. Crystal size tends to decrease with increase in the amount of inclusions but contrast in crystallinity is sometimes displayed by adjacent cherty layers for no apparent reason.

The iron-rich laminae consist of a matrix of intimately associated chert and iron minerals in which are developed spheres of microquartz some 5 to 30 μm (fig. 29.3A). This texture apparently represents unmixing of primary components which may have been deposited in the form of gels. In cherty layers the segregation is sometimes reversed, with iron minerals forming the spheres.

Haematite. This typically occurs as finely disseminated particles in cherty layers or in more concentrated form as iron-rich laminae. Its

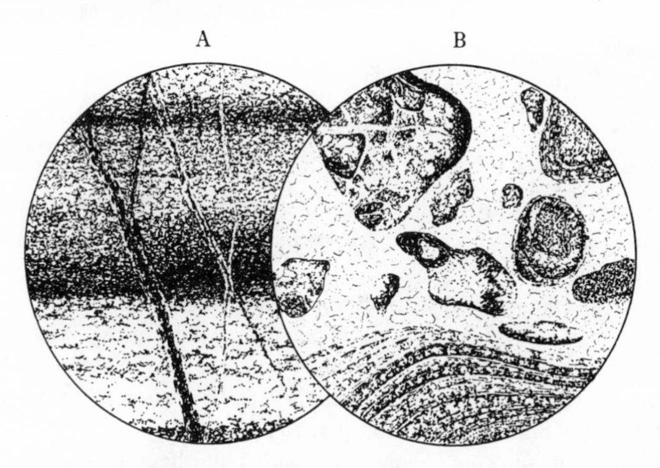

Fig. 29.3. Iron-formation; × 20.

A. Laminated iron-formation; haematite-rich and haematite-poor cherty laminae cut by vein of haematitic chert. Precambrian, South Africa.

B. Granular iron-formation. Granules of haematitic chert and large pisolith set in a matrix of chert. Precambrian, Canada.

finely divided and concordant nature indicates a primary or very early diagenetic origin; mild metamorphism may produce secondary fabrics (fig. 39.2).

Siderite. This forms lozenge-shaped crystals, about 25 μm long, which often contain central inclusions of other minerals. In siderite-rich layers these crystals form closely packed aggregates; more truly crystalline textures are produced by mild metamorphism. Siderite is regarded as a 'primary' mineral, but may be very early diagenetic.

Greenalite. In laminated iron-formation greenalite occurs mostly as cryptocrystalline aggregates interstitial to siderite crystals.

Stilpnomelane. This is usually finely crystalline and fibrous; it forms massive aggregates, networks, and spheres whose textural relationships suggest late diagenetic recrystallisation.

Minnesotaite. This forms radiating aggregates of acicular crystals and is typically associated with sideritic laminae. It often transgresses the bedding and is therefore late diagenetic or metamorphic.

Magnetite. This forms either scattered euhedral crystals or subhedral aggregates which cut across the lamination and are therefore of late diagenetic or metamorphic origin.

Glauconitic rocks

Glauconitic grains are composed of finely crystalline aggregates which display a distinctive speckled aggregate polarisation effect. They include the following types:

Lobate grains, internally structureless and displaying rounded protuberances separated by crudely radial cracks (fig. 29.4A). These are best developed in muddy sediments, where they originate as microconcretions or by replacement of faecal pellets. In sandstones such grains show abrasion and fragmentation (fig. 29.4B).

Flaky grains, showing aggregate extinction and basal cleavage; these apparently originate by alteration of mica flakes.

Internal fossil moulds, including rod-shaped grains derived from the canals of sponge spicules (fig. 29.4A) and internal moulds of other microfossils such as foraminiferans.

Fossil replacements, which affect various shell types, but particularly echinoderm grains.

Glauconitic clays. These display a poorly sorted assemblage of grains in which unabraded lobate types and those of organic association are most conspicuous (fig. 29.4A).

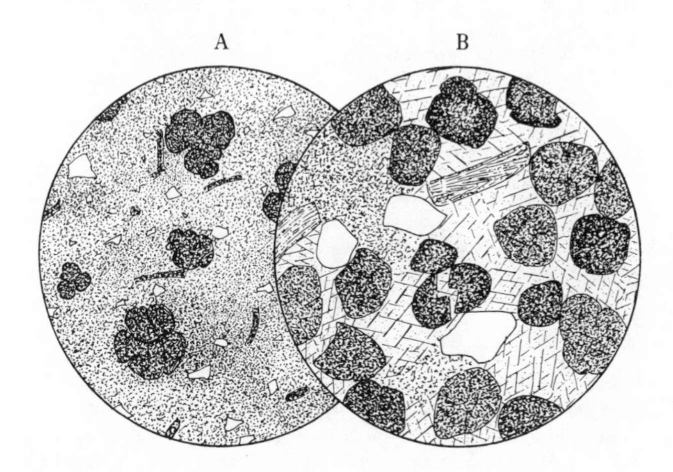

Fig. 29.4. Glauconitic rocks; × 20.

A. Glauconitic clay; irregular lobate glauconite grains and glauconitic sponge spicules in a matrix of detrital clay. Cretaceous, Sussex.

B. Glauconite sand; rounded glauconite grains with scattered quartz grains set in calcite cement with local patches of phosphatised chalk. Cretaceous, Sussex.

Glauconitic sandstones. These are characterised by structureless ovoid glauconite pellets which are associated with terrigenous grains of about the same size. The ovoid grains originate by abrasion of lobate grains. Micaceous and organic grains may also be present. Winnowing of glauconite grains from glauconitic clays can produce almost pure glauconite sands (fig. 29.4B).

Glauconitic limestones. The glauconite of calcarenites is similar to that of glauconitic sandstones except that organic grain types may be more abundant. Micritic varieties more closely resemble terrigenous clays in their glauconite content.

References and notes

(1) James, H. L., 1966. Chemistry of the iron-rich sedimentary rocks. *Prof. Pap. U.S. geol. Surv.* **440**-W, 11.
(2) Carroll, D., 1958. Role of clay minerals in the transportation of iron. *Geochim. cosmochim. Acta.* **14**, 1–28.
(3) Hallam, A., 1975. *Jurassic environments.* London: Cambridge University Press, pp. 39–46.
(4) Knox, R. W. O'B., 1970. Chamosite ooliths from the Winter Gill ironstone (Jurassic) of Yorkshire, England. *J. sedim. Petrol.* **40**, 1216–25.
(5) James, 1966, pp. 47–50.
(6) Hough, J. L., 1958. Fresh-water environment of deposition of Precambrian banded iron formations. *J. sedim. Petrol.* **28**, 414–30.
(7) Trendall, A. F., 1968. Three great basins of Precambrian banded iron formation deposition: a systematic comparison. *Bull. geol. Soc. Am.* **79**, 1527–44.
(8) Takahashi, J., 1939. Synopsis of glauconitisation. *In* Trask, P. D., ed., *Recent marine sediments*, pp. 503–12. Tulsa, Oklahoma; Am. Ass. Petrol. Geol. McRae, S. G., 1972. Glauconite. *Earth Sci. Rev.* **8**, 397–440.
(9) James H. L., 1954. Sedimentary facies of iron-formation. *Econ. Geol.* **49**, 235–93.
(10) LaBerge, G. L., 1964. Development of magnetite in iron-formations of the Lake Superior region. *Econ. Geol.* **59**, 1313–42. Ayres, D. E., 1972. Genesis of iron-bearing minerals in banded iron-formation mesobands in the Dales Gorge Member, Hamersley Group, Western Australia. *Econ. Geol.* **67**, 1214–33.
(11) Huber, N. K., 1958. The environmental control of sedimentary iron minerals. *Econ. Geol.* **53**, 123–40.
(12) Curtis, C. D. & Spears, D. A., 1968. The formation of sedimentary iron minerals. *Econ. Geol.* **63**, 257–70.
(13) Knox, 1970.
(14) Pulfrey, W., 1933. The iron-ore oolites and pisolites of North Wales. *Q. Jl geol. Soc. Lond.* **89**, 401–30.
(15) Taylor, J. H., 1949. Petrology of the Northampton Sand Ironstone Formation. *Mem. geol. Surv. U.K.* p. 5.
(16) Hallimond, A. F., 1925. Iron ores: bedded ores of England and Wales: petrography and chemistry. *Mem. geol. Surv. spec. Rep. Miner. Resour. Gt Br.* **29**, 139 pp. Hayes, A. O., 1915. Wabana iron ore of Newfoundland. *Mem. geol. Surv. Can.* **78**, 163 pp. Taylor, 1949. *Surv. Can.* **78**, 163 pp. Taylor, 1949.
(17) Rastall, R. H. & Hemingway, J. E., 1940. The Yorkshire Dogger: I. The coastal region; petrographical descriptions. *Geol. Mag.* **77**, 257–75.
(18) Dimroth, E., 1968. Sedimentary textures, diagenesis, and sedimentary environment of certain Precambrian ironstones. *N. Jb. Geol. Paläont. Abh.* **130**, 247–74.
(19) Dimroth, E. & Chauvel, J-J., 1973. Petrography of the Sokoman Iron Formation in part of the Central Labrador Trough, Quebec, Canada. *Bull. geol. Soc. Am.* **84**, 111–34. Gross, G. A., 1972. Primary features in cherty iron-formations. *Sediment. Geol.* **7**, 241–61.
(20) Ayres, 1972.

30
Manganiferous rocks

Introduction

Manganiferous sediments occur sporadically in beds of muddy, sandy, calcareous, ferruginous, and cherty facies (1). They form nodules or beds, with one type dominating in any one deposit. Ferromanganese nodules (in which either iron or manganese may dominate) are present on the ocean floors (2) and in freshwater lakes; they are rare in ancient deposits due to the low preservation potential of their associated facies. Several varieties of manganese and iron hydroxides contribute to the nodules. Bedded manganese deposits are known only from ancient deposits; their mineralogy is variously dominated by carbonates, silicates, and oxides.

Origin

Although the geochemistry of iron and manganese is closely similar, it appears that manganese is generally the more mobile in natural environments (3). Ferromanganese nodules associated with rapidly deposited sediments appear to reflect this greater mobility as they are thought to originate by selective leaching of manganese from within the sediment and its deposition in more concentrated form where the pore solutions reach the sediment surface. Nodules associated with areas of very low sedimentation rates may have formed by direct precipitation from sea-water. Ocean-floor nodules are often associated geographically with submarine volcanic deposits and volcanism may enrich bottom waters in manganese, although a direct relation with nodule precipitation has yet to be demonstrated.

Bedded deposits have mostly been interpreted as being of primary origin. Some probably represent deposition in shallow restricted marine environments during phases of intense chemical weathering and low detrital influx (4). Others are associated with slowly deposited pelagic limestones and oceanic lava flows (5); these may originate in the same way as nodules of similar facies association.

Petrography

Manganese-rich sediments contain minerals not usually encountered in sedimentary rocks and some which are not easily identified in transmitted light. They include oxides (pyrolusite, manganite, etc.), the carbonate rhodocrosite, and less commonly manganiferous silicates. The primary mineralogy is uncertain as mineral transformation appears to take place on the sea floor and during very early diagenesis. Some of the older deposits contain manganiferous garnets of metamorphic origin.

Ferromanganese nodules (6)

These are accretionary masses composed of oxides and hydroxides of manganese and iron. Some nodules appear to be devoid of a nucleus, whereas others have grown around a variety of hosts, including lava fragments, limestone fragments, and clasts of skeletal origin. The envelope commonly shows some degree of concentric layering (fig. 30.1A) which consists of an alternation of dark and light layers, enriched in iron and manganese respectively. The lighter layers may also be rich in carbonate.

The envelope structure may be complex due to the presence of colloform or crenulated textures, which resemble small-scale stromatolites. These textures are sometimes clearly diagenetic, where they cut across primary lamination, but in other cases they are primary and represent true micro-stromatolites of probable bacterial origin (fig. 30.1A). Grains which land on the envelope are included within it, so that nodules often contain detrital grains and microplankton (radiolarians, foraminiferans, etc.). Other organic inclusions represent encrustation during growth of the nodules; these include foraminiferans (fig. 30.1A), serpulids, and bryozoans.

The environment of formation of nodules can thus to some extent be inferred from the nature of the nucleus, the presence of encrusting organisms, and the nature of incidental inclusions.

Bedded manganese deposits

These are variously dominated by carbonates, silicates, and oxides. The combined effects of diagenesis and metamorphism have often led to

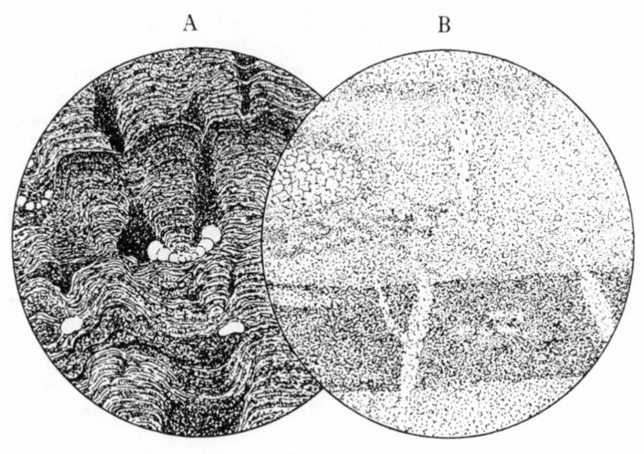

Fig. 30.1.

A. Manganese nodule, showing stromatolitic structure and inclusion of encrusting foraminiferans. Jurassic, Sicily.

B. Manganese carbonate rock. Cambrian, Gwynned.

extensive modification of mineral composition and texture. Deposits dominated by oxides of iron and manganese have probably originated in much the same way as ferromanganese nodules, and the two may even be associated in the field (7). Carbonate deposits (8) appear to have originated as gels or amorphous muds; dehydration, unmixing, and crystallisation have produced textures such as shrinkage cracks, diffusion banding, and concretionary sphaeroids. Colour is influenced by haematite (red), manganese oxide (black), metamorphic garnet (brown), and chlorite (green). Minor amounts of silt and sand are present, together with authigenic microquartz. The purest rocks consist of finely crystalline rhodocrosite (fig. 30.1B).

References and notes

(1) Accounts of various types of manganese deposit can be found in the following:
Varentsov, I. M., 1964. *Sedimentary manganese ores*. Amsterdam: Elsevier.
Symposium on manganese, 20th. Int. geol. Congr., Mexico (1956).

(2) Mero, J. L., 1965. *The mineral resources of the sea*. Amsterdam: Elsevier.

(3) Krauskopf, K. B., 1957. Separation of manganese from iron in sedimentary processes. *Geochim. cosmochim. Acta* **12**, 61–84.

(4) Mohr, P. A., 1964. Genesis of the Cambrian manganese carbonate rocks of North Wales. *J. sedim. Petrol.* **34**, 819–29.

(5) Sorem, R. K. & Gunn, D. W., 1967. Mineralogy of manganese deposits, Olympic Peninsula, Washington. *Econ. Geol.* **62**, 22–56.

(6) Jenkyns, H. C. (in press) Fossil nodules, *in* G. P. Glasby, ed., *Marine manganese deposits*. Amsterdam: Elsevier.

(7) Elderfield, H. *et al.*, 1972. The origin of ferromanganese sediments associated with the Troodos Massif of Cyprus. *Sedimentology* **19**, 1–19.

(8) A petrographical description of a carbonate-rich deposit is given by: Woodland, A. W., 1939. The petrography and petrology of the Lower Cambrian Manganese Ore of West Merionethshire. *Q. Jl geol. Soc. Lond.* **95**, 1–36.
The geochemistry and origin of this deposit is discussed by: Mohr, 1964.

31
Saline rocks

Introduction

The term *saline* is here applied to deposits composed of sulphates and chlorides which have been deposited from concentrated brines. Such deposits are usually interbedded with carbonates (especially dolomite) and often show an intimate association with carbonate minerals under the microscope. Beds consisting of carbonates, sulphates, and chlorides are generally referred to as *evaporites*. The terminology of saline rocks is straightforward as they are named after the dominant minerals.

Origin

Saline deposits originate from the evaporation of brines in standing water or in the interstices of sediments (1). Recent deposits are known from inland lakes, lagoons, and tidal flats (2) but the extent and thickness of many older deposits indicate formation in environments not properly represented at the present day. Thick beds of a soluble mineral such as halite require that the host brine was held within the limits of precipitation of the mineral over a long period. This in turn requires a continuous influx of sea-water and a reflux of other mineral components. These processes may take place in shallow-water environments restricted either by location on extensive shoaling shelves or by the existence of a seaward reef or offshore bar. An alternative mechanism of deep-water precipitation has been proposed for deposits associated with sediments of deep-water facies (3).

Saline minerals are so chemically reactive that primary mineralogy and fabric are usually obscured by diagenesis. At the present day only halite and gypsum (together with calcite and aragonite) are precipitated in any abundance and probably occurred as primary minerals in most ancient evaporite deposits. Other minerals, particularly carnallite, sylvite, polyhalite, and the calcium and magnesium carbonates, may have been primary precipitates in the past. It is possible that under conditions of high concentration alteration of primary minerals could take place at the sediment surface or even in suspension, so that petrographically primary minerals may not represent the true initial precipitate.

Petrography

The petrography of saline deposits concerns crystalline rather than particulate textures. From the mutual relationships of the crystalline constituents it should be possible to determine the order of crystallisation of a given mineral assemblage, but this assemblage may itself represent the culmination of a series of textural and mineralogical modifications. Many saline deposits display complex mineralogy but only the more common minerals are described here as they suffice to illustrate the typical kinds of textural relationship.

Petrography of saline constituents

Gypsum. This occurs both as massive anhedral aggregates and as euhedral porphyroblastic crystals up to several centimetres in length. Inclusions of detrital grains or other saline minerals are common in the latter (fig. 31.1C) and may be arranged zonally.

Anhydrite. This occurs as massive aggregates made up of interlocking subhedral rectangular tablets (fig. 31.1B) or occasionally as more coarsely crystalline aggregates. Porphyroblastic anhydrite forms thick elongate crystals occurring individually or as radiating sheaves or rosettes. Anhydrite commonly contains inclusions.

Halite. Halite occurs as interlocking aggregates in which either the rectangular cleavage or crystal outline can usually be distinguished (fig. 31.1A). Oriented elongate crystals indicate recrystallisation related to plastic flow under the influence of diapiric or tectonic pressures. Halite mostly contains finely divided inclusions; in red-bed facies the halite is often coloured pink or red by inclusions of haematitic clay. Halite may also be rendered cloudy by fluid inclusions which may show growth banding corresponding to variation in rate of crystallisation.

Other minerals encountered in saline deposits include the chlorides sylvite and carnallite, the sulphate polyhalite, and the carbonate magnesite. Silica minerals may also occur.

Petrographical features of 'primary' origin

These include textures resulting from crystallisation or sedimentation at or immediately beneath the sediment surface. Primary fabrics which survive diagenesis usually involve some form of coarse layering, which may involve alternation of saline layers with detrital layers or carbonate layers and in some cases alternation of layers of the same mineral showing different texture.

The best primary crystallisation textures are shown by halite, whose crystallisation takes place both at the water surface and at the sediment surface. The former process produces 'hopper' crystals which may coalesce to form rafts before sinking to the bottom (4). Crystallisation at the sediment–water interface produces more continuous layers which possess a distinctive crystal fabric of vertically elongated and coarsening upward crystals (5). Phases of precipitation of halite are often terminated by detrital influx and associated lowering of salinity is reflected in the truncation of the halite fabric by a solution surface. A well-defined banding sometimes results from the alternation of clear bottom-precipitated halite with surface-precipitated halite which is cloudy due to abundant liquid inclusions (6).

Primary banding is also displayed by deposits in which gypsum or anhydrite alternate with dolomite on a scale of millimetres; the sulphate–carbonate couples are interpreted as varves (7).

Saline minerals are also precipitated interstitially immediately below the sediment surface. In tidal flats of arid areas gypsum often develops as large vertically embedded crystals which are essentially of displacive origin and therefore tend to be inclusion-free. Halite crystals may also grow in surface sediments by displacement and may coalesce to form a crystalline groundmass in which the grains are 'floating'.

In sediments which have undergone burial to depth of a few metres displacive crystallisation gives way to normal interstitial cementation with the development of poikiloblastic texture.

Petrographical features resulting from neomorphism

In many saline rocks it is possible only to establish a sequence of diagenetic events (8). One of the most widespread diagenetic changes is the conversion of gypsum to anhydrite. In finely crystalline anhydrite direct petrographical evidence of a gypsum precursor is lacking, but in coarser fabrics anhydrite pseudomorphs after gypsum may be detected.

Where the gypsum–anhydrite transformation take place in compacted deposits pseudomorphs after gypsum are well preserved (fig. 31.1B), but in soft sediments subsequent compression deforms the pseudomorphs with the formation of irregular 'nodular anhydrite' bodies (9).

Anhydrite replacing gypsum usually occurs as aggregate of fine interlocking crystals showing partial development of rectangular outlines. Recrystallisation leads to the formation of larger tablets or radiating aggregates. Anhydrite may also form at the expense of dolomite and on a small scale by the replacement of halite.

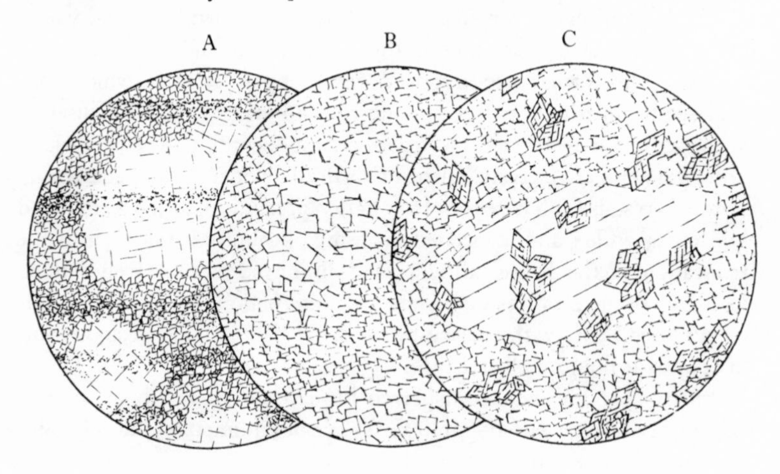

Fig. 31.1. × 20.

A. Replacive halite in laminated dolomite rock.
B. Anhydrite rock, with pseudomorph after gypsum revealed by coarser crystallinity of the anhydrite.
C. Gypsum porphyroblast replacing anhydrite rock; dolomite rhombs within the original rock are retained as inclusions within the gypsum.

Halite commonly occurs as coarsely crystalline aggregates of equidimensional grains, but deformation may produce an anisotropic fabric. The best crystal form tends to be displayed where halite replaces another mineral (fig. 31.1A). Halite crystals sometimes show displacive character even in relatively late diagenetic transformations. This is revealed by the concentration of insoluble crystals or detrital grains around porphyroblast margins and sometimes by distortion of bedding structures.

Anhydritic rocks affected by meteoric water circulation display partial or complete hydration to gypsum. The gypsum often forms large tabular porphyroblasts which may retain relict particles of anhydrite (fig. 31.1c). Later hydration of the matrix produces a porphyroblastic gypsum rock.

The diagenesis of saline rocks sometimes involves minerals which are most familiar in non-evaporative facies; for example calcitisation, accompanied by minor silicification, may cause partial or complete replacement of original sulphate deposits (10).

References and notes

(1) A summary of depositional environments, mineralogy, and geochemistry of saline deposits is given by Stewart, F. H., 1963. Marine evaporites. *Prof. Pap. U.S. geol. Surv.* **440**-Y.
A collection of papers on evaporites (including refs. 2, 4, 8b, 10 below) has been compiled in Kirkland, D. W. & Evans, R., eds., 1973. *Marine evaporites: origin, diagenesis, and geochemistry.* Stroudsberg, Penns.: Dowden, Hutchinson, & Ross.

(2) Kinsman, D. J. J., 1969. Modes of formation, sedimentary associations, and diagnostic features of shallow-water and supratidal evaporites. *Bull. Am. Ass. Petrol. Geol.* **53**, 830–40.

(3) Schmalz, R. F., 1969. Deep-water evaporite deposition: a genetic model. *Bull. Am. Ass. Petrol. Geol.* **53**, 798–823.

(4) Dellwig, L. F., 1955. Origin of the Salina Salt of Michigan. *J. sedim. Petrol.* **25**, 83–110.

(5) Arthurton, R. S., 1973. Experimentally produced halite compared with Triassic layered halite-rock from Cheshire, England. *Sedimentology*. **20**, 145–60.

(6) Dellwig, L. F. & Evans, R. Depositional processes in Salina Salt of Michigan, Ohio, and New York. *Bull. Am. Ass. Petrol. Geol.* **53**, 949–56.

(7) Wardlaw, N. C. & Schwerdtner, W. M., 1966. Halite-anhydrite seasonal layers in the Middle Devonian Prairie Evaporite Formation, Saskatchewan, Canada. *Bull. geol. Soc. Am.* **77**, 331–42.

(8) (a) Detailed descriptions of the petrographical relationships of various evaporite minerals are given by Stewart, F. H., 1949. The petrology of the evaporites of the Eskdale no. 2 boring, east Yorkshire. Part 1. *Min. Mag.* **28**, 621–75. Part 2 (1951) *Min. Mag.* **29**, 445–75. Part 3 (1951) *Min. Mag.* **29**, 557–72.
(b) Petrographical evidence for the time relationships between anhydrite and gypsum are given by Murray, R. C., 1964. Origin and diagenesis of gypsum and anhydrite. *J. sedim. Petrol.* **34**, 512–23. Holliday, D. W., 1970. The petrology of secondary gypsum rocks: a review. *J. sedim. Petrol.* **40**, 734–44.

(9) Kerr, S. D. & Thomson, A., 1963. Origin of nodular and bedded anhydrite in Permian shelf sediments, Texas and New Mexico. *Bull. Am. Ass. Petrol. Geol.* **47**, 1726–32.

(10) West, I. M., 1964. Evaporite diagenesis in the Lower Purbeck Beds of Dorset. *Proc. Yorks. geol. Soc.* **34**, 315–30.

32
Carbonaceous rocks

Introduction

The carbonaceous rocks comprise peats and coals which have originated by the accumulation of plants and plant debris (1). Deposits formed primarily from woody tissue are known as *humic* deposits; in order of increasing rank these are *peat, lignite, bituminous coal*, and *anthracite*. Deposits composed of finely macerated plant debris, spores, and oil-bearing algae are known as *sapropelic* deposits; the principal varieties are *cannel coal* (dominantly macerated plant matter) and *boghead coal* (dominantly algal).

Humic deposits result from the accumulation of vegetable matter in place, whereas the sapropelic deposits result from the settling out of fine plant matter in stagnant water. Early diagenesis of coal involves compaction and dewatering together with bacterial and fungal attack on the organic matter. In later diagenesis the organic matter undergoes jellification and then further dehydration which converts the deposit into lignite and finally coal. Subsequent loss of volatiles and water may cause further enrichment in carbon.

Petrography of bituminous coal constituents

The organic constituents fall into two major groups on the basis of their optical properties: (a) opaque or very slightly translucent red-brown material, and (b) translucent material ranging from golden yellow through reddish yellow to red-brown. The first of these groups includes the *inertinite macerals*; the second group includes both *vitrinite* and *exinite macerals* (table 32.1). Distinction between the latter two maceral types can also be made to some extent on their optical properties in transmitted light, in that the exinite macerals generally display lighter colours. Exceptions occur where colour variation is displayed in different layers of cell walls (vitrinite).

TABLE 32.1

Maceral group	Maceral
VITRINITE	Telinite Collinite
EXINITE	Sporinite Cutinite Resinite Alginite
INERTINITE	Fusinite Micrinite

The colour of macerals in thin section is affected by thickness and, more importantly, coal rank (volatile content). In low rank bituminous coals the exinite macerals range from yellow or golden yellow to orange-yellow, whereas in high rank bituminous coals they range from orange to reddish orange or brownish red. The vitrinite macerals show a more steady increase in colour from orange or orange-yellow to dark red or red-brown; in very high rank bituminous coals (grading into anthracites) the vitrinite and exinite macerals become indistinguishable by colour. The translucent macerals normally display anisotropy, and this increases with increasing rank.

The petrography of coal is here discussed in terms of transmitted light microscopy, although this method of study has now been largely superseded by reflected light microscopy.

Telinite. This maceral consists of the cell wall tissue of wood and bark which has been preserved in vitrinite. The cell structure is defined by an interconnected mesh formed by a thin middle lamella layer (fig. 32.1A); within each cell unit so defined is a discrete cell wall layer which may vary greatly in thickness relative to cell size. At the trilete junction formed by the branches of the middle lamella, a small intercellular space can often be distinguished. The size, shape, and arrangement of cells varies considerably according to the nature of the tissue and the type of plant.

Telinite cell structure shows varying degrees of deformation, from small-scale crumpling of the cell walls to extreme compression with elimination of the cell cavity. The latter is most apparent in thick walled cells which are relatively impervious to fine detritus and impregnating solutions. Thinner walled cells are generally filled with collinite, resinite, micrinite, or clay precipitates which inhibit compactional deformation. Telinite is usually orange to dark red but is occasionally golden yellow.

Collinite. The term collinite is applied to structureless vitrinitic material. Where it occurs as discrete layers it probably represents fragments of wood which have undergone total jellification. Less altered cell tissue displays telinite cell walls enclosing precipitated collinite. The latter is usually featureless but may contain opaque granules representing the original cell contents. Rarely the cell contents are preserved in a shrivelled sac which is embedded in the invading collinite. Collinite is orange to dark red in colour.

Sporinite. Spores include the *megaspores*, with diameters greater than 200 μm and the *miospores* (including microspores and pollen) with diameters less than 200 μm. The megaspores are often conspicuous in thin section, appearing roughly circular in horizontal section and elliptical or compressed linear (fig. 32.1C). The miospores (fig. 32.1C) are less distinctive. Sporinite is yellow or orange, but becomes darker in high rank coals.

Cutinite. Cuticles are derived from the outer layers of stems and leaves. Cutinite is golden yellow or orange in low rank coals. In section it appears as sheets of which one margin (the outer) is smooth and the other serrated due to original contact with underlying cells (fig. 32.1C). More complete plant remains are represented by an entire cuticle which is tightly compressed to form a single sutured sheet from which the internal cellular matter has been lost by decomposition.

Resinite. Resin occurs as isolated bodies roughly oval in vertical section (fig. 32.1C) and circular in horizontal section. It also occurs as cell fillings in telinite. Resinite is yellow in low rank coals; it frequently contains abundant gas-filled cavities.

Alginite. Algal bodies are circular to elliptical in section and typically possess a highly irregular surface. Diameters may exceed 0.5 mm although 0.05 to 0.25 mm is more normal. Alginite is yellow in low rank coals and may display faint cellular or radial internal structure.

Fusinite. This is readily identified by its opacity and well-defined cellular structure (fig. 32.1B). The woody tissue has not undergone jellification and particles show fracture rather than plastic deformation; in some cases fragmentation has reduced the cell structure to a mass of tri-radiate grains. The pores of fusinite may be infiltrated by micrinite or cemented by clay or carbonate cements.

Micrinite. This includes opaque or very dark reddish brown material which consists of internally structureless particles ranging up to about 0.1mm diameter. It commonly forms the matrix of coals (fig. 32.1C) and may also occur as infiltrations of cell and spore cavities.

Inorganic mineral matter. This includes detrital silicates and minerals of authigenic origin. The latter may be intimately associated with the coal macerals indicating that (except in the cementation of fusinite) precipitation occurred before or during jellification. Such *syngenetic* minerals include calcite (as irregular patches), dolomite (as rhombs), siderite (as rhombs or spherulites), pyrite (finely disseminated or as minute spheres), and other sulphides. Minerals deposited after jellification are termed *epigenetic*; they occur in fissures and include medium to coarsely crystalline carbonates and sulphides.

Petrography of coal lithotypes

Vitrain. Vitrain is characterised in hand specimen by its bright clean nature; it is generally closely fractured. In section telinite and collinite are dominant (fig. 32.1A) but thin layers of exinite or inertinite may be present. Vitrain is generally pure, with little visible contamination by inorganic mineral matter.

Fusain. Fusain is dull black and dirty to the touch. If not mineralised it is soft and porous, but cementation by pyrite or carbonates produces a denser and firmer deposit. It is composed of fusinite as fragments, lenses, or thin layers interspersed with minor amounts of other macerals, particularly micrinite (fig. 32.1B). Fragmental fusain is sometimes associated with detrital silicates.

Durain. Durain is grey to brownish black and breaks with a rough fracture with dull or slightly greasy lustre. It consists of exinite macerals, especially spores and resin bodies, set in a micrinite matrix (fig. 32.1C); the latter causes the coal to be relatively opaque.

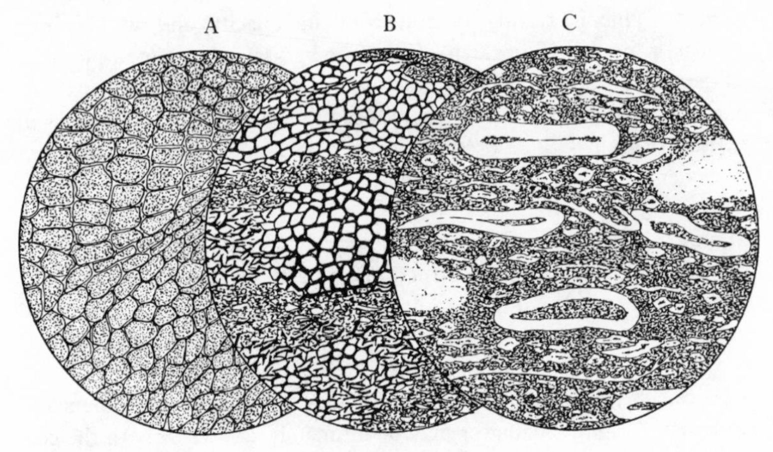

Fig. 32.1. × 40.

A. Vitrain, composed of cellular woody tissue (telinite) filled with structureless collinite.
B. Fusain; fusinite showing local breakdown of structure, in a matrix of micrinite.
C. Durain; megaspores, miospores, resin bodies, and cuticle fragments set in a micrinite matrix.

Clarain. Clarain is characterised by a finely laminated texture (laminae less than 3mm thick). The laminae are made up of bright vitrinite-rich layers alternating with dull exinite or inertinite-rich layers. This alternation imparts lustre to fracture surfaces.

Cannel coal. This is characterised in hand specimen by its dull compact homogeneous appearance and waxy lustre. Sections show it to be finer grained than typical bituminous coal. It is composed of an intimate mixture of vitrinite, exinite, and inertinite macerals. In comparison with durain the groundmass is relatively translucent, indicating a high proportion of vitrinite.

Boghead coal or torbanite. In hand specimen this resembles cannel coal but is distinguished in thin section by an abundance of alginite. Sometimes the algal bodies are so abundant as to form a tightly compressed mass.

References and notes

(1) The following references provide a general coverage of the origin, classification, terminology, and petrography of coals:

Moore, L. R., 1968. Cannel coals, bogheads and oil shales, *in* D. Murchison & T. S. Westoll, eds., *Coal and coal-bearing strata*. Edinburgh: Oliver & Boyd, pp. 19–29.

Raistrick, A. & Marshall, C. E., 1948. *The nature and origin of coal seams*. London: English University Press.

Stach, E. *et al.*, 1975. *Stach's textbook of coal petrology*. Berlin: Gebrüder Borntraeger.

Tomkieff, S. I., 1954. *Coals and bitumens and related fossil carbonaceous substances; nomenclature and classification*. London: Pergamon Press.

III: Metamorphic rocks

33
Metamorphism: deformation and recrystallisation

Introduction (1)

The term 'metamorphism' denotes the production of new minerals, or of new structures or textures (chapter 34) or all three, in pre-existing rock masses. Modern usage excludes weathering and other processes of sub-aerial alteration, and metamorphic processes are considered to embrace the whole range between sedimentary and diagenetic processes at low temperature, and the igneous processes of fusion at high: the boundary definitions are often blurred.

Essentially, however, we are interested in those transformations which are brought about in solid rocks under the influence of high temperature, pressure, and stress. We may thus distinguish *dynamic* metamorphism, due to purely mechanical forces, from *thermal* (or *contact*) metamorphism, due to heat. Such phenomena are usually of local extent. The former, representing the strain and fracture of minerals in response to stress, is developed along planes of movement in rocks cold enough to behave in a relatively brittle fashion, or when strain rates are unusually high. The latter, representing the chemical reconstitution which occurs when a mineral assemblage is rendered unstable by increase in temperature alone, is developed at contact of an igneous body, emplaced, usually at high crustal levels (i.e. low pressure).

The most important class of metamorphic rocks is, however, the product of both heat and mechanical deformation, with confining pressure an important factor. The production of high temperature at high pressure normally requires deep burial and is thus necessarily regional in extent; hence *regional metamorphism*.

Dynamic metamorphism

Pure dynamic metamorphism, i.e. mechanical deformation, produces aggregates of high strain energy and is normally rapidly obliterated by recrystallisation. Most evidence of mechanical strain we see is thus late-stage, sustained when the rock was cooling below temperatures at which rapid recrystallisation could occur.

Strain phenomena

The most common evidence of inelastic strain in the component crystals of a stubborn rock mass is a modification of the optical properties, immediately evident between crossed polars. The crystal shows dark shadows which move across it as the microscope stage is rotated, owing to the variation in the indicatrix orientation from point to point in the strained crystal. These *strain shadows* or undulose *extinctions* are typically seen in quartz, but any 'bent' crystal will show them.

A similar phenomenon results not from external stress, but from internal strain as a crystal inverts to a low-temperature form: the 'wavy' extinction of some microclines is characteristic of this. The strain set up as the triclinic microcline attempts to conform to the 'mould' of the original high-temperature monoclinic sanidine is, however, usually high enough to induce twinning: such inversion twins show impersistent, spindle-shaped lamellae giving the micro-

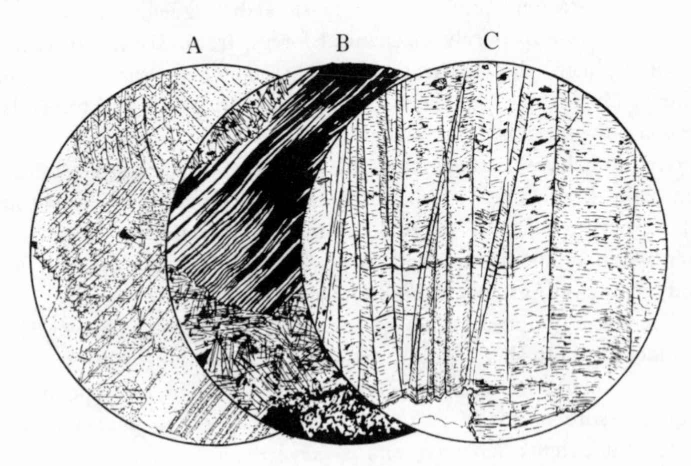

Fig. 33.1. Strain textures; × 35.

A. Stress twinning in calcite. Mont Gendres, Pyrenees.
B. Plagioclase in cataclastic gabbro, the Lizard, Cornwall (crossed polars), showing albite twinning produced by stress.
C. Kink-bands in kyanite of Moine schist, Ross of Mull, Argyll. The excellent cleavage is {100}.

cline 'tartan-twinned' appearance (fig. 33.6B). Analogous strains, set up on cooling or on the release of pressure from intimate intergrowths of minerals having differing expansion coefficients or compressivities, give *strain birefringence*, commonly seen in garnets (fig. 34.1C).

Flexible minerals, such as micas, often show bending of their crystals, or yielding to the stress by twinning or slip. Such secondary twinning can be distinguished from original growth twinning only by its inconstant character and its relation to other strain phenomena (fig. 33.1). Sometimes, in a single crystal the fineness of the twinning is seen to increase with the strain until the crystal has yielded along a crack or granulated vein. Common products of deformation in minerals with a good cleavage are *deformation bands* or *kink bands*, which differ from twin lamellae in having no precise crystallographic relationship, the bounding planes running irregularly across the crystal (fig. 33.1C). Minerals without prominent cleavage or twin planes (e.g. quartz) may show rows of fluid pores marking directions of shearing strain (fig. 33.4A).

Cataclastic structures

The foregoing effects, so often seen as the late-stage products of regional metamorphism or of igneous processes, hardly affect the main appearance or interpretation of the primary structures or textures. The phenomena of internal fracture, crushing, and milling of hard rocks, however, produce distinct rock types of seemingly endless variety in regions of great mechanical disturbance. They may be developed in lesser or greater degree; they may affect some or all of the mineral constituents of a composite rock; they may or may not tend to a parallel arrangement. In the initial stage of cataclasis, fragments or individual crystals (*porphyroclasts*) of the originally crystalline rock are surrounded by borders of finely granular material. As a further stage the fine-grained milled material forms a matrix enclosing strained porphyroclasts; rocks in which the matrix dominates, with a strongly banded or planar structure, are called *mylonites*, or, if a significant proportion of layered minerals (chlorite, muscovite) occurs, *phyllonites*. The constituent grains of the mylonite matrix commonly appear ribbon-like in thin section, suggesting that they have developed as much by ductile deformation as by actual crushing; the planar structure developed is then appropriately called a flow or fluxion structure (fig. 33.2B). Rocks lacking fluxion structure are called *cataclasites*.

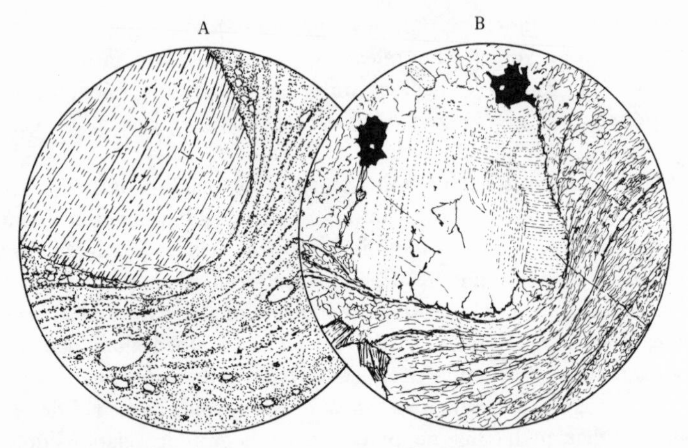

Fig. 33.2. Cataclastic effects; × 35.

A. Mylonitised peridotite, St. Paul's Rocks, Atlantic (Mid-Atlantic Ridge). Porphyroclasts of pyroxene and olivine in a finely comminuted matrix.

B. Crushed granite in shear-zone, Kavaala, Greece. The strained porphyroclast, showing distorted albite and pericline twinning, is of oligoclase. The individual grains of the quartzofeldspathic matrix are elongated, due to ductile deformation.

Shock metamorphism

The term 'shock metamorphism' has been given to deformation under very high strain rates, usually associated with meteorite impact. Quartz and feldspar show multitudes of lamellae and cleavages, often 'decorated' with fluid inclusions: in some examples the shock has converted the mineral to a glass without the intermediate stage of melting. Such alteration is highly dependent on the orientation of the lattice with respect to the applied stress: thus in a single twinned plagioclase one set of twin lamellae may have been rendered isotropic and the other may have had the normal plagioclase optical properties. Very dense phases produced under the transient high pressures of the shock wave passage (e.g. coesite and stishovite from quartz) are usually too fine grained to be identified under the petrographic microscope, but in the Tenham (N.S.W.) meteorite the spinel Ringwoodite can be easily seen as the inversion product of parent olivine (2).

Slates

Slates, in which fissility was thought to have been imposed on largely detrital material, were formerly separated from *phyllites*, in which the mineral content was largely recrystallised. Since, however, recrystallisation clearly commences at the diagenetic, clay stage, the progression *shale → phyllite → schist* is now merely regarded as one of increasing grain size. Nonetheless, the characteristic features of slates can be interpreted mainly as the result of mechanical deformation and are appropriately considered here.

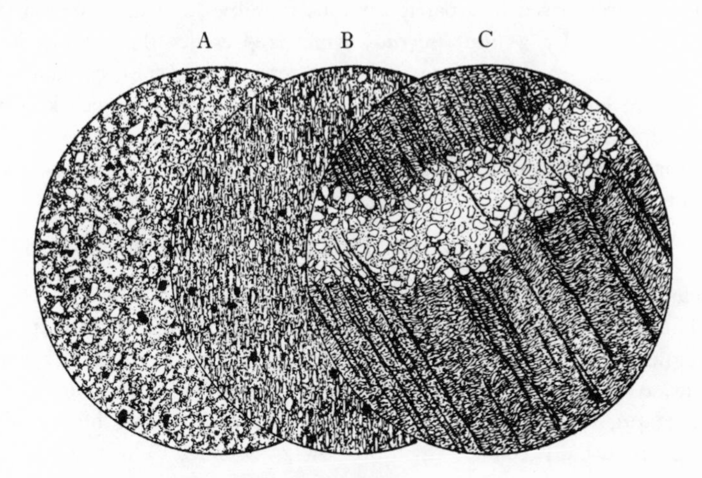

Fig. 33.3. Slates; × 20.

A. Cambrian slate, Moel Tryfaen, Caernarvonshire, cut parallel to cleavage plane. Granules of clastic quartz with muscovite, chlorite, scales of haematite and some calcite.

B. The same, cut transversely to the cleavage, showing more clearly the parallel orientation of mica flakes normal to the maximum principal stress.

C. Skiddaw slate, Bassenfell, Cumberland, with false cleavage (strain-slip); viz. close-set folds passing into faults. This structure is interrupted by a gritty seam in the slate.

Slaty cleavage is produced by the rotation of fine flakes of mica into the plane normal to the direction of maximum compression; the fine grain size ensures that the rock will cleave along any one of innumerable parallel planes (fig. 33.3A, B). Change in the axes of deformation at a later stage produces crenulations of the slaty cleavage known as *false cleavage*, *strain slip*, or *fracture cleavage* (fig. 33.3C); here the rock will only part along well-separated planes of perturbation (3).

Contact and regional metamorphism

When examining metamorphic rocks in the field, a zonal relationship can often be seen as we pass from unmetamorphosed material through *low-grade* metamorphic rocks to the *highest-grade* material at the high-temperature focus of metamorphism. In contact metamorphism, the focus is the contact of the igneous body whose intrusion has produced the temperature gradient; in regionally metamorphosed rocks, the focus represents the channel of high heat flow resultant on deep burial and heat transport by igneous material and by juvenile waters. Any two zones, characterised in a particular rock type by differing mineralogy, are separated by a line (*isograd*) which represents the intersection between the ground surface and the pressure/temperature plane of univariant equilibrium for a chemical reaction (p. 373). A well-known isograd, found in argillaceous rocks of moderate grade, separates a zone with assemblages containing muscovite and quartz from one with sillimanite–microline assemblages; it represents the reaction

$$\text{muscovite} + \text{quartz} = \text{silimanite} + \text{microcline} + H_2O$$

To an approximation the isograd can be regarded as an isotherm (4).

The earliest studies (long preceding microscopic petrography) recognised increasing grade from increasing grain size in hand specimens. Although many factors control the size of mineral grains (e.g. fig. 33.4B), temperature is such a dominant control that high-grade rocks are usually coarser than the low-grade ones. The coarsening

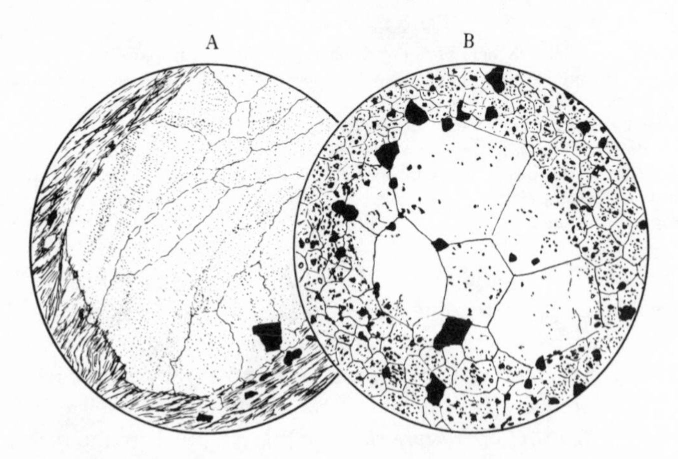

Fig. 33.4. Recrystallisation of quartz; × 40.

A. Deformed quartz in clastic pebble, Glen Esk, Kincardineshire. Elongate subgrains show strain shadows and deformation lamellae, decorated with fluid pores, at a high angle to them. Grain boundaries serrated.

B. Annealed quartzite, Big Rock, New Mexico, with prevalence of 120° grain boundary intersections (granoblastic texture). Grain size is finer where magnetite inclusions have inhibited grain boundary migration.

represents the attempt by crystals to reduce the proportion of surface area to volume. Since contiguous grains are unlikely to have precisely the same orientation, atoms at the intermediate interface will be either too close or too distant from their neighbours in the opposing crystal; thus they will have a higher energy than atoms regularly arrayed within the crystal. *Surface free energy* is thus associated with the grain boundaries and crystallisation tends to reduce this to a minimum.

The ordinary process of recrystallisation, mainly involving reduction of strain- and surface-energy, and uncomplicated by chemical reaction, is best illustrated by monominerallic rocks, such as quartzites, composed of minerals stable throughout most metamorphic régimes. If the quartz grains have been strained, migration of dislocations (5) within the strained crystal produces a network of dislocation 'walls' which separate unstrained 'subgrains' differing only slightly in orientation one from the other (fig. 33.4A). In such aggregates, however, the proportion of grain surface area to volume is high, and further recrystallisation will tend to reduce the surface area by the growth of the larger grains at the expense of the smaller ones; if applied stress favours the growth of crystals of a particular orientation, a preferred orientation (6) will result. The final product with minimal surface area would be a single crystal; in general, however, recrystallisation stops at the stage of the simple *mosaic* texture, the individual grains, as seen in thin section, showing polygonal outlines with any three grain boundaries meeting at equal angles (7). The process is similar to that described by metallurgists as 'recovery' and 'annealing'.

The simple annealed mosaic texture may also be found where minerals of similar structural frameworks – feldspars, scapolite, cordierite – are associated with quartz. The presence of minerals with faces of strongly differing surface energy, however, results in the imposition by one mineral of its dominant crystal form on another. The $\{001\}$ form of mica, for example, corresponding to the orderly array of atoms in the phyllosilicate sheet, has a much lower surface energy than any other form, either in mica or in other minerals, and is thus almost always developed. Similarly, garnet and hornblende will tend to develop $\{110\}$, regardless of the orientation of neighbouring framework minerals. The interpretation of such relations thus differs fundamentally from that of igneous rocks, in which euhedral character is determined by order of crystallisation.

In the unstressed environment of many contact aureoles, the growth of micas or amphiboles in random array gives the *decussate* texture characteristic of the massive contact hornfels (fig. 33.5C). The decussate texture may, however, be absent when contact metamorphism is superimposed on rocks which have previously developed a slaty cleavage; here new growth on pre-existing micas mimics their preferred orientation giving a *mimetic* fissility.

In regional metamorphism, and in those contact aureoles imposed on a stressed environment, orientation of mica cleavages normal to the direction of stress gives the *schistosity* of the crystalline schists. It is uncertain to what extent the production of schistosity results from mechanical rotation (as in slaty cleavage) of seed crystals and to what extent growth of favourably oriented grains at the expense of those less favourably oriented. In stressed rocks the amphiboles commonly lie with their prisms parallel to the fold axes to give *lineation*. The term 'schist' is retained for rocks in which the planar or linear elements are regularly distributed throughout the rock; however, because with increasing temperature the proportion of hydrous minerals decreases markedly, the segregation of micas and amphiboles to form discrete *foliae* separated by lenticles of quartzofeldspathic and other material, results in schistosity giving way to the foliation of a gneiss.

The foregoing phenomena, involving the solution of unfavourably oriented minerals, the growth of favourably oriented minerals, and the reduction in the area of grain boundaries, require diffusion of material within the rock and are temperature-dependent. An additional kinetic factor is raised when, with increasing metamorphic grade, one set of minerals becomes thermodynamically unstable with respect to another. Two or more minerals unstable together are termed *incompatible*, and the attainment of equilibrium must then involve two steps: the nucleation of new minerals, and the diffusive transfer of material to them. The rate of the reaction is the rate of the slower of the steps.

The textures of many rocks are strongly dependent upon nucleation rates. A mineral which nucleates readily will form large numbers of small crystals, whereas a mineral slow to nucleate will form relatively small numbers of large crystals, which in a finer grained matrix appear as *porphyroblasts*. Nucleation difficulties can also affect the manner in which the reaction product forms: often nucleation occurs most readily on other minerals which have similar lattice arrays in one plane or another (*epitaxis*), e.g. oriented growths of sillimanite, garnet or magnetite on mica.

We must recognise here an important difference between regional and contact metamorphism. In regional metamorphism the rates of burial and hence of heating are slow in comparison with the rates of mineral reaction. Regional metamorphism is thus truly *progressive*, the mineral assemblage recrystallising continuously to keep pace with increasing temperature and pressure. Each high-grade assemblage has passed through the lower grade stages. Occasionally relicts of previous stages may be preserved, usually as inclusions in another mineral, but adjustment is commonly complete. The only notable exception involves igneous rocks (particularly gabbros and dolerites) which, being initially coarse grained and dry, often remain unaltered at low grades and at high grades transform directly to the appropriate assemblage.

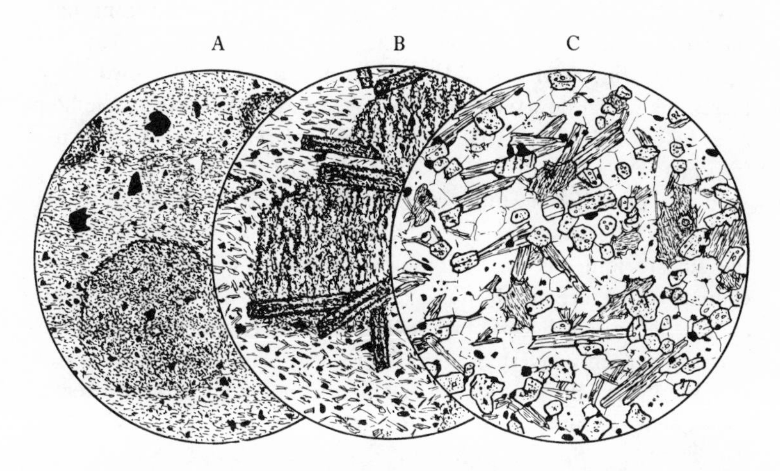

Fig. 33.5. Contact metamorphism in aureole of Insch Gabbro, Aberdeenshire; × 35.

A. Spotted slate, 2 km from contact. Porphyroblasts of cordierite in a fine-grained micaceous groundmass.
B. Knotted mica schist, 1·5 km from contact. Porphyroblasts of cordierite and chloritoid in a schistose groundmass.
C. Hornfels, 300 m from contact. Approximately equigranular aggregate of cordierite, andalusite, orthoclase, quartz and biotite, the biotite showing decussate (disoriented) texture.

In contact metamorphism, however, the relatively rapid emplacement of the igneous body ensures that the rate of heating is commensurate with that of reaction. Contact metamorphic zones thus often represent the direct crystallisation of original rock to final assemblages.

The dependence of rock texture upon nucleation may be illustrated in a general way by classic accounts of the contact metamorphism of argillaceous slates, in which each stage has grown from the original slaty aggregate. Below a certain temperature, reaction rates are too slow for evidence of reaction to be seen. The first indication of recrystallisation is the appearance of 'spots' in the rock. These represent the growth of porphyroblasts either of cordierite from the decomposition of chlorite, or, if the rock contained an original excess of kaolinite, of andalusite (8); they are crowded with fine-grained inclusions of graphite or of iron oxides. The main rock matrix of muscovite, biotite, quartz, plagioclase and graphite is largely inherited from the original slate and is still too fine grained to be readily resolved under the petrographic microscope. The second stage, that of the knotted mica schists, shows a much coarser matrix. Newly crystallised micas, quartz and feldspar are now obvious: they still possess, however, an orientation derived from that of the parent slate. The porphyroblasts of the newly nucleated minerals again have dimensions considerably greater, but are less crowded with inclusions which now tend to be arranged in patterns to give the hourglass habit of chloritoid (fig. 33.5B) or the *chiastolite* habit of andalusite (fig. 33.10A).

In the third, or hornfels, stage, high nucleation rates consequent on high temperatures have completely obliterated epitaxial effects and the silicate minerals are of roughly equivalent grain size. The mica content is much reduced, muscovite having reacted with quartz to give andalusite and orthoclase, while the more magnesian component of biotite has been converted to cordierite and orthoclase; but the iron-rich biotite which remains has the poorly oriented, *decussate* texture consequent on nucleation and growth in an unstressed environment (fig. 33.5C).

The foregoing, rather idealised, account illustrates a specialised example of the metamorphic process. The fineness of grain of the starting product ensures that reacting minerals are close and that the distance traversed by diffusing ions is small. Furthermore, as the temperature increases the initial aggregate becomes less and less stable with respect to the equilibrium assemblage and so adequate energy differentials exist to drive reaction and recrystallisation.

Mineral reactions in coarse-grained rocks require the diffusion of constituents over rather greater distances; here diffusion rates may govern the rate and extent of reaction. Measurements on metals show 'self-diffusion' within crystal lattices to be a much slower process than diffusion along grain boundaries; this applies *a fortiori* to silicate minerals in which the bonds are stronger and stoichiometry is more important. Grain boundary diffusion in rocks, at least at intermediate grades, is probably aided by the presence of a *fluid phase*: since in a rock under pressure the mass of solids greatly exceeds that of the pore spaces, such fluid is strongly dependent on the composition of the surrounding rock, i.e. is predominantly water in dehydrating rocks, and carbon dioxide in calcareous rocks (9). The process of a metamorphic reaction may thus be pictured as the solution of a mineral in its surrounding fluid; concentration (strictly activity) gradients in the fluid connecting incompatible minerals then provide the potential for transfer of material, by diffusion, to the newly nucleating product minerals. If the incompatible minerals are outside the range or 'sphere' of diffusion, reaction will not occur, and they must be considered to belong to a separate chemical system. The extent of the 'sphere of diffusion' increases exponentially with temperature but the relative diffusivity of elements is also important. Thus aluminium often appears to diffuse more slowly than silicon or magnesium – where, for example, the incompatible phases spinel and quartz find themselves within the same sphere of diffusion, the cordierite reaction product forms not around quartz, but around the spinel.

The clearest cut examples of mineral reaction are seen in *polymetamorphic rocks*, in which well-crystallised assemblages have been rendered unstable under physical conditions differing greatly from those of their original formation. The coarseness of grain, the lack of volatiles and the consequent slow rates of diffusion ensure preservation not only of

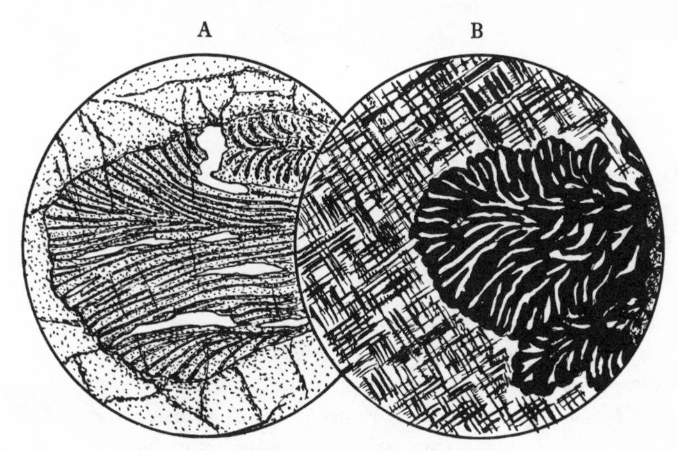

Fig. 33.6. Symplectites; × 50.

A. Dactylitic (finger-like) intergrowth of hypersthene, spinel, and cordierite forming as low-pressure breakdown product of garnet; in granulite from Gangallarpalli, Madras.

B. Myrmekite, Ardveen, Donegal (crossed polars). Plagioclase-quartz symplectite growing from twinned microcline; the reaction involves ionic exchange:

$$\underset{\text{microcline}}{4\ KAlSi_3O_8} + Ca^{2+} + 2\ Na^+ \rightarrow \underset{\text{plagioclase}}{(2\ NaAlSi_3O_8 + CaAl_2Si_2O_8)} + \underset{\text{quartz}}{4\ SiO_2} + 4\ K^+$$

original textures but also quite frequently of relicts of the original assemblages; both reactants and products are readily identified. Simplest of interpretation is the pseudomorphing of a mineral by its polymorph, the new mineral nucleating within the old and growing at its expense. Another simple texture (symplecite) (10) may result from the decomposition of a mineral to two or more products. Fig. 33.6A illustrates the breakdown reaction:

$$\underset{\text{garnet}}{5(MgFe)_3Al_2Si_3O_{12}} \rightarrow$$
$$\underset{\text{hypersthene}}{10(MgFe)SiO_3} + \underset{\text{spinel}}{3(MgFe)Al_2O_4} + \underset{\text{cordierite}}{(MgFe)_2Al_4Si_5O_{18}}$$

Nucleation of the decomposition products on garnet grain boundaries was succeeded by growth of cuspate aggregates of intimately associated rod-like or plate-like crystals. The intimacy of intergrowth ensured the minimum distance of diffusion necessary for any element, the radius of the interface between garnet and intergrowth remaining more or less constant from the position of initial nucleation.

Examples of more complex reactions are found in many metamorphosed gabbros in which on cooling and/or increasing pressure the original magmatic olivine–labradorite assemblage has become unstable. A typical reaction involves the formation of hypersthene, diopside, andesine and spinel. At the pressures concerned the albite component of the plagioclase plays a passive part, and hence we may represent the reaction by the equation:

$$\underset{\text{anorthite}}{CaAl_2Si_2O_8} + \underset{\text{olivine}}{2(Mg, Fe)_2SiO_4} =$$
$$\underset{\text{diopside}}{Ca(Mg, Fe)Si_2O_6} + \underset{\text{hypersthene}}{2(Mg, Fe)SiO_3} + \underset{\text{spinel}}{MgAl_2O_4}$$

Characteristically the products form a *corona* surrounding the reactant olivine, but the precise disposition varies widely in different cases. In the example of fig. 33.7B, the hypersthene replaces original olivine, and the plagioclase adjacent to it is replaced by diopside and spinel. The remainder of the original labradorite plagioclase has its form and orientation preserved by andesine, clouded with fine-grained spinel. We can thus deduce diffusion of calcium from the plagioclase interior to the boundary with olivine to form diopside; and complementary diffusion of (Mg, Fe) from the olivine to the diopside corona and into the plagioclase giving the pervasive spinel clouding. The clouding process involved mainly diffusion within the plagioclase, and clearly the preservation of

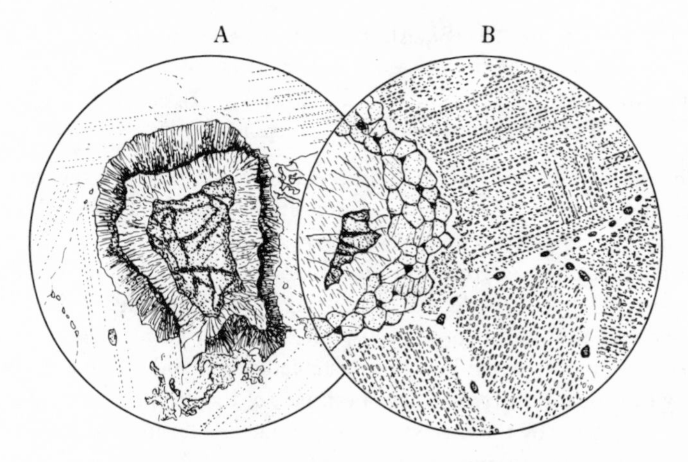

Fig. 33.7. Corona textures, in metagabbros; × 35.

A. Risor, S. Norway. An inner, fibrous corona of hypersthene (replacing olivine) is surrounded by a corona of diopside and spinel (replacing plagioclase). Some hornblende occurs in the diopside corona.

B. Barton Garnet Mine, Gore Mountain, New York. Alteration at higher temperature than in A has allowed a larger sphere of diffusion for Mg, Fe, Ca. Some spinel occurs in the granular diopside corona, but most forms a 'clouding' of feldspar; note the accentuation of albite and pericline twin-lamellae. Many coronas have an extra garnet rim (*see* Fig. 37.1B); hornblende and biotite occur.

electrostatic balance in the feldspar as it became more albite-rich must have involved delicate diffusive controls. In the evolution of the assemblage, aluminium is the only element not to have diffused in substantial amount, and its low diffusivity may have controlled the way in which the reaction occurred. Certainly the 'clouding', with its innumerable tiny grains, has produced a situation of very high surface energy and in the figure it can be seen that subsequent grain boundary diffusion has begun to clear the plagioclase, the fine spinel coalescing into larger, more stable grains (11).

Fig. 33.8 illustrates polymetamorphism. A kyanite–biotite schist originally metamorphosed during the Caledonian regional metamorphism (486 million years) was remetamorphosed within a low-pressure aureole at the contact of a late-Caledonian 'newer' granite (406 million years). Two reactions are evident. One is the simple inversion of kyanite to its low-pressure polymorph, andalusite. The other is a more complex reaction of kyanite and biotite to produce cordierite, orthoclase, and spinel, which may be represented by the simplified equation:

$$\underset{\text{regional biotite}}{(OH)_8\,K_4\,Mg_4\,Fe_5\,Al_2\,(Si_{10}Al_6O_{40})} + \underset{\text{regional kyanite}}{7\,Al_2\,SiO_5} \rightarrow$$

$$\underset{\text{cordierite}}{Mg_2\,Al_4\,Si_5\,O_{18}} + \underset{\text{orthoclase}}{4KAlSi_3O_8} + \underset{\text{spinel}}{Fe_5\,Mg_2\,Al_{14}\,O_{28}} + 4H_2O$$

The spinel occupies the volume previously occupied by the reacted kyanite; whereas the orthoclase–cordierite intergrowth replaces biotite. From the simplified equation it follows that (Mg, Fe) diffused into kyanite, Si diffused from kyanite to biotite, and Al and K remained approximately immobile. An interesting sidelight on fig. 33.8A is the absence of the spinel rim where kyanite has inverted to andalusite: the free-energy differential between unstable kyanite and stable andalusite may thus be interpreted to have been part of the driving energy for the reaction; when by the inversion of kyanite to andalusite this differential disappeared, the reaction stopped.

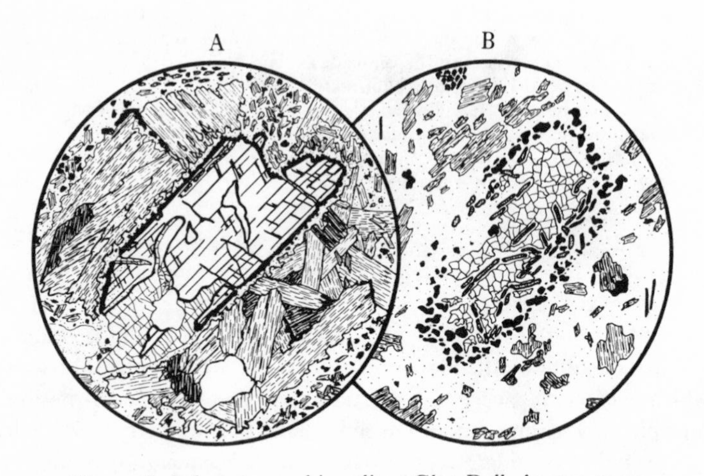

Fig. 33.8. Polymetamorphic pelites, Glen Doll, Angus; × 32.

A. Reaction rims of spinel, and of (cordierite + orthoclase) surrounding kyanite against biotite. Note the capricious inversion of kyanite to andalusite (lower relief) and the absence of the spinel rim in those parts.

B. A more recrystallised stage. All kyanite now inverted to andalusite; note the development of corundum within the pseudomorph due to loss of silica to the matrix.

Examples such as fig. 33.8 are, regrettably, rare. In most progressively metamorphosed rocks the reactant minerals are not in contact but are separated by other minerals, not chemically concerned in the reaction, which must at least in part dissolve to make way for product minerals and recrystallise to fill the voids left by dissolution of reactant minerals.

The resulting textures may thus not directly indicate the chemical reaction which was occurring, and their successful interpretation may depend on independent knowledge of that reaction – deduced, for example, in the field from the position of the rock in a progressive metamorphic sequence. A simple illustrative case involves the sillimanite isograd, representing in many low-pressure regions the polymorphic inversion of andalusite to sillimanite. The obvious mechanism for this reaction, involving minimal diffusion of Al and Si, would seem to be by nucleation of andalusite in sillimanite. Pseudomorphs are, however, uncommon. At temperatures close to those of the univariant equilibrium between the two phases, energy differentials are insufficient to promote nucleation of sillimanite within andalusite. More energetically favourable is the nucleation of sillimanite elsewhere and the transfer of material to these from dissolving andalusite.

An interesting mechanism for such transfer has been suggested by D. M. Carmichael. Textures of dissolving andalusite often suggest replacement by muscovite (fig. 33.9A), while sillimanite develops in pools of quartz embaying muscovite (fig. 33.9B). A reaction representing fig. 33.9A might be:

$$3 \text{ andalusite} + 3 \text{ quartz} + 2K^+ + 3H_2O \rightarrow 2 \text{ muscovite} + 2H^+$$

whereas the texture of fig. 33.9B might be represented by

$$2 \text{ muscovite} + 2H^+ \rightarrow 3 \text{ sillimanite} + 3 \text{ quartz} + 2K^+ + 3H_2O$$

The two reactions added together cancel out to the inversion reaction

$$3 \text{ andalusite} \rightarrow 3 \text{ sillimanite}$$

This mechanism, however, involves immobile aluminium and thus the aluminium which formerly constituted andalusite does not now constitute sillimanite. The driving force for the reaction was the free-energy differential between unstable andalusite and stable sillimanite.

We might regard the volumes containing the two disparate textures as 'domains' outside the diffusion range of aluminium but within the sphere of diffusion of potassium, silicon, water, and hydrogen. Other more complex coupled reactions may be suggested by biotite–sillimanite intergrowths and various replacement textures. Every such texture must, of course, be examined very carefully in the light of its own immediate environment and its geological position. Textures suggesting replacement of, for example, andalusite by muscovite at distances far removed from complementary textures must be regarded as belonging to a much enlarged sphere of diffusion for potassium (12).

The intergrowth of two minerals is thus often to be regarded as the

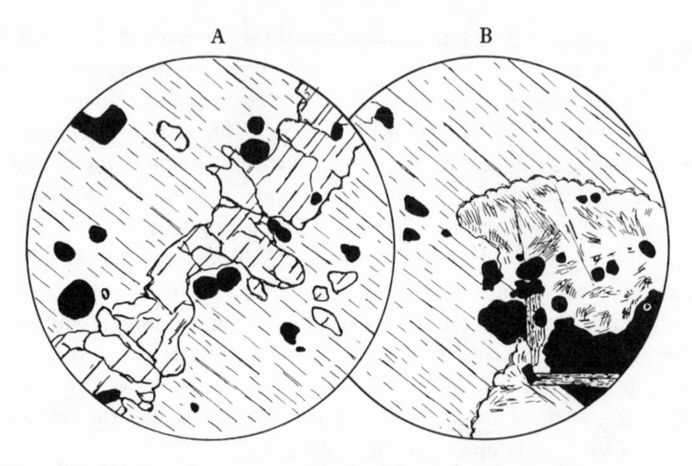

Fig. 33.9. Replacement textures, Buck of the Cabrach, Banffshire; × 50.
A. 'Outliers' of andalusite in optical continuity with the main crystal suggest the replacement of andalusite by muscovite. Opaque inclusions are megnetite.
B. Embayment of muscovite by quartz-sillimanite aggregate suggests the breakdown of muscovite to sillimanite + quartz. Opaque inclusions magnetite. A and B separated by 5 mm in actual rock.

product of some reaction, whether exsolution, decomposition, or replacement. Nonetheless, we can recognise a large group of intergrowths resulting from the passive incorporation of matrix materials into the body of a growing porphyroblast.

A familiar case is that of the *chiastolite* cross-form characteristically developed by andalusite but common enough in other minerals (e.g. staurolite, scapolite), which arises as a combination of passive incorporation and of cooperative recrystallisation. The chiastolite crystals grew substantially by accretion to the $\{110\}$ and $\{001\}$ forms. To achieve the elongate prism, the $\{001\}$ faces clearly grew at a greater rate than those of the $\{110\}$ form; the more rapidly growing faces incorporated matrix material which is seen as the cloudy central area of the chiastolite. On the more slowly growing $\{110\}$ faces, matrix material was able to dissolve and reprecipitate, partly as the clouding in the re-entrant angles and partly as long spindlets of quartz oriented normal to the growing face. Such spindlets are similar to the tubes of air trapped normal to the freezing interface in ice cubes, and their growth is energetically advantageous since the tubelet form involves both a minimum diffusion distance and necessitates no further nucleation (fig. 33.10).

The concentration of graphite at the borders and along the re-entrant angles of andalusite and other minerals was considered by Harker (13) to

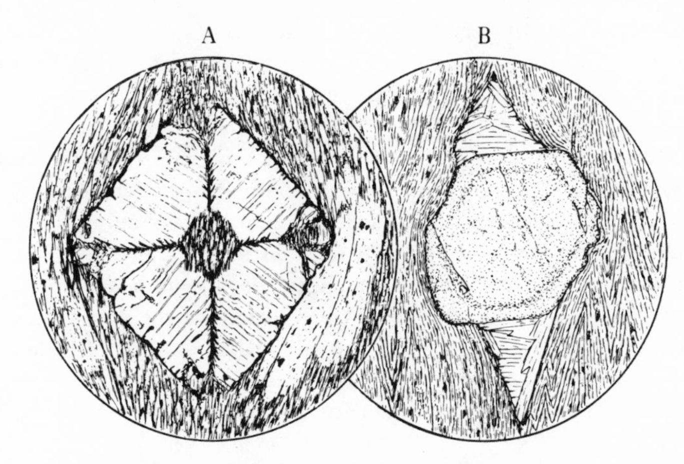

Fig. 33.10. Porphyroblastic textures; × 25.

A. Andalusite (chiastolite) cut normal to the prism; schist in aureole of Skiddaw granite. The pre-existing schistosity is preserved in the central sector which represents the rapid growth along [001]. Spindlets of quartz run parallel to the direction of growth of the {110} sectors. Mica concentrated at edges. Cumberland.

B. Fine-grained graphite-mica schistose matrix 'flattened' on garnet, Boharm, Banffshire. Note coarse-grained biotite in 'pressure shadow' and also growth stage revealed by density of inclusions in garnet.

represent the 'thrusting outward' of the growing crystal in the directions normal to {110}, and the consequent 'brushing aside' of foreign material. Whether or not the force of crystallisation of minerals is sufficient to enable them, in this graphic phrase, to 'thrust aside' the matrix of a rock under pressure is still in dispute. *Bowed* textures suggesting deformation of the matrix by a growing crystal may be interpreted in the opposite way, viz. deformation of the matrix about a pre-existing porphyroblast. The ultimate example of deformation of matrix about a pre-existing porphyroblast is seen in those slates which contain hard unyielding authigenic crystals of magnetite or pyrite. The matrix may be 'flattened', and quartz or chlorite dissolved from the region of higher pressure where the matrix is squeezed against the crystal is redeposited in the *pressure shadow* at the side. Where rotation of the crystal has occurred, growth of fibrous quartz (*pressure fringes*) normal to the faces may plot accurately the stages of detachment between crystal and matrix (fig. 33.11A) (14).

Many textures analogous to pressure shadows and fringes undoubtedly occur in the crystalline schists. The growth of a crystal, however, probably involves both flattening and thrusting to varying degrees.

Room for the growing crystal was made by the solution of more soluble phases (e.g. quartz); the concentration and disturbance of less soluble layered minerals then gives the impression of their having been 'thrust aside'.

Other porphyroblasts (occasionally of the same mineral in the same rock) adopt a seemingly less forceful mode of growth, replacing the matrix without disturbance of delicate sedimentary features such as bedding (fig. 39.2C) or fossils (fig. 36.9). A distinctive growth mechanism involves extension of the crystal as fingers along grain boundaries, the matrix minerals being incorporated or replaced as the porphyroblast enlarges (fig. 33.11B). A stage in the eventual elimination of included minerals by such a growing crystal is represented by the *poikiloblast*, in which the volumes of crystal and inclusions are comparable (fig. 33.11C).

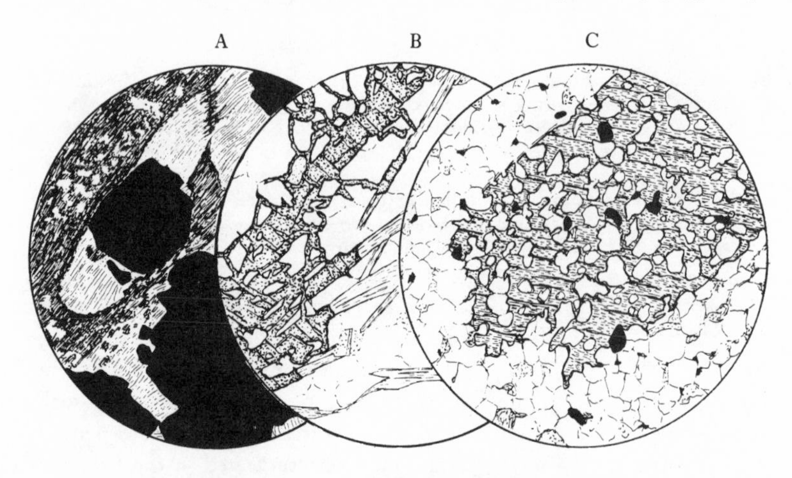

Fig. 33.11. Growth textures.

A. Pressure fringes in slate, Snowdon, Wales. Relative movement between pyrite crystals and matrix has produced voids which as they open have been progressively filled by fibrous quartz. × 40.

B. Skeletal growth of garnet along boundaries between quartz grains, and by replacement of muscovite. Pigeon Island, Newfoundland. × 27.

C. Portion of hornblende poikiloblast in quartz-calcite-muscovite schist, Glen Roy, Invernesshire. × 18.

Remanent inclusions so formed may thus preserve to a greater or lesser degree the fabric of a pre-existing matrix. Sigmoidal trails of inclusions ('helicitic structure') may be thus inherited; however, s-shaped arrangements may also reflect relative movement between porphyroblast and matrix during growth (fig. 33.12B). Examples of 'rotated', 'rolled' or 'snowball' porphyroblasts are found in many minerals, especially staurolite, andalusite, and garnet; spectacular instances may suggest angles of rotation of greater than 360° (15).

The evidence of textures suggesting the growth of minerals pre-, syn-, or post-deformation has been used in the erection of elaborate chronologies of metamorphic rock evolution. Although current understanding of crystal growth and of the metamorphic environment is perhaps insufficient to interpret all examples unequivocally, the chronological information to be gained from the correct interpretation of mineral textures is certainly considerable (16).

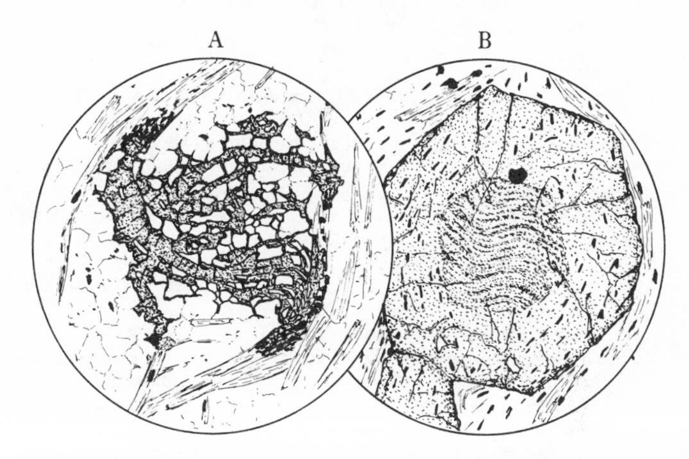

Fig. 33.12. Rotation textures in garnet, Pigeon Island, Newfoundland; × 3.5.

A. Garnet growth by infilling of tension cracks and infiltration of quartz grain boundaries.
B. Rotated core with sigmoidal graphite trails surrounded by post-tectonic outgrowth preserving ilmenite inclusions of replaced matrix.

Retrogressive metamorphism

Although mineral textures may preserve a record of the path of progressive metamorphism, the mineralogy of a rock as we see it generally represents adjustment to the highest temperatures undergone. Preservation of the climactic mineralogy is aided by the hysteresis introduced by the inevitable lag between temperature drop and consequent reaction. Furthermore high-grade rocks are impoverished in the volatiles expelled by prograde dehydration or decarbonation reactions, which now must diffuse back into the rock if readjustment to falling temperatures is to occur. *Retrogressive* reactions, although common, are thus seldom complete enough to obscure the high-temperature mineralogy.

References and notes

(1) Three excellent texts comprehensively treating the mineralogical aspects of metamorphism are:
Miyashiro, A., 1973. *Metamorphism and metamorphic belts.* London: Allen and Unwin.
Turner, F. J., 1968. *Metamorphic petrology : mineralogical and field aspects.* New York: McGraw-Hill.
Winkler, H. G. F., 1967. *Petrogenesis of metamorphic rocks.* Berlin: Springer (2nd ed.).
Textural aspects are comprehensively considered in Spry, A. 1969. *Metamorphic textures.* London: Pergamon.
As far as possible, metamorphic references will be confined to these.

(2) Spry, 1969, p. 247.

(3) Ramsay, J. G., 1967. *Folding and fracturing of rocks.* New York: McGraw-Hill, p. 177 ff.

(4) At moderate pressures (5–10 kilobars) the entropy difference ΔS of a devolatilisation reaction is very much greater than the volume change ΔV, so that the slope $\Delta P/\Delta T = \Delta S/\Delta V$ is virtually isothermal. This does not apply for solid–solid reactions (ΔV and ΔS comparable) or where mixed gases are involved. See Verhoogen, J., Turner, F. J., Weiss, L. E., Wahrhaftig, C. & Fyfe, W. S., 1970. *The Earth.* New York: Holt, Rinehart and Winston, p. 547 ff. Miyashiro, 1973, chapter 2.

(5) Spry, 1969, p. 29 ff.

(6) Verhoogen *et al.*, 1970, pp. 530–2.

(7) In three dimensions the mosaic texture is analogous to foam films and 120° triple junctions are only ideally seen when the plane of the section is normal to the grain boundary planes. Spry, 1969, p. 40. The term *granoblastic* (blastos = 'sprout') is also used for this type of texture. 'Blast' or 'blasto' as suffix or prefix is used for specifically metamorphic textures.

(8) Simplified equations are:

$$\underset{\text{kaolinite}}{8Al_2Si_2O_5(OH)_4} + \underset{\text{chlorite}}{2(Mg,Fe)_5Al\,(Al, Si_3)\,O_{10}\,(OH_8)} + \underset{\text{quartz}}{3SiO_2} \rightarrow$$
$$\underset{\text{cordierite}}{5\,(Mg, Fe)_2\,Al_3\,(Al, Si_5)\,O_{18}} + 20\,H_2O \qquad \text{(a)}$$

$$\underset{\text{kaolinite}}{Al_2Si_2O_5\,(OH)_4} \rightarrow \underset{\text{andalusite}}{Al_2SiO_5} + \underset{\text{quartz}}{SiO_2} + 2H_2O \qquad \text{(b)}$$

Where albite feldspar is present, these are metastable reactions, i.e. they do not represent the lowest energy state. The reactions stable relative to (b) would then be

$$\underset{\text{kaolinite}}{Al_2Si_2O_5(OH)_4} + \underset{\text{albite}}{NaAlSi_3O_8} \rightarrow$$
$$\underset{\text{paragonite}}{NaAl_2\,(Al, Si_3)\,O_{10}(OH)_2} + \underset{\text{quartz}}{2SiO_2} + 2H_2O \qquad \text{(c)}$$

and at a higher temperature

$$\underset{\text{paragonite}}{NaAl_2\,(Al, Si_3)O_{10}\,(OH)_2} + \underset{\text{quartz}}{SiO_2} \rightarrow \underset{\text{andalusite}}{Al_2SiO_5} + \underset{\text{albite}}{NaAlSi_3O_8} + H_2O$$

However, at low pressures the equilibrium temperature of (c) is so low, and the reactions consequently so slow, that the paragonite step is missed out and kaolinite dehydrates directly by reaction (a).

(9) The presence of graphite may introduce methane and carbon monoxides: pyrite and pyrrhotite give oxides of sulphur. Miyashiro, 1973, pp. 53–4. Fluid inclusion analyses suggest chlorine also to be significant.

(10) A texture analogous to *pearlite* of metallurgy in which austenite (Fe-C alloy) has undergone eutectoid decomposition to ferrite (bodycentred Fe) and cementite (Fe_3C). Reed-Hill, R. E., 1964. *Physical metallurgy principles*. Princeton: Van Nostrand, p. 444 ff.

(11) The precise interpretation of the coronite structure depends on the location of the original olivine–plagioclase boundary. In some zones the direction of diffusion may also be ambiguous – hypersthene may form from olivine *either* by accession of Si *or* by loss of Mg. The latter, however, involves the smaller change in volume. The variety of corona types may be seen in Whitney, P. R. & McLelland, J. M., 1973. Origin of coronas in metagabbros of the Adirondack Mts., N.Y. *Contr. Min. Petrol.* **39**, 81–98. Griffin, W. L. & Heier, K. S., 1973. Petrological implications of some corona structures. *Lithos* **6**, 315–35.

(12) See metasomatism. Carmichael, D. M., 1969. On the mechanism of prograde metamorphic reactions in quartz-bearing pelitic rocks. *Contr. Mineral. Petrol.* **20**, 244–67.

(13) Harker, A., 1939. *Metamorphism*. London: Methuen, pp. 42–3. Spry, 1969, p. 176.

(14) Spry, 1969, p. 240 ff.

(15) Rosenfeld, J. L., 1970. Rotated garnets in metamorphic rocks. *Bull. Geol. Soc. America Special Paper* **129**. Spry, 1969, p. 253. *op. cit.*, p. 253.

(16) A variety of textures suggesting pre-, syn-, and post-tectonic crystallisation is illustrated by Spry, 1969, figs. 58–61.

34
Metasomatism and melting

Metasomatism

Where the magnitude of the sphere of diffusion is less than that of a thin section or of a hand specimen, the metamorphic reaction is considered to have been *isochemical*, involving no essential change in rock chemistry. In many rocks the assumption of isochemical metamorphism seems justified, although we may find developed on the centimetre scale *segregations* or *metamorphic differentiation* into contrasting layers, e.g. hornblende- and feldspar-rich layers in a gneiss. Many such inhomogeneities result purely from surface energy difference – thus quartz segregations tend to form, especially from highly strained aggregates, since quartz disseminated in a fine-grained mica-schist has a higher surface free energy than the same volume concentrated in a coarse-grained, monomineralic segregation.

Migration on a greater scale is referred to as *metasomatism*; the distinction is somewhat arbitrary and the concept is certainly in no way different from that of the diffusive interchanges discussed earlier. The enstatite zone of fig. 33.7 could, for example, be regarded as a mini-metasomatic system since its composition is dependent on the diffusive interaction of the two incompatible phases olivine and plagioclase which we may define as lying outside the system. In the conventional usage of the term, however, a metasomatic system is several orders of magnitude greater than such a reaction rim and the source of the chemical gradients is therefore less obvious. Furthermore the magnitude of mass transfer operative in some metasomatic systems requires postulation of movement of metasomatic fluids. We may thus regard metasomatism as the chemical gain or loss suffered by a rock in its attempt to reach equilibrium with large quantities of fluid flowing through it (1).

The main problem with metasomatism is the initial one of deciding whether or not chemical change has in fact occurred. During the progressive metamorphism of clay-rich sediments, for example, large quantities of water are expelled and migrate from higher grade to lower grade down the temperature gradient; it is certain that in many cases this migration is accompanied by movement of other constituents, especially of more soluble ions such as alkalis. Rocks resulting from equilibration with such fluids, however, differ from the isochemical assemblages only in the proportions of mica to feldspar or to aluminium silicates; detailed and sophisticated analytical work is then necessary to establish the nature of chemical change.

The best criterion of metasomatism is the preservation in a metasomatic rock of a texture clearly inherited from a rock of originally different composition (figs. 34.1A, B). Such textures, preserving the original form and size of pre-existing minerals, may also be a guide to the relative profit and loss during the metasomatic transaction. Thus the common observation that minerals surrounding pseudomorphs of serpentine after olivine are undistorted has suggested that the replacement took place with negligible change in volume; the metasomatic reaction must then have involved loss of Mg and Si as well as gain of water, as in the constant volume (for solids) equation (2)

$$5Mg_2SiO_4 + 4H_2O + 6H^+ \rightarrow$$
olivine: 218 cm³
$$2Mg_3Si_2O_5(OH)_4 + 4Mg^{2+} + Si^{4+} + 6OH^-$$
serpentine: 220 cm³

The most readily identified and distinctive examples of metasomatism come from the environment of contact metamorphism. Here the energy source is clearly recognised as the intrusive body, and chemical changes are often zoned along the intrusive contact. Furthermore the metasomatic episode often clearly postdates the initial intrusion and contact metamorphism. Examples already given (p. 32, under the heading of pneumatolysis) are of the Cornish tourmalinisation and greisen-formation which affects indifferently both country rock and the marginal parts of the intrusion (fig. 3.6A); the extensive kaolinisation of granite which produced the Cornish china-clay deposits is often also considered to have been a late-magmatic hydrothermal event. The formation of kaolinite from, say, orthoclase requires a high ratio of hydrogen to potassium in the metasomatic fluid, as suggested by the simple reaction

$$2KAlSi_3O_8 + 11H^+ \rightarrow Al_2Si_2O_5(OH)_4 + 2K^+ + 4Si^{4+} + 7(OH)^-$$
orthoclase — in soln. — kaolinite — in soln.

Clearly with lower values of $^aH^+$: ($^aK^aSi^aOH$) the product might be not kaolinite, but muscovite; such alkali equilibria, grouped under the heading of ***alkali metasomatism***, are readily reproduced experimentally and have importance in all aspects of metamorphic study including wall rock alteration associated with ore deposition (3).

Such alkali exchanges are readily conceived in terms of reaction between crystallising magma and surrounding country rock by the agency of migrating fluids. Many contact metasomatic deposits, however, display the principle of polarity: that is, they are enriched in chemical components strikingly deficient in the supposed progenitor igneous body. Outstanding as such metasomatic products are the iron- and magnesia-enriched *skarns*, developed in calcareous rocks at many granite contacts. Here the CaFe garnet andradite, with axinite, epidote, and hedenbergite, form common associates with sulphide minerals (sphalerite, chalcopyrite) and oxides (magnetite, hematite) (4) (fig. 34.1C). Fluorine- and chlorine-bearing minerals (humite, chondrodite, scapolite) are common skarn associates.

Albite-rich *adinoles*, often associated with tourmaline, are often developed at dolerite contacts. Such rocks may show beautiful pseudomorphism of country rock slates and even of fossils contained in them (5). Another sodium-rich type is *fenite*, an orthoclase–arfvedsonite–aegyrine rock characteristically developed at carbonatite contacts but also found in the country rock surrounding many alkaline complexes and occasionally of granites. The fenitisation of granitic country rock of the Oldonyo Dili carbonatite in Tanzania involved the conversion of original microline to orthoclase and of biotite to soda-amphibole; more altered types contain greater quantities of soda-pyroxene, often as veins traversing the rock (6). As in the skarns, the presence here of highly ferriferous minerals indicates the oxidising character of the environment (fig. 34.1B).

Rodingites, lime-rich, grossularite or prehnite-bearing rocks, develop at the contacts of ultramafic intrusions from calcium liberated by the serpentinisation of periodotite pyroxene. Rodingite as originally defined from the contact of the Dun Mountain intrusive of Nelson, New Zealand, is a metagabbro with hydrogrossularite replacing the plagioclase; but other country rocks (e.g. slates) show comparable alteration. Rodingites superimposed on contact aureoles may in fact conceal evidence of high-temperature emplacement of dunitic and peridotite intrusives (7).

Even in the contact environment, however, identification of metasomatic products may be difficult. Hornfels consisting predominantly of anthophyllite and cordierite were formerly considered to result from contact metasomatic alteration (either by addition of Mg, Fe or by loss

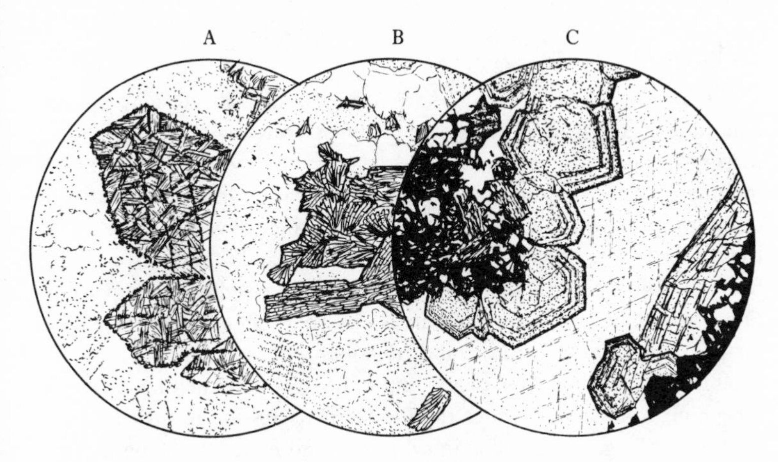

Fig. 34.1. Metasomatic rocks; × 40.

A. Vogesite dyke, metasomatised at potassic granite contact, Catacol, Arran. Euhedral hornblendes replaced by biotite, sphene granules outlining original cleavages.
B. Fenite, Holla near Fen, Oslo, Norway. Original granite gneiss with biotite replaced pseudomorphously by an aegyrine-acmite pyroxene.
C. Skarn at granite contact, Tregullan, Cornwall. Chalcopyrite rimmed by andradite garnet and axinite set in a matrix of calcite. Zones of birefringent garnet are outlined by fine-grained inclusions. Only carefully established field relationships can identify this rock as a metasomatic product.

of Ca, K, Na) of dacitic or basaltic extrusives, mainly on the ground that the curious chemistry could not be the product of any pre-metamorphic weathering process. Recent studies have, however, shown that basalts in the suboceanic environment may undergo low-grade metasomatism to chlorite–quartz aggregates (cf. p. 424) which have precisely the Mg, Al-rich, alkali-poor chemistry required to give, on subsequent thermal metamorphism, a rock with substantial content of cordierite and anthophyllite. The likelihood that any arbitrary assemblage of chlorite and quartz would be in the precise proportions required to produce *only* cordierite and anthophyllite is small; invariably therefore we see the associates quartz or, where chlorite was in excess, spinel (8).

An interesting consequence of metasomatism as the adjustment of minerals to an externally imposed fluid of arbitrary constitution is the production of rather simple mineralogies (9). Thus a metasomatic vein cutting a polymineralic rock will commonly consist of one mineral only (fig. 39.1C).

Melting

At high temperatures in metamorphism a liquid silicate may appear, and here the processes of metamorphism and magmatism begin to merge. Metamorphic rocks, however, rarely show evidence of wholesale melting – rather a liquid of low melting point coexists with a greater or lesser proportion of solid mineral phases, which give cohesion to the rock and allow preservation of metasedimentary and other compositional features. Distortion of the solid framework may, however, expel portions of the liquid: 'rheomorphic' veins so produced have been described from both contact and regional environments. From the geological viewpoint, the most important agent in melting is water. The high solubility of water in silicate melts results in a decrease in volume as the gas enters the liquid phase, and hence the slope

$$\triangle T/\triangle P = \triangle V/\triangle S$$

of the melting curve is negative, the increase in pressure decreasing the melting temperature. Albite melts in the atmosphere at 1115 °C; under anhydrous pressure of 3 kilobars this is raised to 1193 °C; but under a water pressure of 3 kilobars the melting point is reduced to 825 °C, the saturated melt containing some 8 per cent by weight of water. It is thus due to water that melting processes are brought within the temperature range of metamorphism within the earth's crust; however, lack of water sufficient to saturate the liquid limits the degree of melting in most metamorphic environments (10).

Silicate liquids with the lowest melting temperatures are, in the broad sense, granitic, corresponding in composition to mixtures of quartz, alkali feldspar, and sodic plagioclase, the exact proportion being dependent upon the water pressure (11). At any temperature and water pressure, the greatest proportion of liquid will be developed in rocks with the highest proportion (actual or normative) of these three constituents, and so evidence of fusion in metamorphic rocks is most commonly confined to arkoses, sandstones, greywackes and pelites.

The curve produced by plotting in terms of P and T the experimentally determined points of first melting has a form similar to that of the univariant melting curve, in, say, the system albite-water, and is known as a *minimum melting curve*. Minimum melting curves for various quartzofeldspathic rocks are not, in fact, very different; at 2 kilobars water pressure, for example, the range of minimum melting temperatures from granite to shale is some 670–750 °C. The composition of the melt and the proportion of liquid to solid phase existing at the minimum temperature will, however, vary widely with rock and mineral composition.

The clearest case of melting in metamorphism is found in the buchites, products of local fusion of metasediments at the contacts of high-level basic intrusives and, occasionally, of burning coal seams; in these, rapid cooling (quenching) has preserved the liquid as a glass.

The fused arkose at the contacts of Tertiary periodotite sills on Soay (Hebrides) are typical. The original sediment is predominantly quartz, with lesser amounts of alkali feldspar, plagioclase, and detrital sericite. Liquid initially formed at the contacts between quartz and the alkali bearing minerals; rounded relicts of the refractory residue, generally quartz, are abundantly found (see fig. 34.2A). The reluctance of quartz/tridymite inversion is well illustrated by the confinement of tridymite to the edges of, and cracks within, the relict quartz grains, and it is possible that in these cases alkali from the melt promoted metastable tridymite formation. Certainly the presence of tridymite (and, more rarely, cristobalite) in buchites cannot be taken as unequivocal evidence

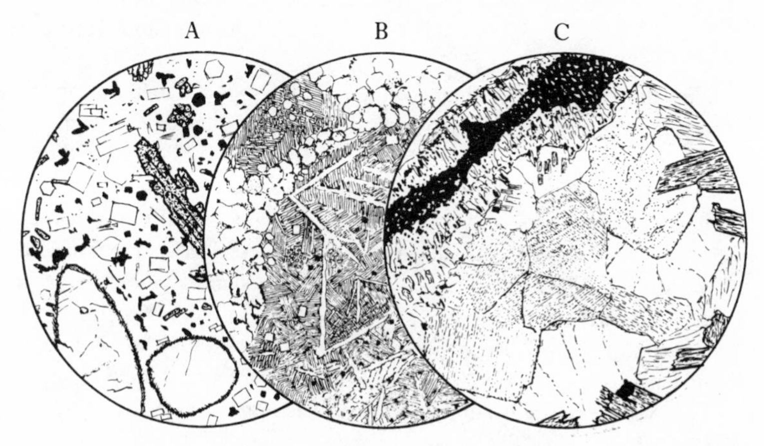

Fig. 34.2. Melted rocks; × 40.

A. Buchite (here a fused ashy Devonian sandstone) at contact of Tertiary dolerite, Tieveragh, Co. Antrim. Residual quartz grains (with marginal zone of inversion to tridymite) in a glassy matrix with euhedral crystals of magnetite, enstatite and cordierite.

B. Fused Torridonian sandstone, Tertiary gabbro contact, Isle of Rum. Residual sandstone fragments in granophyric matrix. Tridymite phenocrysts now inverted to quartz.

C. Fused micaceous schist at contact of Caledonian diorite, Glen Clova, Angus. The refractory residue, here silica-deficient, is seen as lenses of almost opaque spinel surrounded by cordierite-spinel symplectite (melanosome). The melted portion has crystallised to a quartz-feldspar aggregate (= leucosome).

that the inversion temperatures (at 1 atmosphere, 870 °C and 1470 °C respectively) had been exceeded. Despite such suggestions of disequilibrium, however, the glasses are often surprisingly homogeneous, reflecting not only the narrow range of melt compositions at temperatures little above the minimum melting point, but also the high diffusivity of alkalis in the presence of a silicate liquid.

Where cooling has been slower the liquid has crystallised, typically to an aggregate in the intimate granophyric intergrowths suggestive of simultaneous crystallisation (fig. 34.2B). Tridymite has then crystallised as coarse prisms subsequently inverted to quartz, the lath-like forms being composed of quartz grains in random orientation. At deeper levels the fused portion of shales and sandstones crystallises to aggregates texturally indistinguishable from plutonic igneous rocks. Good examples are found at the contacts of the 'newer' Caledonian gabbro and granite intrusives of N.E. Scotland, where liquids from the partial fusion of argillaceous metasediment have crystallised to quartzofeldspathic aggregates, often with garnet or cordierite. The more refractory remainder from the fusion of mica-rich rock is silica-deficient, and occurs as clots or lenses of spinel, basic feldspar, sillimanite and cordierite (fig. 34.2C).

With the even more gradual cooling found in regional metamorphism, annealing and recrystallisation may destroy the characteristic igneous textures, rendering recognition of formerly fused rocks very difficult. Characteristic of argillites and psammites metamorphosed at high grades is the presence of quartzofeldspathic veins, varying in thickness from the centimetre to the metre order, of coarser grain size than that of the surrounding rock. The material between such *leucosomes* is often enriched in mafic and aluminous minerals, especially biotite, sillimanite and garnet (melanosome). Such *migmatitic* rocks may arise from a homogeneous metasediment by the segregation of a low-melting liquid fraction whose crystallisation on cooling has left the leucosome (12).

References and notes

(1) A comprehensive account of metasomatic changes, with a chemical classification of metasomatism, is given by Turner, F. J. & Verhoogen, J., 1960. *Igneous and metamorphic petrology*. New York: McGraw-Hill (2nd ed.), pp. 561–86.

(2) Turner & Verhoogen, 1960, p. 318. Serpentinisation of olivine without removal of Mg and Si involves almost 100 per cent increase in volume. Where serpentines are foliated, however, constant volume alteration cannot be proved, and volume increase by serpentinisation has been invoked as a tectonic mechanism.

(3) Meyer, C. & Henley, J. J., 1967. Wall rock alteration, *in* Barnes, H. L., ed., *Geochemistry of hydrothermal ore deposits*. New York: Holt, Rinehart and Winston, pp. 166–235.

(4) *Skarn* is an old Swedish mining term, and many economic deposits are associated with this metasomatic environment. Cf. Park, C. F. & MacDiarmid, R. A., 1970. *Ore deposits*. San Francisco: Freeman, chapter 11.

(5) Agrell, S. O., 1939. The adinoles of Dinas Head, Cornwall. *Min. Mag.* **25**, 305–36.

(6) McKie, D., 1966. Fenitisation *in* Tuttle, O. F. & Gittins, J. *Carbonatites*, New York: Wiley.

(7) Challis, G. A., 1965. High-temperature contact metamorphism at the Red Hills ultramafic intrusion. *J. Petrol.* **6**, 395–419.

(8) Vallance, T. G., 1967. Mafic rock alteration and isochemical development of cordierite anthophyllite rocks. *J. Petrol.* **8**, 84–96. Chinner, G. A. & Fox, J. S., 1974. The origin of cordierite–anthophyllite rocks in the Lands End Aureole. *Geol. Mag.* **111**, 1–12.

(9) At fixed pressure and temperature, an equilibrium constant K exists for the serpentinisation reaction of page 365, olivine and serpentine can only coexist stably when the gas phase is of that unique constitution

$$K = \frac{({}^{a}H_2O)_4\,({}^{a}H)^6}{({}^{a}Mg)^4\,({}^{a}Si)\,({}^{a}OH)^6}$$

The likelihood that any arbitrarily imposed fluid phase would be of precisely that composition is negligible; i.e. in an equilibrium metasomatic system either olivine or serpentine will be stable, but not both. In practice this ideal situation is complicated by kinetic factors, especially the limited permeability of rocks under pressure which allows the solid assemblages partly to buffer the fluid. See Thompson, J. B., 1959. Local equilibrium in metasomatic processes, *in* Abelson, P. H., ed., *Researches in Geochemistry*. New York: Wiley, pp. 427–57. Barton, P. & Skinner, B. J., 1967, *in* Barnes, H., ed., *Origin of hydrothermal ore-deposits*, pp. 243–6.

(10) However water-saturated melts will tend to freeze when the pressure is released. Most anatectic granites which stay liquid until emplacement at high level were probably derived by dry melting at higher temperatures, i.e. deeper, often subcrustal, levels. See Cann, J. R., 1970. Upward movement of magma. *Geol. Mag.* **107**, 335–40.

(11) Tuttle, O. F. & Bowen, N. L., 1958. The origin of granite in the light of experimental studies. *Geol. Soc. America Memoir*.

(12) The leucosome may closely resemble pre-metamorphic structures or products of sub-solidus segregation; sophisticated chemical analysis may be necessary to decide. A thorough account of migmatites is given by Mehnert, K. R., 1968. *Migmatites*. Amsterdam: Elsevier.

35

The facies classification of metamorphic rocks

In attempting a useful yet unencumbered classification of recrystallised rocks, the metamorphic petrologist encounters conflicts familiar in most petrographic nomenclature; however, chemical composition, being usually a product of a pre-metamorphic process, does not assume the importance it has in igneous or sedimentary classification. Rocks may be referred to as 'calcareous' or 'argillaceous' but since the chemical composition is indicated broadly by the minerals it contains, the simplest, non-genetic, terminology merely states the salient mineralogy of the rock, e.g. cordierite–anthophyllite rock. Names descriptive of the appearance or texture – e.g. slate, phyllite, schist, gneiss, hornfels, mentioned in chapter 33 – are commonly suffixed to the mineralogy but these may develop too genetic a connotation; thus granulite, originally a granular ('small-grained') rock, now denotes high-grade rocks regardless of texture and cannot be used unambiguously as a descriptive term.

Many attempts have been made to classify rocks in terms of the pressures and temperatures of recrystallisation (1). Implicit in all these is the assumption that metamorphic rocks as we see them represent a recognisable approach to equilibrium under the physical conditions prevailing when quenching occurred; in the polymetamorphic rocks shown in figs 33.7 and 33.8, for example, we generally assume that the product minerals formed the most stable association under the new conditions. In assessing the equilibrium state of a system the Gibbs' phase rule (2) is often used. The number (degrees of freedom or variance, ν) of externally imposed variables which may be changed in a system of C chemical components and ϕ phases without losing or gaining a phase is given by the relation:

$$\phi + \nu = C + 2$$

Since for practical purposes at this stage the only external variables we must consider are those of pressures and temperature, the usage of the phase rule is well illustrated by a P/T plot such as that of the system Al_2SiO_5 (fig. 35.1). Only one chemical species (the combination Al_2SiO_5)

is necessary to specify the composition of the polymorphs andalusite, sillimanite, and kyanite. The system thus has one component, and three phases ($\phi = 1 + 2-3$) may coexist at an invariant point. Two phases may coexist along each of three univariant lines which emanate from the invariant point ($\phi = 1 + 2 - 2$), and which separate three divariant fields in each of which one phase only is stable.

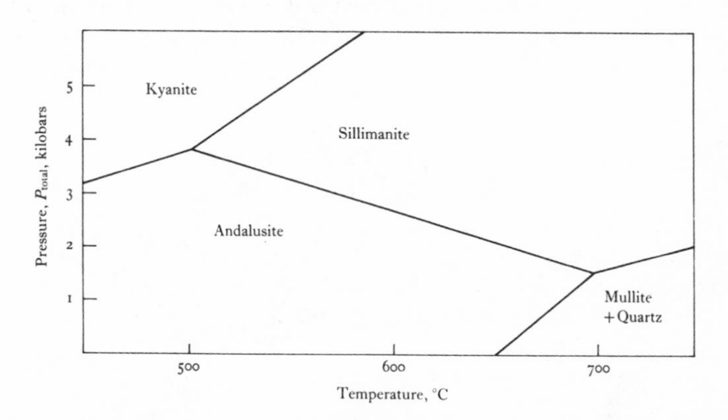

Fig. 35.1.
Stability fields of the Al_2SiO_5 polymorphs in terms of the variables pressure and temperature. The precise positions of the invariant points are in doubt.

Possible difficulties in the use of the phase rule are foreshadowed in the low-pressure, high-temperature portion of the diagram, where mullite (approximately $Al_6Si_2O_{13}$) appears. The species Al_2SiO_5 is no longer adequate to specify the composition of all phases present; the combination of two species is now required and the system therefore has two components. Whether we identify the components as $Al_6Si_2O_{13}$ and SiO_2, or as Al_2O_3 and SiO_2 is immaterial. The divariant space is now occupied by two phases, mullite + quartz ($\phi = 2 - 2 + 2$); along the univariant lines three phases coexist; and at the invariant point the four phases mullite, quartz, andalusite, sillimanite coexist uniquely ($\phi = 2 - 4 + 2$).

Along the univariant lines, heat put into or removed from the system will be used in converting one divariant assemblage into the other, and hence one may specify the lines by the univariant reactions which occur across them, e.g.

$$\text{kyanite} \rightleftharpoons \text{andalusite}$$
$$\text{mullite} + \text{quartz} \rightleftharpoons \text{sillimanite}$$

This provides the surest way of identifying a univariant assemblage

(3). If a chemical reaction can be written between the phases present, an assemblage has a variance of one or less.

A more complex example is that of the three-component system MgO-SiO_2-H_2O, illustrated in fig. 35.2. Along the univariant lines, four phases coexist. The chemical combinations in a three-component system are readily represented on a triangular diagram and on this the chemical reaction specific to each univariant line on the P/T plot may be identified by crossed tielines between pairs of minerals. The divariant regions which the univariant lines separate are characterised by triplet mineral assemblages.

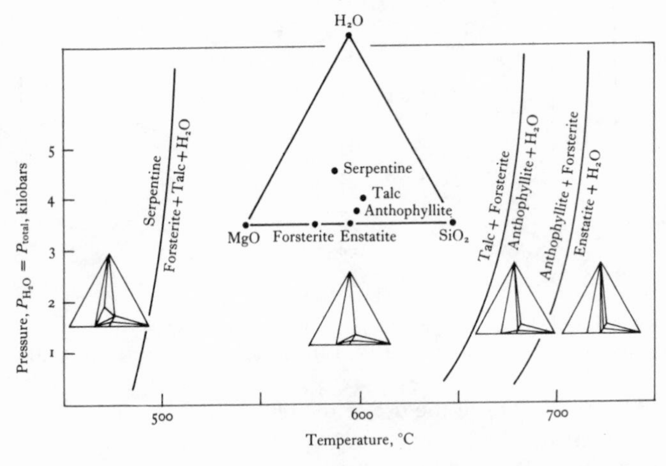

Fig. 35.2.

Univariant reactions in the system MgO–SiO_2–H_2O, after several authors. Inset triangle shows the phases concerned in terms of the chemical components. For simplicity only reactions affecting the serpentine composition are shown. Note the steep slope of these dehydration reactions in contrast to the lower slopes of the solid–solid reactions in fig. 35.7.

The natural occurrence of low-variance assemblages is limited by the fact that temperature and pressure, though often generally dependent upon depth in the earth's crust, are semi-independent variables and are most unlikely to correspond precisely with any invariant point or univariant curve over large volumes of rock. The consequent requirement of two degrees of freedom (divariance) for a regionally developed assemblage reduces the phase rule to $\phi \leqslant C$, a relation demonstrated in 1911 by V. M. Goldschmidt and often known as the mineralogical phase rule. Since, however, this is not a phase rule independent of Gibbs', it might be better to call it Goldschmidt's criterion.

The Goldschmidt criterion is not concerned with the *local* development of univariant or invariant assemblages. Univariant assemblages will outcrop locally along the line of contiguity of two divariant regions, and, representing the intersection of the erosion surface with a plane of univariant P/T conditions, define the *isograd*.

The identification of such univariant lines of outcrop is clearly of great importance in determining the nature of 'fossil' P/T gradients; and an ideal quantitative classification might be based on sets, closely spaced in terms of P and T, of laboratory-determined univariant reactions (the *petrogenetic grid*). Experimental work in the last two decades has brought such a scheme within prospect of realisation. Currently, however, the practical disadvantage of the petrogenetic grid lies in the complex solid solution series which constitute most common metamorphic minerals – particularly substitution of Mg for Fe (and Mn). Reactions involving such minerals are not univariant but divariant, and proceed over a temperature range, involving a continuous modification in the composition of the mineral. Such reactions have been termed (4) *continuous reactions* in contra-distinction to the discontinuous univariant reactions between minerals showing no solid solution. If we ignore the small amount of sodium involved, the reaction in argillaceous rocks of *muscovite* and quartz to give sillimanite, potassium feldspar, and water (p. 348) is a classic illustration of a discontinuous reaction. Iron–magnesium diadochy causes the equivalent reaction between *biotite* and quartz to be continuous. At 250 bars water vapour pressure, the iron biotite, annite, reacts with quartz at 650 °C (5):

$$\underset{\text{annite}}{KFe_3(AlSi_3)O_{10}(OH)_2} + \underset{\text{quartz}}{SiO_2} + \tfrac{1}{3}(O) =$$
$$\underset{\text{orthoclase}}{KAlSi_3O_8} + \underset{\text{fayalite}}{Fe_2SiO_4} + \underset{\text{magnetite}}{\tfrac{1}{3}Fe_3O_4} + H_2O$$

Phlogopite, however, does not react with quartz until 800 °C has been attained:

$$\underset{\text{phlogopite}}{KMg_3(AlSi_3)O_{10}(OH)_2} + \underset{\text{quartz}}{3SiO_2} = \underset{\text{orthoclase}}{KAlSi_3O_8} + \underset{\text{enstatite}}{3MgSiO_3} + H_2O$$

Throughout this temperature interval of 150°, therefore, biotites of intermediate composition are continuously reacting, with the production of anhydrous reactants and smaller amounts of a more magnesian biotite. The temperature at which biotite disappears thus depends rather intimately upon the composition of the rock; the erection and use of a petrogenetic grid on such a basis would require so detailed a knowledge of the mineral chemistry as to render it impracticable for common use by the geologist.

On the basis of the Goldschmidt criterion we may, however, recognise some divariant assemblages characterised by their widespread or regional occurrence, and relate these to each other in terms of pressure and temperature to give a broad genetic classification. By grouping together associations from a variety of rock compositions we may produce a syndrome of assemblages characteristic of a particular divariant interval of P and T. This, the *metamorphic facies* classification, derives from that introduced by P. Eskola soon after the First World War.

Although each metamorphic facies (or strictly mineral facies, since we could use the system also for igneous rocks) is a syndrome of assemblages and is named from one of the characteristic mineral associations, the name can also be used to denote the P/T range over which the syndrome is stable; expressions such as 'the rock crystallised within the amphibolite facies' are accepted by most petrologists.

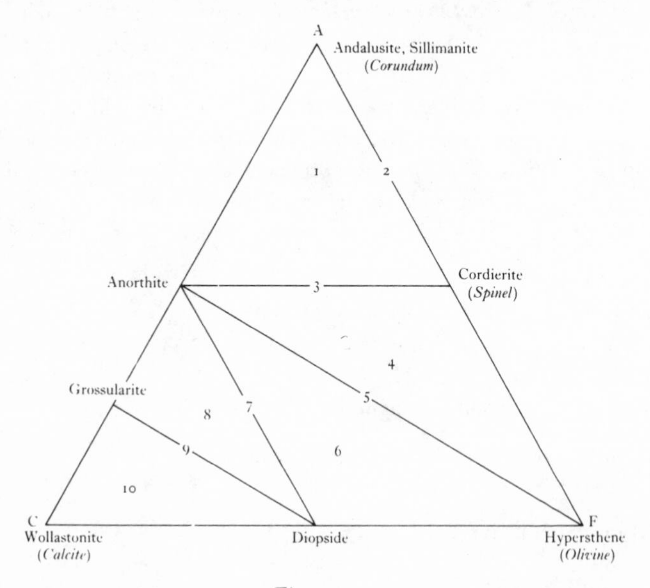

Fig. 35.3.

ACF (A = $Al_2O_3 - (Na_2O + K_2O)$; C = CaO; F = (FeO + MgO + MnO) plot for Eskola's pyroxene hornfels facies. The numbers correspond to the hornfels classes originally recognised by Goldschmidt in the Oslo aureoles. Where SiO_2 is present in excess, quartz is an additional phase; where SiO_2 is deficient, the phases in parenthesis may appear in the appropriate class (*see* figs. 35.4 and 35.5).

The classic illustration of a metamorphic facies is provided by Goldschmidt's description of metasediments in the inner portions of the contact aureoles surrounding the Permian intrusions of the Oslo igneous province. Despite a wide diversity of chemical composition, the assemblages found were relatively simple combinations of a restricted number of minerals. The ACF diagram (6), in which rock and mineral chemistry are projected onto the plane Al_2O_3 – CaO – (Mg, Fe, Mn)O, is commonly used to illustrate this. All assemblages recorded by Goldschmidt correspond to one or other of the three-phase triangles or to a two-phase line in the ACF diagram of fig. 35.3. Extra phases could be accounted for by extra chemical components not represented on the diagram, such as albite, Na_2O; orthoclase, K_2O; apatite, P_2O_5. The presence of quartz represents SiO_2; the presence of spinel, corundum, or olivine represents SiO_2 in deficiency. Univariant assemblages, identified by writing reactions between coexisting minerals (i.e. by crossed tielines on fig. 35.3 such as cordierite-anorthite-diopside-hypersthene), are absent.

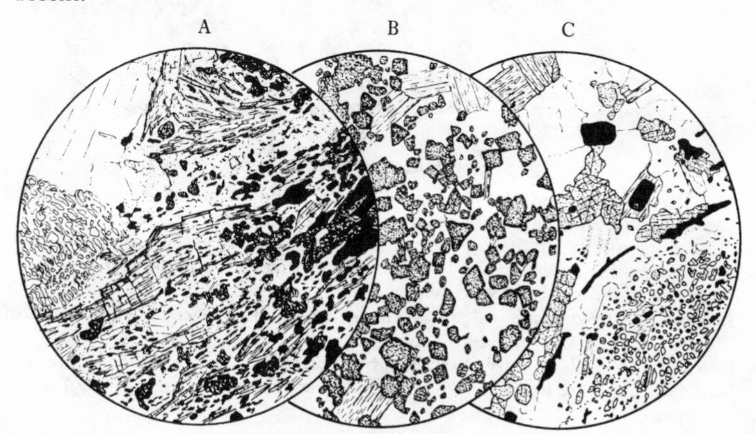

Fig. 35.4. Hornfels, Lochnagar aureole; × 40.

A. Sillimanite-cordierite-spinel-andesine-magnetite-apatite hornfels (pelitic): the silica deficient equivalent of Goldschmidt class 2. The plagioclase shows antiperthitic exsolution of orthoclase; orthoclase, as discrete crystals showing cleavage, also occurs. The preferred orientation of sillimanite fibres is inherited from a precursor mica schist.

B. Spinel-cordierite-andesine-orthoclase hornfels: the silica-deficient equivalent of class 3.

C. Hypersthene-andesine-cordierite-quartz hornfels: the silica-excessive equivalent of class 4. The fine-grained part, containing a green hercynite spinel and no quartz, is the silica-deficient equivalent. Sphere of diffusion for silicon here is of the order of 0.3 mm radius.

All three rocks contain biotite, indicating incomplete dehydration.

The syndrome of divariant assemblages shown in figs. 35.4 and 35.5 is found in many contact aureoles and is known as the pyroxene hornfels facies from the assemblage hypersthene-diopside-plagioclase developed in metabasic hornfels. Figs. 35.4 and 35.5 illustrate six assemblages of this facies as developed in the aureole of the late-Caledonian calc-alkaline Lochnagar complex of Aberdeen and Angus.

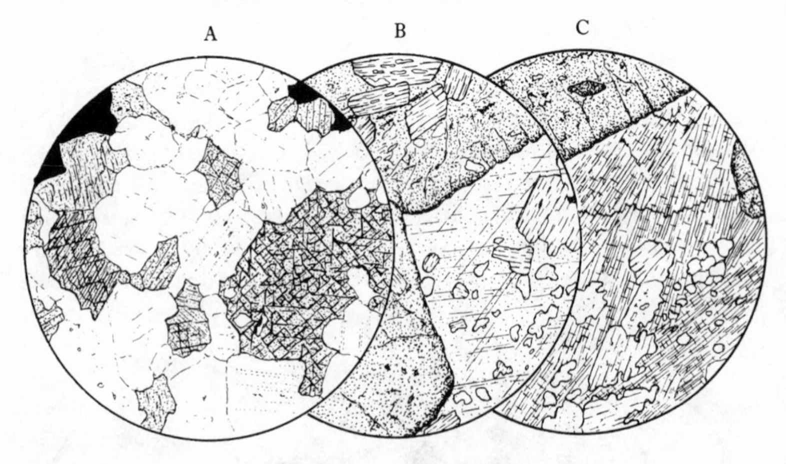

Fig. 35.5. Hornfels, Lochnagar aureole; × 40.

A. Two-pyroxene andesine-magnetite hornfels: the meta-basic equivalent from which the pyroxene hornfels facies is named (class 6). (001) sections of both diopside and hypersthene, showing {110} and {100} cleavages, are illustrated; the diopside crystal, to the left, shows the 'diallage' structure of magnetite crystals outlining the {100} cleavage.

B. Grossularite-diopside-calcite hornfels: the silica-deficient equivalent of class 10, a metamorphosed marl.

C. Grossularite-diopside-wollastonite hornfels: the silica-saturated equivalent of class 10. Note lozenge-shaped crystal of sphene (representing TiO_2) in the garnet.

This simplified treatment of rocks in terms of three essential chemical components works quite well for the low-pressure, high-temperature products of a thermal aureole, in which the ferro-magnesian minerals show virtually complete Mg-Fe diadochy and sodium is always accommodated as the albite molecule of plagioclase. Although less satisfactory in depicting many other facies, the ACF diagram does give a good general idea of the relation between isofacial mineral assemblage and chemical composition. Thus *argillaceous* or *pelitic* rocks, originally rich in clay minerals, occupy the region close to the 'A' apex of the diagram; *calcareous* rocks, originally calcite rich, the 'C' region; and metabasic rocks of basaltic, gabbroic, or tuffaceous origin, the region midway between 'C' and 'F'. The mineral assemblage of chemical compositions intermediate between the main types may also be approximately predicted.

Ideally, an individual metamorphic facies would be characterised by a variety of mineral assemblages for differing rock compositions, which would differ from those of any other facies. It is, however, unlikely that the divariant fields of stability of all isofacial assemblages will coincide precisely; thus in defining the divariant facies fields one particular rock type tends to be given prominence. Eskola had originally proposed a facies nomenclature based mainly on the assemblages of metabasic rocks and current usage closely follows this, rocks being usually assigned to a facies on the basis of metabasic rocks interbedded with them (7).

Difficulties and doubts concerning the significance of Eskola's facies have often been aired since the first enunciation of the concept. Most importantly, the facies involve not only differences in P and T, but also in amounts of volatiles – in particular H_2O and CO_2. The valid use of metamorphic facies to indicate divariant P/T fields of formation must therefore depend on dehydration and decarbonation being functions of P and T, and not variables independent of them. For most rock compositions water is the most important volatile; we may thus distinguish a group of anhydrous facies from those in which hydrous minerals play an essential role.

The anhydrous facies form a series of increasing pressure (decreasing temperature) derived from the lowest P, highest T gabbroic assemblage augite-labradorite. The pyroxene hornfels and granulite facies are characterised in metabasic rocks by the assemblage diopside-hypersthene-plagioclase (= *two pyroxene facies* of some authors), representing breakup of the high-temperature augite solid-solution into two pyroxene phases,

$$\underset{\text{augite}}{Ca(Mg, Fe)Si_2O_6 \, . \, m\,(Mg, Fe)_2Si_2O_6} =$$

$$\underset{\text{diopside}}{Ca(Mg, Fe)Si_2O_6} + \underset{\text{hypersthene}}{2m(Mg, Fe)\,SiO_3} \qquad \text{(a)}$$

Distinction between the low-pressure, dominantly contact metamorphic pyroxene hornfels facies and the regional granulite facies cannot readily be made by metabasic mineralogy alone (compare figs. 35.3 and 37.4). If we equate however the pyroxene hornfelses with the low-pressure granulites we may separate from them an intermediate-pressure granulite (I.P.G. of fig. 35.6) on the basis of the reaction of olivine and anorthite to give two pyroxenes and spinel (p. 354).

$CaAl_2Si_2O_8 + 2(Mg, Fe)_2SiO_4 =$
anorthite olivine (b)
$Ca(Mg, Fe)Si_2O_6 + 2(Mg, Fe)SiO_3 + (Mg, Fe)Al_2O_4$
diopside hypersthene spinel

Increasing pressure within the I.P.G. facies involves a steady decrease in the role of plagioclase as the anorthite and albite molecules, with their high molar volume, are eliminated in favour of accommodation in denser structures. Anorthite is first to go, the remaining plagioclase increasing in albite content as Ca and Al enter the pyroxene as the Tschermak molecule $CaAl_2SiO_6$. This may be represented by equations such as:

$CaAl_2Si_2O_8 + m\ CaMgSi_2O_6 = m\ CaMgSi_2O_6 \, . \, CaAl_2SiO_6 + SiO_2$
anorthite diopside aluminous diopside quartz

When a limiting aluminium content in pyroxene is reached, however, the alumina appears as *garnet*, e.g.

$CaAl_2Si_2O_8 + 2(Mg, Fe)\ SiO_3 = Ca\ (Mg, Fe)_2\ Al_2Si_3O_{12} + SiO_2$ (c)
anorthite hypersthene garnet quartz

The appearance of garnet thus allows a further separation of *high-pressure granulites* in which basic rocks contain an almandine–pyrope rich garnet. Ultimately pressure increase causes complete loss of plagioclase, the albite molecule entering the clinopyroxene by the substitution $CaMg \rightleftharpoons NaAl$ to give the sodic pyroxene *omphacite*, e.g.

$NaAlSi_3O_8 + mCa\ (Mg, Fe)\ Si_2O_6 =$
albite diopside
$mCa\ (Mg, Fe)\ Si_2O_6 \, . \, NaAlSi_2O_6 + SiO_2$ (d)
omphacite quartz

The feldspar-free association garnet–omphacite in basic rocks characterises the *eclogite facies*.

Since the reactions bounding the subdivisions of the granulite facies and eclogite facies involve not only Fe-Mg diadochy but also more complex solid-solutions such as that of alumina in pyroxene, they are clearly continuous reactions and, proceeding over a temperature interval, are developed in different compositional varieties of metabasite at different temperatures. A commonly accepted subdivision, shown in fig. 35.6, is that developed for the quartz–tholeiite chemical composition by Green and Ringwood (8).

The hydrous facies, involving water, figure most prominently in a contact metamorphic or regionally metamorphosed dehydration sequence. The sequence of facies encountered with increasing metamorphic grade in a particular region was called a *facies series* by Miyashiro. A great variety of facies sequences have now been mapped in the field; as an approximation these may be reduced to three, clearly reflecting differing pressure gradients:

(1) Low-pressure facies series (high $\Delta T/\Delta P$) in order of increasing dehydration:
- (a) Greenschist
- (b) Hornblende hornfels
- (c) Pyroxene hornfels (two-pyroxene)

The association of high temperature at low pressure is usually found where an igneous body is emplaced at a high crustal level; and hence this facies series is commonly associated with contact metamorphism (9).

(2) 'Normal' facies series (intermediate $\Delta T/\Delta P$), in order of increasing dehydration:
- (a) Greenschist
- (b) Albite-epidote-amphibolite
- (c) Amphibolite
- (d) Granulite (two-pyroxene)

Such facies, representing simultaneous increase in pressure and temperature, are characteristically found in regionally metamorphosed terrains.

(3) 'High-pressure' facies series (low $\Delta T/\Delta P$):
- (a) Zeolite
- (b) Prehnite-pumpellyite
- (c) Glaucophane (blueschist)
- (d) Greenschist or albite-epidote-amphibolite

The development of high pressure at low temperatures appears to be found where cold material has been rapidly depressed to deep levels, and this is the characteristic facies series of the circum-Pacific and Alpine orogenic belts.

As stated on p. 379, this use of dehydration sequences as indications of geothermal gradients implicitly assumes that progressive dehydration is a product of increasing temperature. Eskola's view was the simplest possible: that a portion of the water driven off from the solids as they

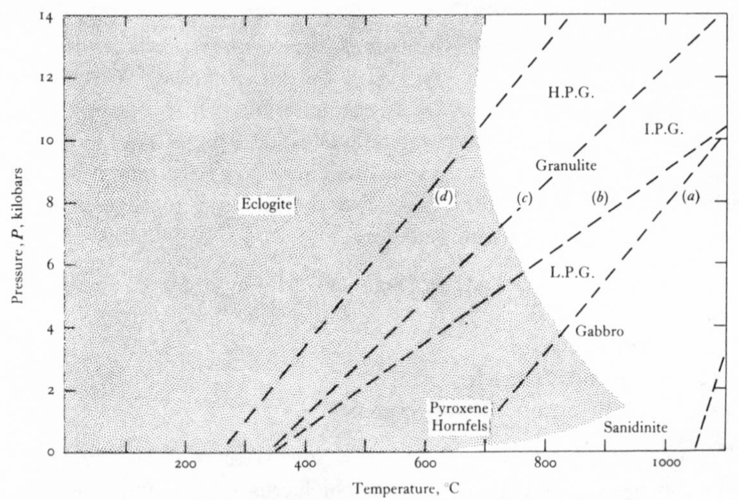

Fig. 35.6. P/T fields of the anhydrous metabasic facies.

The curves (a), (b), (c), (d) correspond to those of the reactions on pp. 379–80 of the text, subdividing the granulite facies (generalised after the work of several authors). The stippled area is blanketed by hydrous facies when

$$P_{H_2O} = P_{total}$$

react to increasing temperature remains as a phase present along grain boundaries. Thus the two-pyroxene facies which occur as the anhydrous culmination of series (1) and (2) were thought to be stable at these high temperatures in the presence of water. Recent experimental work has, however, shown that in an aqueous environment amphibolites begin to melt before dehydration has set in (fig. 35.7). Except at very low ($< c.$ 3 kbar) and at very high pressures ($> c.$ 20 kbar) the anhydrous facies are not stable with water and will revert to hydrous assemblages if exposed to it. Dehydration mechanisms more complex than that envisaged by Eskola must therefore be involved; nonetheless at the present time the correlation of the Miyashiro facies series with geothermal gradients seems to accord well with geological and experimental experience (10).

In treating the metamorphism of rocks, we have the alternatives of considering (a) each rock composition as an isochemical system recrystallising in response to progressive temperature increase along a particular P/T path; or (b) the mineralogy of different rock compositions (the syndrome) within individual facies. Each approach has its advantages and disadvantages; in the necessarily abbreviated and non-rigorous account of the following chapters a compromise will be

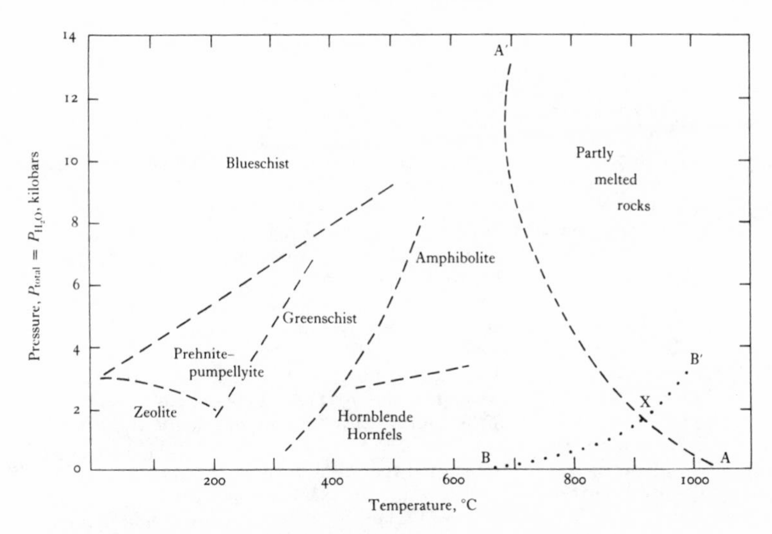

Fig. 35.7. P/T fields of the hydrous metabasic facies.
Epidote amphibolites not differentiated. A-X-A' is minimum melting curve of wet tholeiitic basalt. B-X-B' represents temperatures at which amphiboles begin to dehydrate to pyroxene. At water pressures above that of X, metabasites begin to melt before dehydration to pyroxene occurs.

adopted. Progressive metamorphism will be illustrated by considering three compositional types – basic, argillaceous (pelitic) and calcareous – as isochemical assemblages evolving with increasing temperature under an intermediate P/T gradient. Those facies which reflect more specialised geological conditions – the granulite, eclogite, and blueschist facies – will then be considered as isofacial entities.

References and notes

(1) A history of metamorphic classification is given by Turner, *Metamorphic Petrology*, p. 46.

(2) Accounts of the phase rule may be found in all standard texts on thermodynamics; a useful summary is that of Barton, P. & Skinner, B. J., 1967. Sulfide mineral stabilities *in* Barnes, H. L. ed., *Geochemistry of hydrothermal ore deposits*. New York: Holt, Rinehart and Winston. The use of the phase rule in metasomatic systems is also discussed. The student unfamiliar with the language of thermodynamics may find helpful the derivation of the phase rule given by Findlay, A. R., 1957, *in* Campbell & Smith, eds, *Phase rule*. New York: Dover (9th ed.).
Phase = mineral, if we ignore the fluid phase assumed as an approximation to be present as a component *and* a phase, e.g. water.
Component: the *minimum* number of chemical species required to specify the composition of the phases present is C. Turner

(*Metamorphic Petrology*, p. 56) discusses the ambiguities involved and also examines the assumption that most metamorphic rocks achieved an equilibrium state before quenching.

(3) *Assemblage:* the equilibrium association of minerals in a rock. The term *paragenesis* is also used.

(4) Thompson, J. B., 1957. The graphical analysis of mineral assemblages in pelitic schists. *Amer. Mineral.* **42**, 842–58.

(5) Under appropriate oxidation conditions. The stability of annite is further diminished in highly oxidising or highly reducing conditions; see Wones, D. R. & Eugster, H. P., 1962. Stability relations of the ironbiotite, annite. *J. Petrol.* **3**, 82–125.

(6) Details of the ACF projection and other projections suitable for more specialised compositions are given by Winkler, *Petrogenesis*, chapter 5.

(7) Since basic rocks are so widely distributed the geologist will generally have little difficulty in assigning his rocks to a facies.

(8) Green, D. H. & Ringwood, A. E., 1972. A comparison of recent experimental data on the gabbro-garnet granulite-eclogite transition. *J. Geol.* **80**, 277–88.

(9) This facies series is illustrated for pelitic rocks by fig. 33.5; representative assemblages of the pyroxene hornfels are shown in figs 35.4, 35.5. Eskola recognised a *sanidinite facies* for high temperatures at very low pressures usually developed by rapid heating at volcanic conduits. Except where melting has occurred (fig. 34.2A) the fine-grained products are petrographically difficult and often show strong disequilibrium; they will not be discussed here.

(10) Fry, N. & Fyfe, W. S., 1969. Eclogites and water pressure. *Contr. Mineral. Petrol.* **24**, 1–6. Bryhni, I., Green, D. H. & Fyfe, W. S., 1970. On the occurrence of eclogite in Western Norway. *Contr. Mineral. Petrol.* **26**, 12–19. The possibility of *local* dehydration (near well-developed fissure systems, or in originally dry rocks – e.g. gabbros – or where carbonates are involved) independently emphasises the dictum that metamorphic facies have *regional* significance only.

36

Progressive regional metamorphism

All sequences of increasing metamorphic grade, reflecting P/T gradients themselves resultant upon a particular geological situation, must to a greater or lesser degree be different one from another. Within the intermediate range of P/T gradients, however, sequences of progressive metamorphism show a broad similarity, which allows erection of a generalised scheme of progressive metamorphism for standardised rock compositions (1). We may start with the metabasic compositions, as illustrating the nomenclature of the intermediate, greenschist to granulite, metamorphic facies series (fig. 36.1).

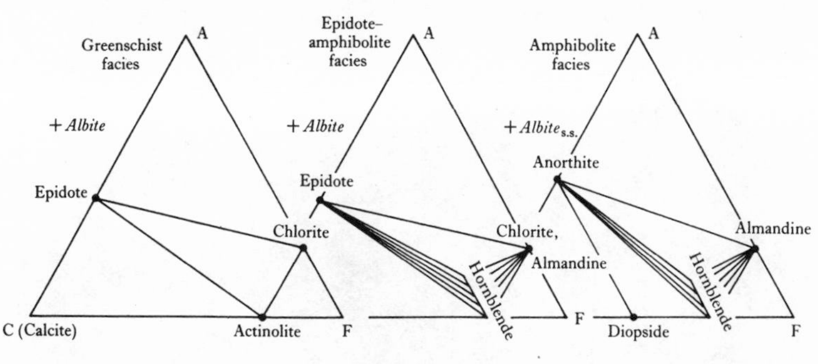

Fig. 36.1.

ACF diagrams illustrating the medium-pressure greenschist amphibolite facies sequence for metabasic rocks only.

Basic rocks

Basic rocks, whether tuffaceous aggregates, lavas, or coarse-grained gabbros, consist originally of aggregates of high-temperature, mainly anhydrous minerals, and accession of water is necessary to form the hydrated minerals characteristic of the lower-grade facies. Less pervious rocks in lowly stressed situations may thus remain unaltered for long periods in a variety of facies conditions. The lowest temperature alteration, involving the development of zeolites (*zeolite facies* – see chapter 39), is generally not pervasive, being confined to amygdales or to glassy portions of a rock. *Greenschist facies* alteration forms the lowest-grade metamorphism commonly seen and readily recognised under the petrographic microscope. Even so, basic rocks within the greenschist facies commonly retain extensive relicts of their original mineralogy, only the amygdale zeolites having given way to epidote, albite and chlorite. Advent of water produces marginal alteration of the pyroxene to fine fibres of amphibole and chlorite and the replacement of labradorite plagioclase by albite and epidote; restricted mobility of calcium and aluminium may result in the clouding (cf. p. 354) of the plagioclase pseudomorph by epidote and an amphibole of the $Ca_2(Mg, Fe)_5Si_8O_{22}$ series referred to here as actinolite. Even where completely recrystallised, the anisotropic minerals may lack preferred orientation and the resulting albite–epidote–actinolite–chlorite–quartz rock still preserves pseudomorphs of the parent igneous texture (fig. 36.2A). Epidote is commonly zoned, with highly birefringent cores and rims of lowly birefringent (iron-poor) clinozoisite. Calcite, resulting from high partial pressures of carbon dioxide and the consequent partial reaction of actinolite with

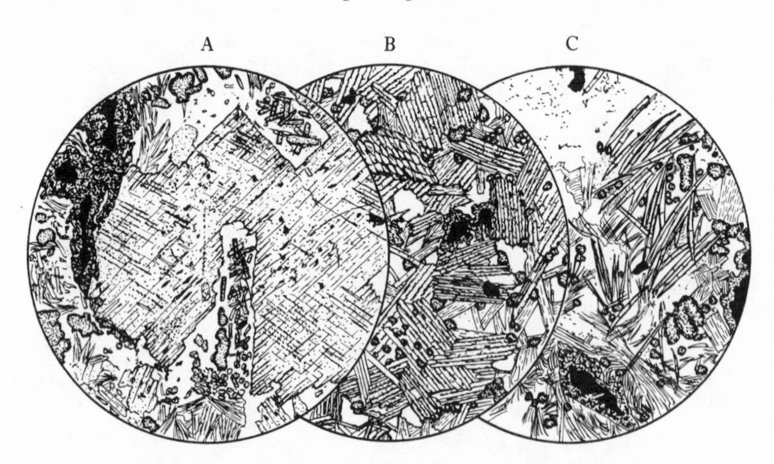

Fig. 36.2. Greenschist facies metabasic rocks; × 40.

A. Albite-epidote-actinolite-chlorite-sphene rock, Loch Fyne, Argyll. Original igneous pyroxene replaced by amphibole preserves the ophitic texture: note the epidote-albite pseudomorphs after basic plagioclase.

B. A slightly higher grade than A, with aggregates of two amphiboles (actinolite and hornblende) intergrown. Some biotite present.

C. A stilpnomelane-bearing greenschist, Loch Fyne. In all three rocks note sphene rimming original igneous ilmenite.

epidote to give chlorite and carbonate, is a common accessory. Titanium at this grade is contained in sphene, which often mantles relict igneous ilmenite.

More potassic variants may show biotite; in relatively higher-pressure sequences, such as that of the Scottish Dalradian, the layered mineral stilpnomelane may appear (fig. 36.2c).

Increasing metamorphic grade now involves a series of continuous reactions whereby plagioclase, at the expense of epidote, steadily increases in anorthite content; the actinolitic amphibole by reaction with chlorite and albite increases in Al content and is transformed to hornblende.

Both plagioclase and amphibole transitions involve the 'jumping' of a compositional miscibility gap, and rocks may be found containing coexisting actinolite-hornblende, and albite-oligoclase (2). In relatively high-pressure sequence, the amphibole transition is achieved first; the production of hornblende independently of plagioclase may be represented by the equations (3)

(a) $NaAlSi_3O_8 + Ca_2(Mg, Fe)_5\,Si_8O_{22}(OH)_2 \rightarrow$
albite actinolite
$NaCa_2(Mg, Fe)_5(AlSi_7)O_{22}(OH)_2 + 4SiO_2$
'edenite' hornblende s.s. quartz

(b) $Ca_2(Mg, Fe)_5Si_8O_{22}(OH_2. + 7(Mg, Fe)_{10}Al_2Al_2Si_6O_{20}(OH)_{16} +$
actinolite chlorite
$24Ca_2Al_3Si_3O_{12}(OH) + 28SiO_2 \rightarrow$
'epidote' quartz
$25Ca_2(Mg, Fe)_3Al_2(Al_2Si_6)O_{22}(OH)_2 + 44H_2O$
'tschermak' hornblende s.s.

The amphibole substitutions involved here are the 'edenite' substitution Si = NaAl; and the 'tschermakite' substitution MgSi = AlAl.

Characteristic of the *albite–epidote–amphibolite facies* are albite–epidote–hornblende–chlorite–quartz schists with a blue-green hornblende, occasional green biotite, and abundant sphene. The development of oligoclase at about the same grade at which almandine garnet appears is accompanied by a decrease in the amount of epidote, sphene and chlorite. Concurrently, hornblende deepens in colour as it gains sodium at the expense of the albite content of plagioclase, and titanium from sphene. The characteristic *amphibolite facies* assemblage is hornblende-andesine, with or without garnet and with ilmenite or magnetite as oxide phase; such amphibolites are usually well crystallised, the deep brown/green hornblende showing a linear preferred orientation and the feldspar polygonal, multiply-twinned grains. Even at this grade, however, one occasionally finds intrusives that have been protected from lower-grade alteration by their rigidity and impermeability to water, and have crystallised directly to amphibolite facies assemblages (fig. 36.3A) (4).

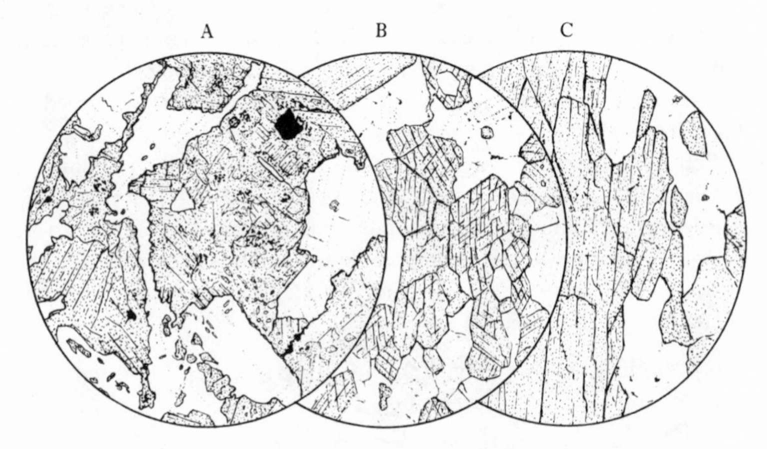

Fig. 36.3. Amphibolites, metamorphosed dyke, Tanunda Creek, S. Australia; × 40.

A. Interior of dyke. Pyroxene transformed to amphibole aggregate, but igneous texture preserved.
B. Margin of dyke. Completely recrystallised, annealed hornblende-andesine aggregate, the amphibole showing pronounced lineation. (Cut perpendicular to lineation).
C. As B, cut parallel to lineation.

The progress of metamorphism to the amphibolite stage has produced a steady increase in the proportion of amphibole to plagioclase, as progressively more of the feldspar constituent has entered the increasingly complex hornblende molecule. The ultimate end, a feldspar-free *hornblendite*, has in fact been achieved in the laboratory by crystallisation of basalt under $P_{H_2O} = P_{total}$. In natural sequences, however, dehydration of hornblende to pyroxene begins before all feldspar has disappeared. The breakdown of hornblende can be represented approximately by the simple reactions

$Ca_2(Mg_4Al)(AlSi_7)O_{22}(OH)_2 \rightarrow$
tschermakitic hornblende
$CaAl_2Si_2O_8 + CaMgSi_2O_6 + 3MgSiO_3 + H_2O$
anorthite diopside hypersthene

and

$NaCa_2Mg_5(Si_7Al)O_{22}(OH)_2 + SiO_2 \rightarrow$
edenitic hornblende quartz
$NaAlSi_3O_8 + 2CaMgSi_2O_6 + 3MgSiO_3 + H_2O$
albite diopside hypersthene

These, with Mg-Fe substitution, produce a complex continuous reaction and hornblende–pyroxene associations persist over a wide temperature range. In many regions grain size may increase notably at this stage, with a general straightening of grain boundaries as seen in thin section, and diminution of the inclusion content of crystals (5). The increasingly equidimensional character of the minerals results in the loss of the strongly schistose appearance of the lower-grade amphibolites, although the segregation of dark and light minerals into bands may give a foliated structure. Such rocks, technically gneisses, are more commonly given the facies appellation of *granulite*. The ideal granulite facies assemblage, hypersthene–diopside labradorite, will be discussed in chapter 37.

Argillaceous (pelitic) rocks

The argillaceous, highly aluminous rocks begin as detrital aggregates of clay minerals with varying amounts of clastic quartz, feldspar, or tourmaline (chapter 23). The processes whereby these environmental-accidental aggregates move towards equilibrium are normally considered under the heading of diagenesis – as for example the reaction of montmorillonite with detrital K feldspar and kaolinite to form mixed-layered complexes of illite and chlorite (6). By the stage at which minerals are coarse enough to be resolved by the ordinary petrographic microscope, the rock is a slate, or phyllite, and will contain muscovite (often phengite-rich), chlorite and quartz, with some sodium-rich plagioclase and highly refringent rutile. Opaque minerals are graphite, recrystallised from organic carbon, or magnetite or hematite: these are *greenschist facies* assemblages.

The sequence of dehydration reactions in the progressive metamorphism of an individual group of argillaceous rocks is intimately dependent on the precise P/T path followed, in addition to its sensitivity to the precise chemical composition. The classic sequence first studied by Barrow (7) in the Scottish Dalradian, although by no means a standard succession, illustrates the general principles involved. Barrow described a zonal sequence characterised by the successive incoming of biotite, garnet, staurolite, kyanite, and sillimanite in successively more dehydrated assemblages. Since the low-calcium pelitic compositions are crammed along the AF perimeter of the ACF diagram, this projection is here of little use. A projection AKF (fig. 36.4) gives a somewhat better indication of the reactions involved. At low grades the mineralogy is dominated by the hydrous, layered phases chlorite, biotite, and musco-

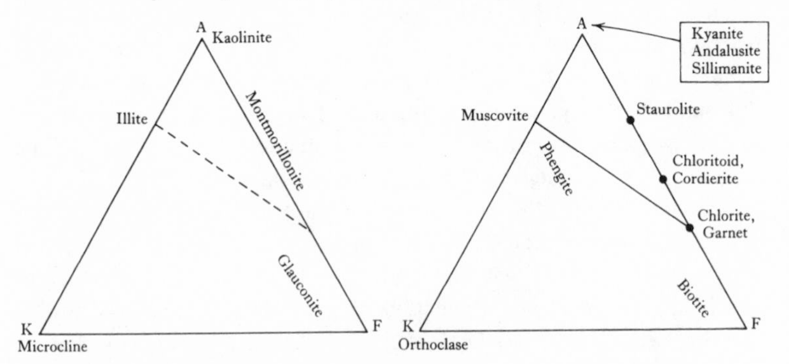

Fig. 36.4.

Representation of argillaceous (pelitic) mineral compositions in terms of A′ (Al_2O_3 – (K_2O + Na_2O + CaO)) K (K_2O) and F (FeO + Fe_2O_3 + MgO + MnO). FeO and Fe_2O_3 are grouped together since during metamorphism ferric iron is normally reduced by graphite to ferrous (however, *see* fig. 36.7). Rock compositions originally rich in illite and glauconite will not develop the peraluminous minerals chloritoid, staurolite, Al_2SiO_5 polymorphs until the muscovite-chlorite barrier has been breached. Rocks lying to the A′ side of the muscovite-chlorite tieline develop peraluminous minerals at much lower grades (p. 39).

vite, and progressive dehydration reactions between these minerals allow the successive appearance of the less hydrous, more aluminium-rich minerals which plot closer to the 'A' apex of the diagram.

Biotite arises by reaction between muscovite and chlorite (8). Biotite contains a higher ratio of Fe to Mg than coexisting chlorite; hence as the reaction proceeds the steadily diminishing chlorite content increases in Mg:Fe. Rutile participates in the reaction; with the substitution of Ti for Fe in the octahedral sites of mica the transmitted colour of biotite changes from green to brown. Excess rutile reacts with magnetite to produce ilmenite. The resulting chlorite–biotite–muscovite–quartz–plagioclase–ilmenite schist is one of the most common of pelitic rock types. The incoming of the iron-rich phase garnet further diminishes chlorite; early-formed garnets may concentrate the manganese and calcium contents of the rocks, but as with increasing grade the garnet content increases, manganese and calcium are diluted and the mineral becomes dominantly an almandine/pyrope solid-solution. Garnets at this stage frequently show skeletal growth and replacement of mica and chlorite flakes, but coalescence soon produces the typical porphyroblastic habit in a granoblastic aggregate of quartz, plagioclase, chlorite, biotite, and muscovite (fig. 36.5B). Such assemblages correspond to the lower range of the *amphibolite facies*.

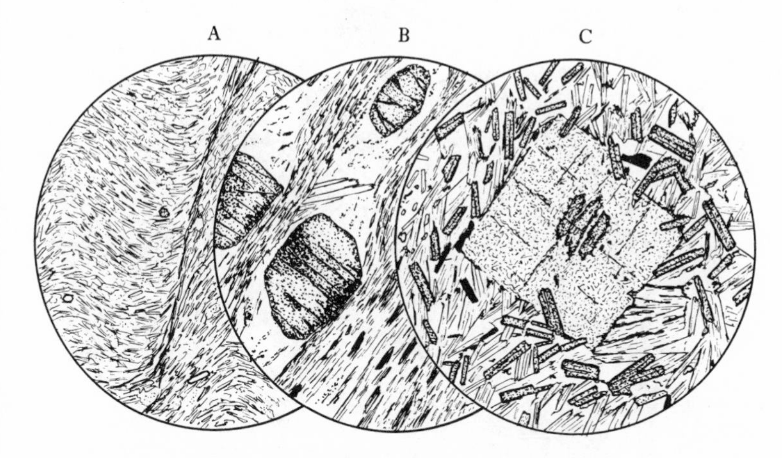

Fig. 36.5. Barrovian peltic schists; × 45.

A. Chlorite-muscovite-albite-graphite schist (chlorite zone), Loch Awe, Argyll.
B. Garnet-chlorite-muscovite-biotite-oligoclase schist, Blair Atholl, Perthshire.
C. Muscovite-chlorite-chloritoid-staurolite-graphite schist, Bannock Hill, Kincardineshire. The staurolite crystal is now largely replaced by fine-grained muscovite (shimmer aggregate) due to late-stage retrogressive metasomatism.

Rocks in which chlorite still persists next show *staurolite*, by reaction of chlorite, garnet and muscovite to give staurolite and biotite; in the Dalradian this continuous reaction finally exhausts the chlorite content of the rock (fig. 36.6A).

The staurolite zone in Scotland is not wide, the incoming of staurolite being followed immediately by that of *kyanite* (fig. 36.6B) as staurolite and muscovite become incompatible (9).

Staurolite and kyanite may, however, arise in another way. Rocks with higher proportions of original kaolinite in general develop the more aluminium-rich silicates at lower grades. Where kaolinite content exceeds that required to produce the lower greenschist facies assemblage muscovite-chlorite-albite, the sodium mica *paragonite* (10) appears,

$$\underset{\text{kaolinite}}{Al_2Si_2O_5(OH)_4} + \underset{\text{albite}}{NaAlSi_3O_8} = \underset{\text{paragonite}}{NaAl_2(AlSi_3)O_{10}(OH)_2} + \underset{\text{quartz}}{2SiO_2} + H_2O$$

Such paragonite, not readily differentiated from muscovite in thin section, disappears within the epidote amphibolite facies by reaction with chlorite, to produce the iron-rich aluminosilicate *chloritoid* (fig. 35.5C) and albite. Even more aluminous rocks, in which the original kaolinite content exceeded the albite available to produce paragonite, develop *kyanite* at comparable grades, by direct dehydration (11). We

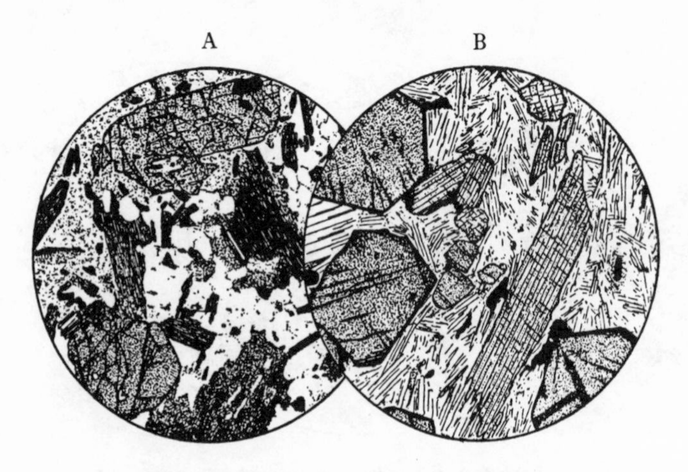

Fig. 36.6. Barrovian pelitic schists; × 20.

A. Staurolite-garnet-mica-oligoclase-quartz schist, Glen Clova, Angus. The biotite is partly oxidised to chlorite-magnetite pseudomorphs during retrogressive alteration.

B. Kyanite-garnet-mica schist, Glen Fernait, Perthshire: rich in muscovite with little biotite. A fractured crystal of tourmaline occurs in the centre.

thus find kyanite–chloritoid–garnet–muscovite assemblages and, in such peraluminous rocks, the development of staurolite by reactions such as

$$\text{kyanite} + \text{chloritoid} + \text{quartz} = \text{staurolite} + \text{chlorite} + H_2O$$

follows the appearance of kyanite (12).

Such amphibolite facies rocks, at the equivalent of the Barrow kyanite grade, are still essentially schistose, with a homogeneous fabric of muscovite, biotite, quartz and albite/oligoclase, and porphyroblasts of pink garnet and prismatic kyanite; opaques are graphite and ilmenite or magnetite. Small amounts of yellow staurolite, soon to react to give more kyanite, are often present. In common with most regional metamorphic rocks, these schists show few signs of the reactions whereby they have recrystallised. Notable exceptions are found in garnet and in tourmaline, which, once formed, seem reluctant to recrystallise and frequently show marked compositional zoning with the preservation of an earlier formed centre. Garnet, in particular, may contain *'armoured relicts'* of minerals which, immolated within the core, have been preserved from reaction with rock matrix. In rocks of high ferric/ferrous ratio, hematite is a prominent constituent (fig. 36.7).

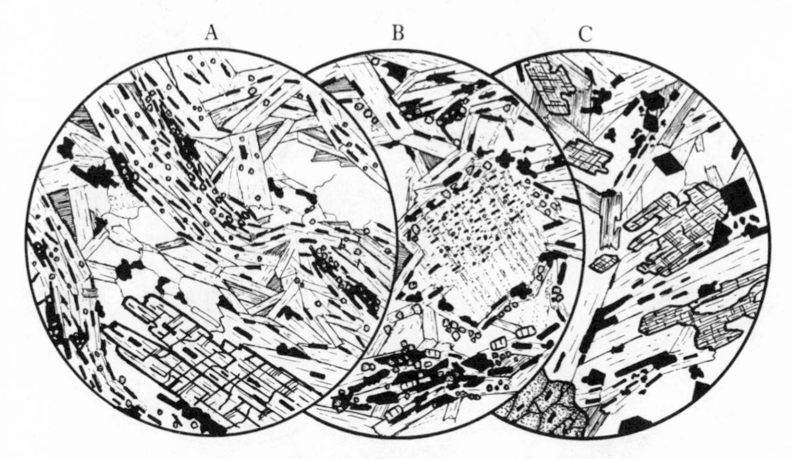

Fig. 36.7. Oxide-rich pelitic schists, Glen Clova, Angus; × 25.

A. Kyanite-muscovite-biotite-spessartite-oligoclase-quartz-hematite schist. Ferric oxide in argillaceous rocks is normally reduced during diagenesis by organic carbon. Where the carbon quotient is insufficient, hematite persists to high metamorphic grade.
B. As in A, with prisms of tourmaline. Note smaller grain size of hematite and garnet included in oligoclase (albite twinned).
C. As in A, with in addition some magnetite (more equant grains).

The final index mineral of the Barrovian sequence is sillimanite. Although the reaction represented is certainly the inversion of kyanite to its high-temperature polymorph, the mechanism took other paths than direct pseudomorphism (as in the andalusite–sillimanite transition discussed on pp. 357–8). Sillimanite, in the form of finely fibrous fibrolite, occurs as mats and sheaves intimately admixed with biotite or quartz; the process is by no means homogeneous, and substantial volumes of kyanite–garnet assemblage may remain untouched. The incoming of sillimanite in Scotland is usually accompanied by a general increase in grain size, with a coarsening and segregation of quartzofeldspathic from micaceous material so that the rocks are more properly termed gneisses (fig. 36.8). The matrix coarsening probably resulted from increased spheres of diffusion as the rocks came within the temperatures of melting, and pools of liquid appeared. Much of the coarser quartzofeldspathic lenticular material probably represents liquid retrogressively crystallised to coarse aggregates as the temperatures again dropped below those of the 'granite' minimum solidus. Sillimanite gneisses from this stage may still show residual staurolite which has survived by its concentration of the zinc content of the rock.

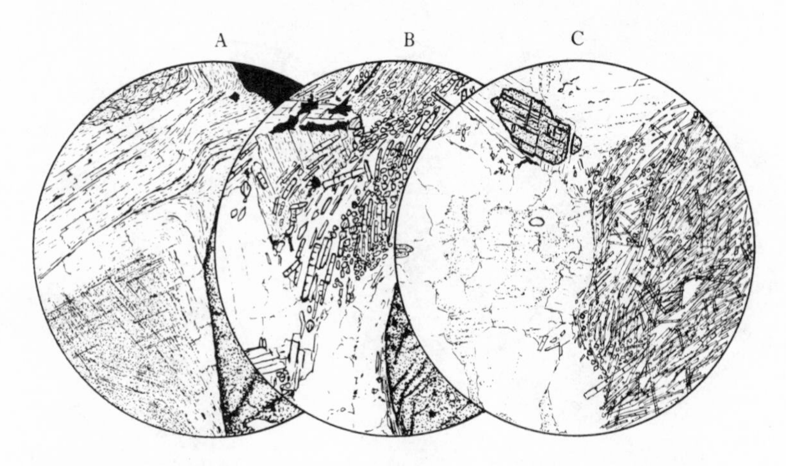

Fig. 36.8. Gneisses of Barrow's sillimanite zone, Glen Clova, Angus; × 40.

A. Fine fibrolitic sillimanite intergrown with biotite; the garnet to the right is 5 mm in diameter. Opaque mineral magnetite.
B. Coarse sillimanite with biotite, garnet, and coarse graphite (opaque). Muscovite and quartz coexist.
C. Matted sillimanite in quartz surrounds a quartz-oligoclase-muscovite lens which probably represents a low-melting patch liquified during the metamorphic climax. Note relict kyanite enclosed in muscovite.

All samples were collected within a radius of 5 metres.

Reactions proceeding at higher grades involve the rapid reduction in the remanent mica content. Muscovite decomposes by reaction with quartz to give sillimanite + K feldspar (p. 348). Sillimanite is thus generated at two distinct stages (often called the first and second sillimanite isograds respectively); at temperatures above the second isograd, K feldspar, as microcline or orthoclase, is stable with sillimanite, and muscovite is found only in the absence of quartz. Such assemblages are transitional to the granulite facies.

Argillaceous rocks by definition are calcium-poor, but with admixtures of tuffaceous or carbonate material, transitions to calcareous assemblages occur. Within the amphibolite facies calcium is normally

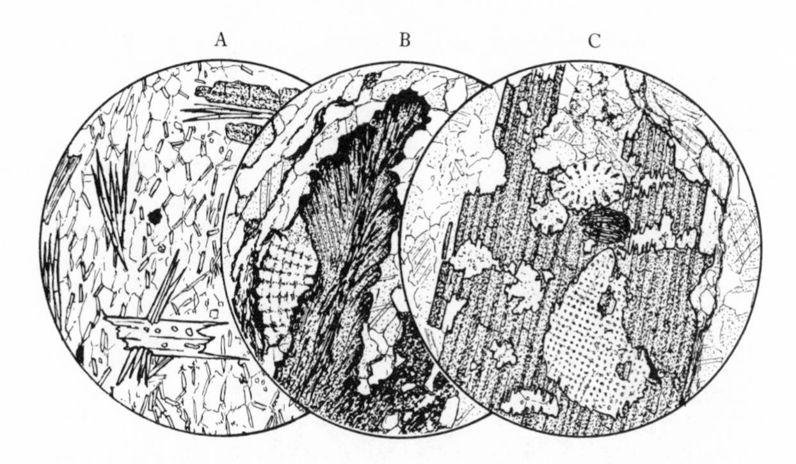

Fig. 36.9. Lime-rich pelites (Triassic Bundnerschiefer), Tessin; × 65.

A. Muscovite-biotite-chlorite-margarite-chloritoid-quartz schist. Note characteristic bladed habit of margarite with the mica (001) cleavage poorly developed.

B. Garnet (almandine-rich) calcite-chlorite-quartz schist. The garnet shows growth lenticles of quartz and graphite and encloses an echinoid ambulacral plate.

C. Chloritoid-calcite-quartz-chlorite. The chloritoid shows intricate polysynthetic twinning and encloses several echinoid fragments which persist as single crystals of calcite.

contained in garnet and plagioclase, but epidote and zoisite may appear; epidote-bearing staurolite- and chloritoid-parageneses are common in the Triassic schists of the Lepontine Alps. The lime-mica margarite (= kyanite + zoisite) similarly occurs abundantly in certain graphite-rich argillaceous horizons in the Alps and in the Dalradian (fig. 36.9). Curious variants include parageneses in which an aluminous hornblende, or calcite or dolomite, coexist with chloritoid or with kyanite (13).

Low-pressure sequences

Although the Barrow (kyanite–sillimanite) sequence gives a fair representation of progressive metamorphism of pelites, the P/T paths of many metamorphic belts are of low enough slope to intersect the Al_2SiO_5 univariant lines at pressures below those of the kyanite stability field. Andalusite is now developed and the lower pressures involved favour the framework silicate *cordierite* rather than the denser equivalent garnet. At the north-eastern extremity of the Scottish Dalradian outcrop such a lower pressure (andalusite–sillimanite) sequence is developed. Amphibolite facies representatives are andalusite–staurolite–biotite–muscovite schists; rocks originally rich in chlorite contain cordierite. Garnet is only found as a spessartite (manganese-rich) almandine.

The appearance of sillimanite, at first as fine fibres associated with micas, doubtless involves the dissolution of andalusite by complex reaction paths such as those previously discussed and illustrated (p. 357; fig. 33.9). In low-pressure sequences, however, temperatures of muscovite–quartz incompatibility are achieved soon after the sillimanite field is entered. Commonest sillimanite-bearing assemblages thus normally contain microcline, with biotite, cordierite, quartz, and plagioclase. The equivalents in N.E. Scotland, being transitional to the higher-pressure Barrow sequence, contain in addition an almandine-rich garnet (fig. 36.10). In much lower-pressure sequences, as in most contact aureoles, garnet is lacking and microcline appears, by the incompatibility of muscovite and quartz, within the andalusite stability field. Rocks originally lacking in excess chlorite do not develop cordierite

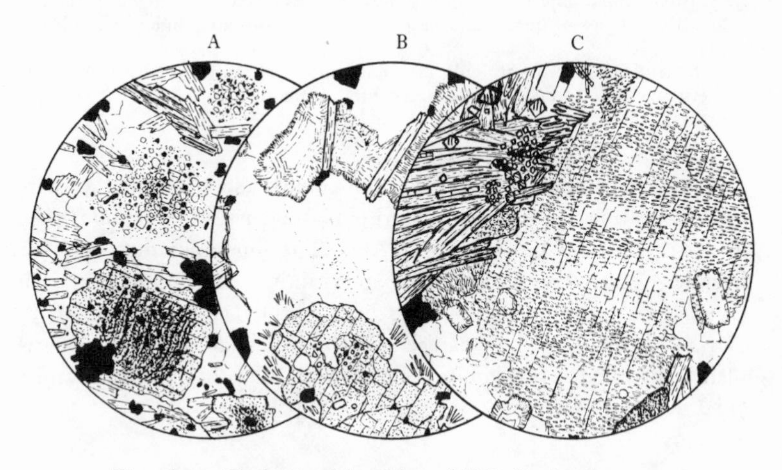

Fig. 36.10. 'Buchan' peltic schists of N.E. Scotland; × 30.

A. Cordierite-andalusite-magnetite schist, Coreen Hills, with biotite, muscovite, and quartz. Note the poikiloblastic habit of cordierite and the andalusite cored by sigmoidal inclusion trails suggesting rotation at an early stage of growth.

B. Andalusite-cordierite-magnetite-muscovite-biotite gneiss, Buck of the Cabrach. Cordierite is altered to a characteristic lowly birefringent serpentinous material, here faintly coloured in shades of green and yellow. Some sillimanite fibres in quartz (cf. fig. 33.9); muscovite and quartz still coexist.

C. Sillimanite-biotite-microcline-cordierite-quartz gneiss, Inverurie. Perthitic exsolution in K feldspar: note its coexistence with sillimanite.

until a grade higher than that at which microcline appears; the reaction here is

$$\text{biotite} + \text{sillimanite} \rightarrow \text{cordierite} + \text{K feldspar} + H_2O$$

Retrogressive reactions

Occasional retrogressive alteration indicates that high-grade pelitic rocks are sometimes susceptible to widespread permeation by water. 'High-grade' minerals are most readily attacked: sillimanite, staurolite and kyanite being often altered to *'shimmer aggregates'* of fine-grained muscovite. Garnet and biotite show low-temperature alteration to chlorite. The distinction between such secondary chlorite and chlorite of primary origin has long been a petrographic puzzle. In general, however, chlorite secondary from high-grade biotite contains needles of rutile precipitated by the alteration of the titaniferous host. Chlorite retrogressive after garnet inherits the Fe:Mg ratio of its parent and is thus bright green in colour with the anomalous blue interference colours of iron-rich chlorite, in contradistinction to the lower absorption and anomalous brown interference colours of the more magnesian prograde chlorite.

Calcareous rocks

The foregoing account has emphasised the role of water in determining the metamorphic facies attained by basic and argillaceous rock compositions. Rocks containing substantial amounts of carbonates fit with some difficulty into such a scheme, for the high proportions of CO_2 in a gas phase buffered by carbonate assemblages produce local environments differing greatly from those of adjacent argillaceous and basic rocks. Nonetheless some generalisations may be made. Calcite-bearing marbles persist over the whole range of metamorphic conditions, excepting only the higher-pressure range of the glaucophane schist facies, where the dense $CaCO_3$ polymorph *aragonite* (p. 415) is stable. Dolomites are likewise stable except at high temperatures in the relatively low pressure pyroxene hornfels facies, where, by *dedolomitisation*, periclase appears:

$$\underset{\text{dolomite}}{CaMg(CO_3)_2} = \underset{\text{calcite}}{CaCO_3} + \underset{\text{periclase}}{MgO} + CO_2$$

Periclase on cooling is very susceptible to alteration; the retrogressive product, however, is not a carbonate (dolomite or magnesite) but is invariably the hydrate brucite, $Mg(OH)_2$. Clearly prograde expulsion of CO_2 was followed by retrograde ingress of a water-rich gas.

Where quartz is present in the original sediment, the metamorphic products are more varied. The classic analysis of progressive metamorphism in such rocks is that of Bowen (14): his scheme, somewhat modified by later experimental work, shows a sequence of increasingly decarbonated assemblages which also involve initial hydration and subsequent progressive dehydration. These may be summarised by the minerals which progressively develop in favourable compositions, viz. talc, tremolite (greenschist facies), diopside, forsterite (amphibolite facies), wollastonite. The appearance of wollastonite by the reaction

$$\underset{\text{calcite}}{CaCO_3} + \underset{\text{quartz}}{SiO_2} = \underset{\text{wollastonite}}{CaSiO_3} + CO_2$$

marks the 'high-water mark' of ordinary contact or regional progressive metamorphism, occurring as it does with the pyroxene hornfels and (rarely) the granulite facies. The development of talc and tremolite is clearly dependent on a moderate proportion of $H_2O:CO_2$ in the gas phase: where the partial pressure of CO_2 is greater than some 90 per cent, stages involving these minerals are omitted (15)

'Impure' calcareous rocks

The presence of argillaceous (Fe and Al) impurity introduces micas, chlorite, and plagioclase into low-grade calcareous rocks. With fluids of high $CO_2:H_2O$ ratio such rocks retain to high grades their carbonates, which may coexist with minerals as aluminous as chloritoid, staurolite, or kyanite (p. 395). Commonly, however, calcite and dolomite disappear from marly compositions within the greenschist facies, complex decarbonation reactions producing epidote, zoisite, and actinolite,

$$\underset{\text{anorthite}}{3CaAl_2Si_2O_8} + \underset{\text{calcite}}{CaCO_3} + H_2O = \underset{\text{zoisite}}{2Ca_2Al_3Si_3O_{12}(OH)} + CO_2$$

Although assemblages with epidote-actinolite-quartz-plagioclase may superficially resemble metabasic assemblages, the more calcareous character and lower water pressure (due often to CO_2 from interbedded, more carbonate-rich sediments) ensure that in calcsilicates dehydration reactions are more advanced than in isograde basic rocks. The lime-garnet grossularite

$$\underset{\text{zoisite}}{4Ca_2Al_3Si_3O_{12}(OH)} + \underset{\text{quartz}}{SiO_2} = \underset{\text{anorthite}}{5CaAl_2Si_2O_8} + \underset{\text{grossularite}}{Ca_3Al_2Si_3O_{12}} + 2H_2O$$

is an abundant member of epidote–amphibolite facies assemblages (garnet-epidote-hornblende-quartz); zoning towards more andraditic (Ca, Fe^{3+}) composition is often seen and where strain birefringence is developed the garnet may resemble the related tetragonal mineral idocrase (vesuvianite). Within the amphibolite facies amphibole diminishes rapidly and calcsilicate assemblages are usually dominated by pyroxene (diploside–hedenbergite solid-solution) with grossularite, quartz, plagioclase, and some epidote and hornblende. In some regions a chlorine-rich gas phase (from NaCl trapped in the original sediment) has allowed the formation of *scapolite*. When rich in the $3NaAlSi_3O_8.NaCl$ molecule the lowly birefringent scapolite closely resembles albite, and unequivocal uniaxial figures may be hard to obtain; however, the calcium-rich scapolites $3CaAl_2Si_2O_8.CaCO_3$ have a birefringence high enough for confusion to be difficult. Finely banded pyroxene–scapolite rocks are a persistent and beautiful feature of many amphibolite facies terrains (fig. 36.11).

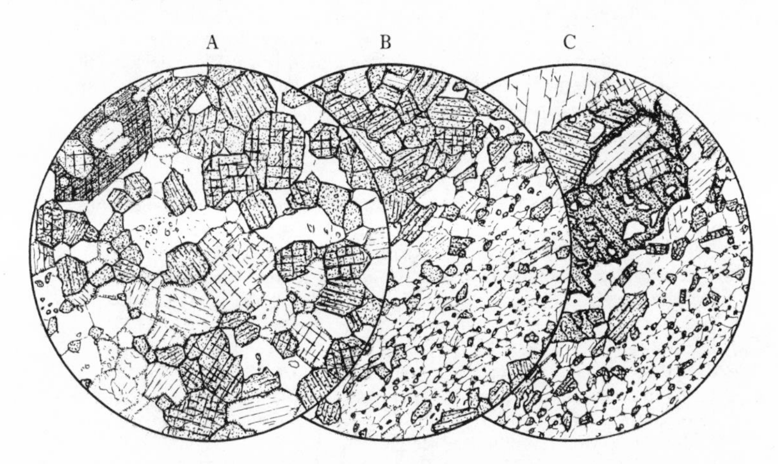

Fig. 36.11. Pyroxene-scapolite rocks, Tanunda Creek, S. Australia; × 35.

A. Pyroxene (diopside-hedenbergite) scapolite-microcline-albite rock, with hornblende and tiny grains of sphene.

B. As A, with calcite, metasedimentary compositional banding being accentuated by grain size difference between pyroxene-rich and pyroxene-poor bands.

C. A banded variety showing grossularite garnet and epidote in the coarse band with pyroxene, scapolite, and microcline.

References and notes

(1) In more detailed consideration of metamorphism, such generalisation may be unsatisfactory. The diversity in detail of metamorphic facies series is comprehensively treated in both Miyashiro's and Turner's books.

(2) The feldspar 'peristerite' gap, see Cooper, A. F., 1972. Progressive metamorphism of metabasic rocks from the Haast Schist group of New Zealand. *J. Petrol.* **13**, 457–92. Graham, C. M., 1974. Metabasite amphiboles of the Scottish Dalradian. *Contrib. Min. Pet.* **47**, 165–85.

(3) Lower pressures favour the less dense plagioclase solid-solutions rather than epidote, and the andesine–actinolite assemblage may occur. Miyashiro, *Metamorphism*, p. 249. The production of more anorthite-rich plagioclase *with* hornblende may involve such reactions as

$$\underset{\text{chlorite}}{3(Mg,Fe)_{10}Al_2Si_6Al_2O_{20}(OH)_{16}} + \underset{\text{'epidote'}}{12Ca_2Al_3Si_3O_{12}(OH)} + \underset{\text{quartz}}{14SiO_2} =$$

$$\underset{\text{'tschermak' hornblende s.s.}}{10Ca_2(Mg,Fe)_3Al_2(Al_2Si_6)O_{22}(OH)_2} + \underset{\text{anorthite}}{4CaAl_2Si_2O_8} + 20H_2O$$

(4) Compositional variants at this grade include cummingtonite-bearing types and, more rarely, amphibolites with chloritoid, staurolite, and kyanite. The occurrence of such highly aluminous minerals with hornblende is not well understood: one would normally expect kyanite and hornblende to react to give, for example, plagioclase–cummingtonite–garnet associations.

(5) Binns, R. A., 1964. Zones of progressive regional metamorphism in the Willyama complex, Broken Hill district, New South Wales. *J. Geol. Soc. Australia* **11**, 283–330.

(6) Frey, M., 1974. Alpine metamorphism of pelitic and marly rocks of the Central Alps. *Schweiz. Min. u. Pet. Mitt.* **54**, 489–506.

(7) See map and description in Miyashiro, *Metamorphism*, p. 186 ff.

(8) The complex character of biotite isograd reactions has been illustrated by Mather, J. D., 1970. The biotite isograd and the lower greenschist facies in the Dalradian rocks of Scotland. *J. Petrol.* **11**, 253–75.

(9) Staurolite + muscovite = biotite + kyanite + H_2O, a continuous dehydration reaction.

(10) Not generally distinguishable from muscovite in thin section. Paragonite is not common in the Scottish succession; its occurrence is, however, well documented in the closely related Taconic metamorphic belt of New England. Thompson, J. B. & Norton, S. A., 1968. Palaeozoic regional metamorphism in New England and adjacent areas *and* Albee, A. L., 1968. Metamorphic zones in Northern Vermont, chapters 24 and 25 *in* Zen *et al.* ed., *Studies of Appalachian geology*. New York: Wiley.

(11) The dehydration of kaolinite in quartz-bearing rocks involves another layer silicate mineral, pyrophyllite, $Al_2Si_4O_{10}(OH)_2$, thus

$$\underset{\text{kaolinite}}{Al_2Si_2O_5(OH)_4} + \underset{\text{quartz}}{2SiO_2} = \underset{\text{pyrophyllite}}{Al_2Si_4O_{10}(OH)_2} + H_2O$$

and

$$\underset{\text{pyrophyllite}}{Al_2Si_4O_{10}(OH)_2} = Al_2SiO_5 + 3SiO_2 + H_2O$$

Pyrophyllite, however, has not yet been widely recognised as a rock-forming mineral. (Miyashiro, *Metamorphism*, p. 200–1.)

(12) The necessity of identifying the precise reaction involved when mapping an isograd is hereby emphasised. For the variety of possible reactions in pelites see Albee, A. L., 1965. A petrogenetic grid for the Fe-Mg silicates of pelitic schists. *Amer. J. Sci.* **263**, 512–36.
(13) See note 4. The inhibition of expected reaction between calcite and aluminium rich phases to produce epidotes or lime-rich plagioclase must result from partial pressures of CO_2 higher than those normally to be expected during the dehydration of argillaceous rocks.
(14) Bowen, N. L. (1940) first showed the systematic handling of devolatilisation reactions. For a detailed critique in the light of modern experimental work see Turner, *Metamorphism*, p. 131 ff.
(15) Metz, P. W. & Trommsdorff, V., 1968. On phase equilibria in metamorphosed siliceous dolomites. *Contr. Mineral. Petrol.* **18**, 305–9.

37
Granulites and eclogites

The granulites

Experimental studies at moderate pressures (p. 382) show that, in the presence of water as a phase ($P_{H_2O} = P_{total}$), amphibolites transform with increasing temperature to feldspar-free hornblendites, and then begin to melt without the intervention of the anhydrous granulite assemblage diopside-hypersthene-plagioclase. The granulite facies thus appears to be a water-deficient facies: in its P/T field of stability the two-pyroxene plagioclase assemblage would either melt or be converted to hydrous assemblages if water were available.

The cause of such water deficiency is not always obvious. Occasionally, local water deficiency within the amphibolite facies may be found where gabbroic dykes intruded into wet sediments have been subsequently metamorphosed: the margin of the dyke may recrystallise to an amphibolite, but the centre, kept anhydrous by the slow rate of inward diffusion of water, has recrystallised to granulite assemblages (fig. 37.1A) (1).

The most significant development of the granulite facies is however on a regional scale, in the deeply-eroded Archaean complexes of all continents. Many of the rock types represented (e.g. pelites) were clearly once water-rich sediments and it may be that in these very old (2900–2700 m.a.) rocks successive episodes of polymetamorphism have driven out all water.

In the case of the central European granulites exposed in the classic granulitgebirge of Saxony, textural relicts suggest that granulite metamorphism was preceded by an episode of crystallisation under high P_{H_2O} (probably amphibolite facies) accompanied by melting. Removal of the low-melting granite fraction with its high content of dissolved water may here have left the remaining rock sufficiently desiccated to give granulite facies assemblages on subsequent metamorphism. Certainly granulite facies and amphilobolite facies rocks are often closely intermingled, and examples where a regular progression may be traced on the ground from amphibolite facies to granulite facies (Adirondacks, peninsular India, and Broken Hill, Australia) seem to be relatively uncommon (2).

The term 'granulite' originally referred to the distinctive texture of quartzofeldspathic rocks of the Saxon granulitgebirge. Long platy crystals of quartz, in thin section resembling ribbons, are separated by aggregates of finely-granular quartz and feldspar (fig. 37.3B). This texture, probably a product of deformation analogous to the mylonitic, is also found in basement complexes in Quebec, India and Central Australia; it is, however, common nowhere and is rarely found in rock types other than the quartzofeldspathic. Granulites, in the modern usage, show as wide a variety of textural types as rocks of any other facies (figs. 37.2, 37.3). A characteristic type is well annealed, the coarse equigranular aggregates showing in thin section a preponderance of straight (fig. 37.5A) or gently curved grain boundaries (3).

Basic and ultramafic granulites

Most basic granulites have hornblende in addition to the andesine or

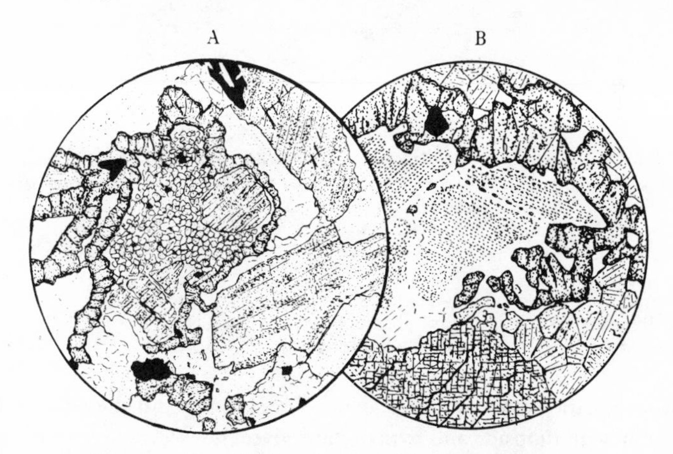

Fig. 37.1. Localised analogues of the granulite facies.

A. The dry interior of a diabase dyke metamorphosed within the (wet) amphibolite facies, Bakersville, N.C. (× 10). Two reactions may be inferred: (i) the breakdown of igneous augite (oriented relicts in the centre of field) to a hypersthene-diopside aggregate – reaction (a) of fig. 35.6; (ii) the reaction of hypersthene with plagioclase to give a garnet corona – reaction (c) of fig. 35.6.

B. Metagabbro relict within garnet-amphibolite, Barton Garnet Mine, New York (× 35). This gabbro, illustrated in fig. 33.7, shows the additional development of garnet between hypersthene and plagioclase. The coarse pyroxene cut normal to [001] is of diopside.

labradorite plagioclase–hypersthene–diopside–magnetite–ilmenite assemblage, indicating incomplete dehydration. The reaction of hypersthene with plagioclase to give garnet in rocks of quartz–tholeiite composition was used by Green and Ringwood (p. 380) to separate the intermediate-pressure granulites (hypersthene–plagioclase) from the garnetiferous high-pressure granulites; however, compositional variants (e.g. more Al-rich rocks) may contain garnet throughout a wider range of the granulite facies.

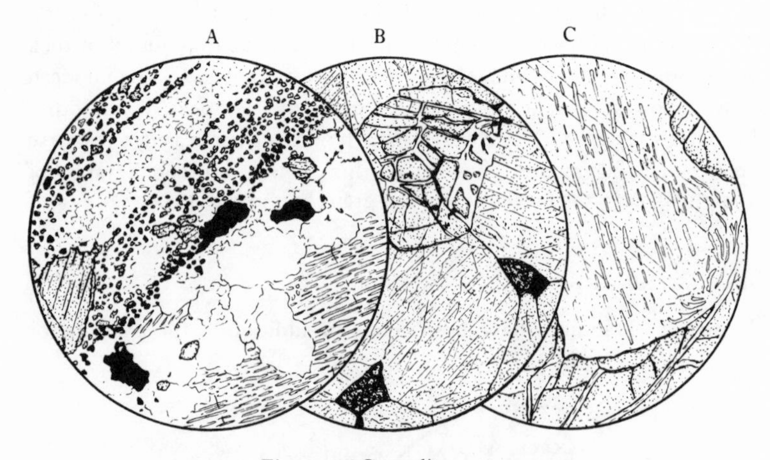

Fig. 37.2. Granulites; × 30.

A. Mylonitic pyroxene syenite, near Harrisville, Adirondacks, New York. A crushed aggregate of diopside and feldspar with porphyroclasts of pyroxene, perthitic orthoclase, oligoclase, and magnetite.
B. Ultrabasic granulite, Scourie, Scotland. A partly serpentinised olivine crystal in a well-annealed aggregate of hypersthene and diopside. Some magnetite and hercynite spinel.
C. Forsterite marble, Meddecombra Estate, Sri Lanka. Calcite shows exsolution of dolomite both as oriented platelets and as symplectite; the {1011} cleavage is not developed in the dolomite platelets.

In ultramafic compositions plagioclase is lacking and the occurrence of olivine with diopside and hypersthene gives the *lherzolite* assemblage; spinel is characteristically present (4).

Acid and intermediate granulites

The granite mineralogy quartz–two feldspar–mica (hornblende) is stable throughout most of the metamorphic range; however, in the granulite facies we find the anhydrous equivalent, quartz-orthoclase-plagioclase-hypersthene, known as *charnockite* (5). Characteristic of charnockite hypersthenes is vivid pleochroism; the presence of mesoperthite, an intergrowth of approximately equal amounts of exsolved sodium and potassium feldspar, indicates that the original homogeneous crystal crystallised at temperatures higher than the crest of the alkali feldspar solvus (fig. 37·3A,C). Other less regular types – 'hair' and 'flame' perthites – probably represent longer-distance migration of alkali ions during exsolution. More aluminous charnockites (e.g. produced from muscovite granites) may contain garnet.

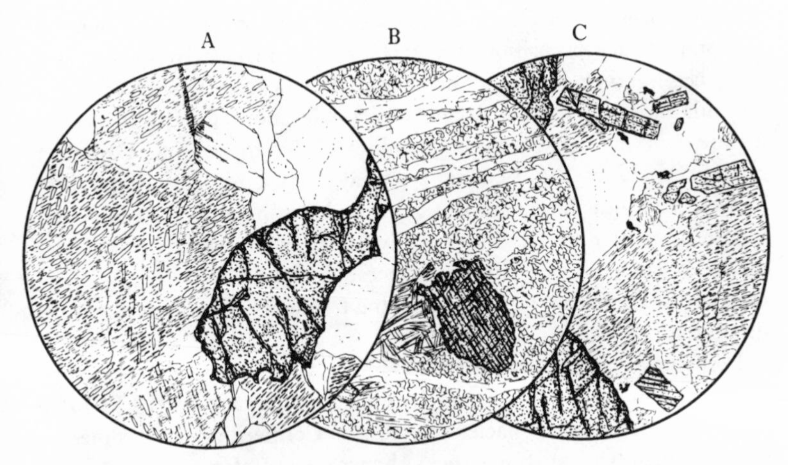

Fig. 37·3. Pelitic granulites; × 35.

A. Charnockite, Madras. (From Job Charnock's tombstone in Calcutta.) This variety is poor in hypersthene and shows garnet with mesoperthite and quartz.
B. Kyanite granulite, Geringswalde, Saxony. Note the ribbons of clear quartz in the fine-grained granulated matrix of perthitic feldspar (the original granulite texture).
C. Sillimanite granulite, Vizagapatam, Madras. Euhedral sillimanite prisms in with perthitic feldspar, quartz, and garnet. In all three drawings the intricate inter-growths of feldspars with differing refractive index give the appearance of moderate relief greater than that normally associated with feldspar.

Charnockites, though extensively developed in peninsular India, are not always common in other regions of granulite facies metamorphism. Intermediate granulites, superficially similar to charnockites but enriched in hypersthene, or plagioclase, or quartz, are more characteristic. A predominant type in the Scourie region of the Scottish Lewisian is a quartz-andesine antiperthite-augite rock with hypersthene and ilmenomagnetite (6).

An important associate of the granulite facies is *anorthosite*, a predominantly feldspar (An_{40-65}) rock with subordinate hypersthene and augite; together with associated norites and syenites, these outcrop over large areas (1000–2000 km^2) of the Adirondacks, Quebec, and southern Norway. Characterised by coarse grain size and abundance of cataclastic and mylonitic textures, these appear to be the products of magma evolution and intrusion under granulite facies conditions (7).

Pelitic rocks

The anhydrous equivalent of argillaceous compositions, representing the complete decomposition of micas (with quartz), is quartz-orthoclase-plagioclase-garnet-Al_2SiO_5. The aluminium silicate polymorph present depends on the precise pressure of crystallisation: usually it is sillimanite, but in high-pressure granulites (as in the central European granulites and in the Lewisian) it may be kyanite (fig. 37.3). The replacement of the dominant low-grade hydrous minerals muscovite and biotite by their granulite facies equivalents, kyanite-garnet-orthoclase, places a strain on the silica resources of rocks not originally over-endowed with quartz, and results in the frequent appearance of corundum or spinel (8).

In the low-pressure range of the granulite facies, the degree of pyrope solid-solution in garnet is insufficient for the rock ratio of Mg:Fe to be satisfied in the granulite facies assemblage. Cordierite then becomes an essential member (9). Such rocks, however, are often still partly hydrated and sillimanite–cordierite–garnet–microline–microperthite–plagioclase–biotite gneisses are common in the region of the amphibolite facies/granulite facies prograde transition.

Pelitic granulites relatively deficient in alumina approach the charnockite assemblage by the development of hypersthene, normally prohibited from association with kyanite or sillimanite by the intervention of cordierite- or garnet-bearing assemblages. In more magnesian rocks, however, the instability of cordierite at pressures not yet high enough to stabilise the pyrope garnet removes this prohibition; kyanite–orthopyroxene associations are thus found in the high pressure range of the granulite facies. Other interesting assemblages permitted by the instability of cordierite contain sapphirine and kornerupine (10).

Calcareous assemblages

The characteristic granulite facies assemblages have reached the forsterite–diopside (p. 398) stage of decarbonation. Occasionally the fluo-

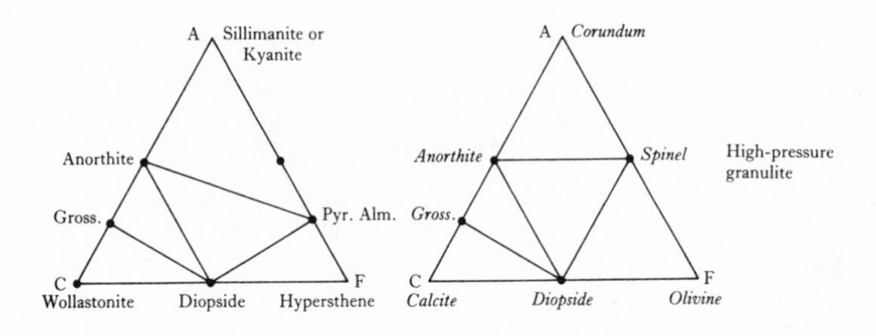

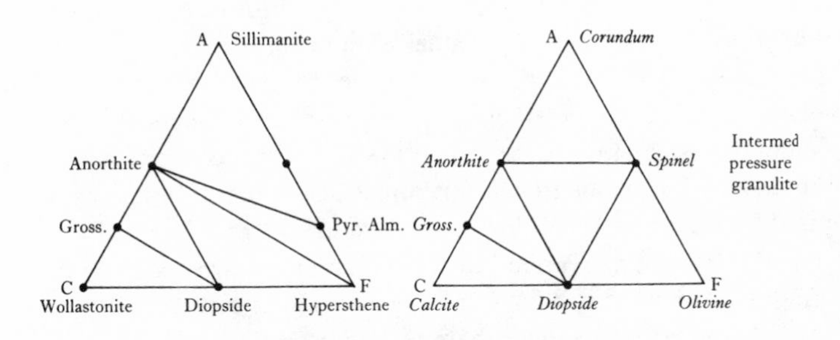

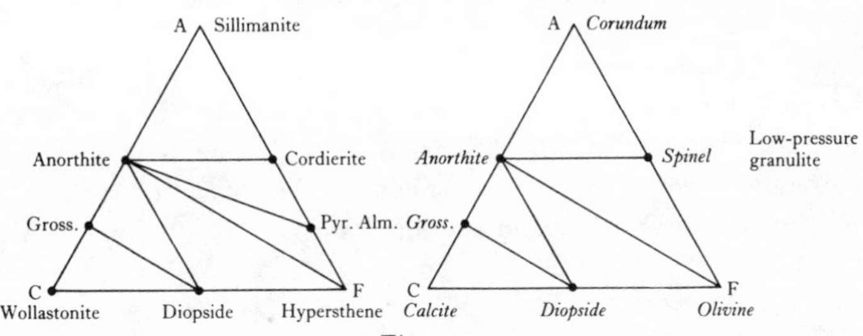

Fig. 37.4.

ACF diagrams illustrating typical assemblages in the low, intermediate, and high pressure echelons of the granulite facies. Silica-deficient assemblages on the right: phases incompatible with SiO_2 in italics. Solid-solution in garnets and pyroxenes ignored. (Modified from Miyashiro.)

silicate chondrodite is present, the silicates forming anhedral grains in a matrix of calcite which frequently shows oriented exsolution of dolomite (fig. 37.2C). Locally under conditions of very low P_{CO_2}, however, wollastonite may appear; examples are recorded from Ceylon, India and from the Adirondacks of New York State, where a granulite facies wollastonite deposit is quarried commercially as raw material for the ceramic industry. Slightly argillaceous limestones develop grossular garnet and a fassaite pyroxene showing some solid-solution of alumina as the 'tschermakite' molecule. Silica-deficient aluminous marbles show spinel or corundum.

Iron-formations

The iron-formations uniquely developed in Archaean terrains (chapter 26) must naturally figure in even a brief account of the granulite facies. The oxide-rich types – magnetite cherts and hematite cherts – recrystallise to coarse-grained, well-annealed magnetite quartzites and hematite (specularite)-quartzite, oxidation or reduction of the iron minerals being minimal. The silicate-rich original sedimentary types contain the anhydrous equivalents of the iron chlorite greenalite and the iron talc minnesotaite, viz. hypersthene with quartz and magnetite. Where dehydration is incomplete, the iron amphibole grunerite is found. The dominant carbonate of iron-formation, siderite, decomposes by reactions with quartz at much lower temperatures than does calcite. Consequently within the granulite facies original carbonate types are distinguished with difficulty from original silicate-rich rocks; thus the type *eulysite*, a garnet–hedenbergite–hypersthene–fayalite–magnetite–quartz rock with varying amounts of amphibole, may have originated either as a carbonate chert or as a dominantly siliceous sediment.

Eulysites normally contain minerals with substantial substitution of manganese: where manganese is present in excess of iron, spessartite garnet is characteristically developed with the oxide braunite, the pyroxenoids pyroxmangite and rhodonite $MnSiO_3$ (fig. 39.2A), and manganiferous pyroxenes and amphiboles (11).

Retrogressive effects

The high temperatures of crystallisation of granulite facies rocks ensure that quenching of assemblages is difficult: retrogressive effects are thus common. Exsolution in feldspars, in magnesian calcite, and in pyroxenes has been mentioned. In addition, numerous corona effects may often be seen, and symplectitic decomposition (fig. 33.6A). The kyanite of the high-pressure pelitic granulites is frequently inverted to sillimanite, and the garnet replaced by cordierite-bearing products. An interesting scapolite-bearing suite of granulites found as inclusions in 'kimberlite' pipes at Delegate, New South Wales, show retrogressive effects due to temperature drop at virtually constant pressure; reaction rims of garnet have formed between plagioclase and aluminous pyroxene (fig. 37.5A), and exsolution of the 'tschermakite' molecule from pyroxene has given

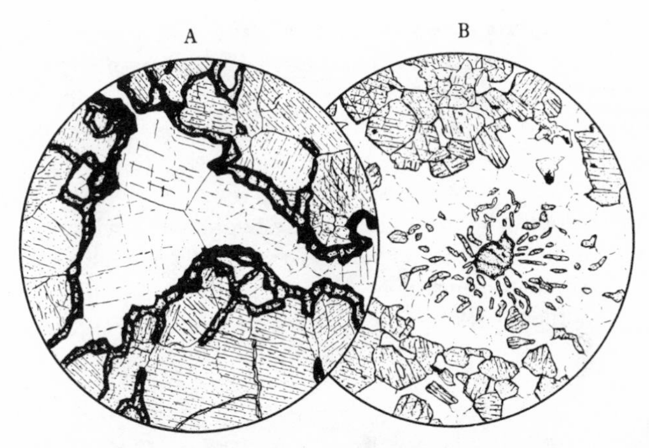

Fig. 37.5. Reaction textures in basic granulites.

A. Pyroxene granulite, inclusion in 'kimberlite' pipe, Delegate, N.S.W. A garnet rim has developed between hypersthene and plagioclase, by reaction (c) of fig. 35.6. Note annealed texture in plagioclase. × 17.

B. The opposite reaction in granulite from the Saxon Granulitgebirge, Hartmannsdorff, Saxony. The garnet in this high-pressure granulite has partly decomposed to a hypersthene-plagioclase aggregate. × 27.

lamellae of garnet (12). The diapirically emplaced peridotite of the Lizard in Cornwall shows recrystallisation of the original assemblage of olivine, aluminous pyroxenes, and spinel (spinel lherzolite, p. 404) to the lower-pressure assemblage olivine-diopside-hypersthene-plagioclase-chromite (13).

Where water is involved, it can be less certain that the effects were retrogressive; hydration may have occurred without change of temperature and pressure. The hydration products – amphibole and mica – are however common around anhydrous primary granulite minerals: their fine grain and zonal relation to the parent mineral normally serves to differentiate them from primary hydrous minerals.

The eclogites

With increasing pressure, the garnet-two pyroxene-plagioclase assemblage of the high-pressure granulites progressively diminishes in plagioclase as sodium and aluminium is accommodated in the pyroxene by the jadeite substitution Ca^{2}–$(Mg, Fe)^{2+} \rightarrow Na.Al^{3+}$ (p. 380). On the final disappearance of plagioclase from metabasic composition we have the garnet–omphacite assemblage of *eclogite* (fig. 37.6).

The garnet–omphacite association is stable in anhydrous metabasic composition over a very wide field of P and T. Unlike the granulites, however, the eclogites do not occur on regional scales of outcrop; rather they form relatively small masses enclosed in rocks of different type. Three main environments may be recognised (14): (a) with blueschist facies rocks; (b) with amphibolite facies rocks; (c) as inclusions in kimberlite and related pipes. Class (a) clearly represents anhydrous recrystallisation within the dominantly hydrous blueschist facies, and will be considered in chapter 38. The origin of (b) and (c) is not quite so certain. Group (c), inclusions brought up from depth in kimberlite pipes, may represent in part samples of mantle material. They show a

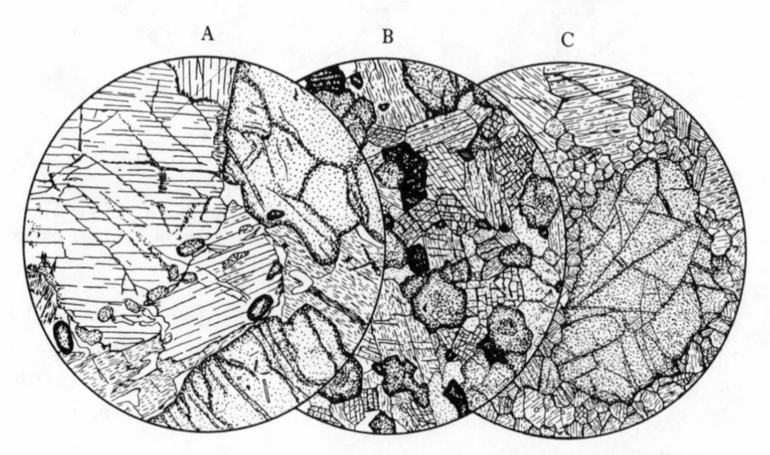

Fig. 37.6. Caledonian eclogitic rocks.

A. Eclogite, Loch Duich, Invernesshire. × 20. Omphacitic pyroxene, hornblende, and garnet with rutile and quartz.

B. Eclogite, Nautsdal, W. Norway. × 37. In addition to euhedral amphibole prisms and rutile, abundant apatite is present.

C. Garnet lherzolite, Sunmore, W. Norway. × 16. Porphyroclasts of garnet and pyroxene set in a cataclastic (subsequently somewhat annealed) aggregate of olivine and pyroxene.

variety of cataclastic textures but the classic *griquaites* from the southern African pipes often have well annealed crystals of green omphacite and pink garnet. Compositional variants show hypersthene (alumina deficient), kyanite (alumina in excess), or corundum.

Eclogites rich enough in calcium to develop a grossularite garnet are characteristic of the Yakutia diamond field in Siberia where richly chromiferous kyanites also are found. Carbon as *diamond* is not commonly seen in thin section.

Kimberlite eclogites rarely contain primary hydrous minerals, excepting only phlogopite mica (15). Group (b) eclogites – those associated with amphibolites – do, however, habitually contain amphiboles and micas whose chemistry (16) and euhedral habit are consistent with crystallisation as equilibrium partners of the garnet–omphacite assemblage.

Ingress of water following the main episode of recrystallisation, or accompanying temperature retrogression, produces secondary amphiboles or micas readily distinguished from 'primary' products – thus, in fig. 37.7A, the secondary products amphibole-plagioclase have developed as a finer-grained aggregate along omphacite grain boundaries.

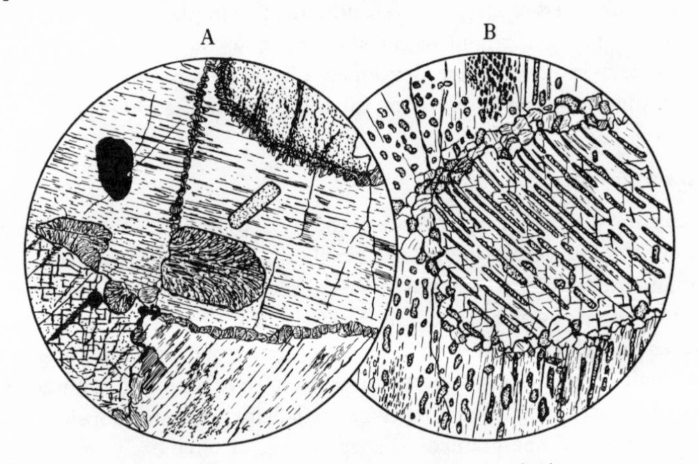

Fig. 37.7. Decompression textures in Lewisian eclogites; × 35.

A. Omphacite with symplectitic breakdown aggregates of diopside and plagioclase. Note also growth of secondary hornblende along grain boundaries. Totaig, Loch Duich.

B. 'Tschermak' molecule $CaAl_2SiO_6$ exsolved from aluminous clino-pyroxene crystallises as platelets of garnet parallel to (100) – note central crystal with (110) cleavages. Along grain boundaries the aggregate has recrystallised to discrete grains of diopside and garnet. Balvraid near Glenelg.

Fig. 37.7 also illustrates the anhydrous retrogression of omphacite to the low-pressure equivalent diopside-albite-quartz: in some eclogite-associated pyroxenites containing pyroxene rich in the 'tschermakite' molecule, decompression is accompanied by exsolution of Ca and Al in the form of garnet (fig. 37.7B). While in some cases such retrogressive changes may have accompanied the emplacement of the eclogite as an allochthonous slice thrust into a foreign environment, the autochthonous character of many amphibolite-associated eclogites may be inferred from their intimacy of interbanding with clearly metasedimentary material.

Eclogites at Glenelg, Invernesshire, may occur in layers as narrow as 2 cm in thickness, grading into a more feldspathic garnet–omphacite–plagioclase–quartz gneiss; they are concentrated along horizons of a dolomite–calcite (forsterite–diopside) marble and are reminiscent of marly meta-sediments. Those central European granulites outcropping in Polish Silesia show transitions to eclogitic rocks apparently interbedded with metasedimentary dolomite-rich layers and pelite. The pelitic eclogite facies assemblage developed here is garnet-kyanite-orthoclase-plagioclase-quartz, i.e. indistinguishable from that of the high-pressure granulite facies; normally, however, only the eclogites themselves show the anhydrous assemblage. A well-described example is that of Puerto Cabello, Venezuela, where basalt sills and flows in a Mesozoic meta-sedimentary sequence have developed a garnet–omphacite association within the prevailing epidote–amphibolite recrystallisation (17).

Ultramafic rocks

Interesting ultramafic equivalents of the type (b) eclogites are the garnet peridotites and garnet lherzolites which are found as masses amongst amphibolite facies gneisses and schists of the Caledonian and Hercynian in most countries of Europe. On the coast of western Norway striking layered complexes with dunite, eclogite, peridotite, and high-pressure pyroxenite occur. Cataclastic textures are often dominant, with coarse prophyroclasts of garnet, bright green jadeitic chrome diopside, and olivine, set in a fine-grained crushed matrix of olivine, pyroxene, and amphibole (fig. 37.6c); many such peridotites seem not to be autochthonous but to represent (? mantle-derived) material tectonically emplaced in its present position (18).

References and notes

(1) Wilcox, R. E. & Poldervaart, A., 1958. Metadolerite dike swarm in Bakersville, Roan Mountain area, N.C. *Bull. Geol. Soc. America* **69**, 1323–68.

(2) Winkler, *Petrogenesis*, p. 139. Turner, *Metamorphic petrology*, pp. 322–8. Fluid inclusions of granulites are unusually high in CO_2; it has been suggested that in deep crustal metamorphism mantle derived CO_2 may cause desiccation.

(3) A discussion of granulite facies textures is given by Collerson, K. D., 1974. Descriptive microstructure for high-grade metamorphic tectonites. *Geol. Mag.* **111**, 313–18.

(4) O'Hara, M. J., 1967. Mineral facies in Ultrabasic rocks *in* Wyllie, ed. *Ultramafic and related rocks*. New York: Wiley.

(5) The naming of charnockite from Job Charnock's tombstone in

Calcutta is notorious. A description of the type area near Madras where the stone was quarried is given by Subramaniam, A. P., 1959. Charnockites of the type near Madras. *Amer. J. Sci.* **257**, 321–53.

(6) Such rocks are remarkably poor in Li, Na, K, Rb relative to their SiO_2 content and do not correspond to any described igneous suite. Possibly they represent a refractory residue after abstraction of low-melting granitic liquid to which the granulite desiccation can also be attributed; see Fyfe, W. S., 1973. The granulite facies, partial melting, and the Archean crust. *Phil. Trans. R. Soc. London* **273**, 457–61.

(7) With anorthosites, charnockites, lherzolites and many peridotites, we have the difficulty of differentiating between igneous rocks crystallised (and annealed) under granulite facies conditions, and supracrustal rocks recrystallised within the granulite facies. These rocks are thus mentioned under both igneous and metamorphic 'labels'.

(8) Cf. the reaction muscovite + quartz = kyanite + orthoclase + H_2O.

(9) See Miyashiro, *Metamorphism*, pp. 211–14.

(10) Seifert, F., 1975. Boron-free Kornerupine – a high-pressure phase. *Amer. J. Sci.* **275**, 57-87.

(11) Such a cursory account does little justice to the iron-formations, which of course in preCambrian terrains can be seen in all stages of dehydration, through a wide range of metamorphic facies. See Klein, C., 1973. Changes in mineral assemblages with metamorphism of some banded preCambrian iron-formations. *Econ. Geol.* **68**, 1075–88.

(12) Lovering, J. E. and White, A. J. R., 1969. Granulitic and eclogitic inclusions from basic pipes at Delegate, Australia. *Contr. Mineral. Petrol.* **21**, 9–52.

(13) Followed by extensive serpentinisation. Green, D. H., 1964. The petrogenesis of the high-temperature peridotite intrusion in the Lizard area, Cornwall. *J. Petrol.* **5**, 131–88.

(14) Other subdivisions have been made. See Miyashiro, *Metamorphism*, pp. 316–24, who also discusses the strong differences in mineral chemistry between the eclogite groups.

(15) Anhydrous breakdown products of phlogopite involve K feldspar or leucite, high molecular volume framework silicates which have no high pressure equivalent. Phlogopite is thus uniquc amongst hydrous minerals in having a stability field not limited at very high P_{H_2O} (except by melting).

(16) For example, amphiboles with compositions intermediate between hornblende and glaucophane. Binns, R. A., 1967. Barroisite-bearing eclogite from Naustdal, Norway, *J. Petrol.* **8**, 349–71.

(17) Morgan, B. A., 1970. Petrology and mineralogy of eclogite and garnet amphibolite from Puerto Cabello, Venezuela. *J. Petrol.* **11**, 101–45.

(18) The relation between garnet peridotite (eclogite facies) and spinel lherzolite (granulite facies) may be expressed by the equation: fosterite + Ca pyrope = enstatite + diopside + spinel. O'Hara, M. J., Richardson, S. W. & Wilson, G., 1971. Garnet peridotite stability and occurrence in crust and mantle. *Contr. Mineral. Petrol.* **32**, 48–68.

38
The blueschist (glaucophane-schist) facies

Basic rocks crystallised under the gradients of low $\Delta T/\Delta P$ characteristic of (although not confined to) post-Palaeozoic orogenic belts, such as the Circumpacific and Alpine (1), typically contain amphiboles rich in the glaucophane (Na, Al^{3+}) through crossite to the riebeckite (Na, Fe^{3+}) molecules: the resulting often spectacularly blue schists provide an appropriate name for this high-pressure facies.

The Ca $\rightleftharpoons$ NaAl substitution (2) whereby

glaucophane, $Na_2(Mg, Fe)_3Al_2(Si_4O_{11})_2(OH)_2$, is derived from actinolite, $Ca_2(Mg, Fe)_5(Si_4O_{11})_2(OH)_2$, permits such amphiboles to accommodate all of the potential albite content of the metabasite. In other rock compositions a similar role is played by jadeite-rich pyroxenes. The potential anorthite content is represented by the hydrous lime-silicates, lawsonite (= anorthite + water) and the epidote-like mineral *pumpellyite* (very approximately equivalent to epidote + actinolite). Basic assemblages of blueschist facies rocks are thus ideally (if not always practically) feldspar free (3).

The textures and appearance of blueschists are extremely variable. Due perhaps to the brevity of the recrystallisation period and to the low temperatures (~300 °C) involved, many blueschists are excessively fine-grained and massive, preserving pre-existing textures, and may hardly merit the name 'schist' (figs. 38.1, 38.2, 38.3). Rapid variation in space from such fine-grained rocks to coarsely crystalline schists with well-defined lineation or schistosity is however common (4). The appearance in thin section of basic compositions is dominated by the high proportion of faintly pleochroic glaucophane; pink porphyroblasts of an almandine-rich garnet and coarse lawsonite prisms are typical. Other minerals encountered include a pyroxene – commonly omphacite but sometimes grading towards jadeite or aegyrine in composition – stilpnomelane, chlorite, albite, and sphene. Although epidote characterises the higher temperature, and the denser minerals lawsonite and pumpellyite the higher pressure blueschists, varying atomic substitution enables all of

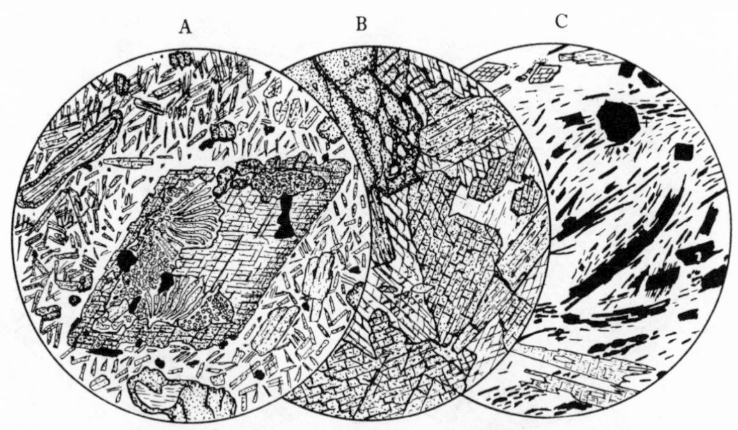

Fig. 38.1. Alpine blueschist facies rocks; × 18.

A. Polymetamorphic metabasalt, Spitz Flue, Zermatt. A glaucophane crystal is rimmed by hornblende in optical continuity reflecting waning pressures during Alpine (Cretaceous) metamorphism. Subsequent Lepontine (Eocene) epidote-amphibolite metamorphism is reflected in the matrix of epidote, hornblende, and clear albite; paragonite is rimmed by epidote, and the interior of the glaucophane has partly decayed to an hornblende-albite symplectite.

B. Glaucophane-eclogite, Syros, Cyclades. Euhedral omphacite and glaucophane with garnet and some inclusions of rutile.

C. Metachert, Bethe Ghinivert Mine, near Sestriere. Stilpnomelane and glaucophane with magnetite and hematite in quartz matrix. Note the cross-fractures which help to distinguish the layered mineral, stilpnomelane, from its relative, biotite. The flaky opaque mineral is the rare iron silicate, Deerite.

these related minerals to occur together and in fine-grained rocks they may be difficult to differentiate. The three show moderate to high relief ($n \sim 1.65$–1.8) but in sections of normal thickness epidote grains are readily identified by their maximum second-order green interference colours and by the strong compositional zoning shown towards the less iron-rich, more lowly birefringent clinozoisite. Lawsonite, being unique in composition, shows interference colours uniform throughout a grain; orthorhombic prisms with $\{001\}$ and $\{110\}$ developed are characteristic. Pumpellyite, often patchily green in thin section, is monoclinic with low (+ve) 2V (40°).

Carbonate minerals are commonly calcite in the low pressure range and, at high pressure, aragonite. Lamellar twinning on sections of low birefringence, and its biaxial character, normally serve to distinguish the orthorhombic aragonite from the calcite-type carbonates. In coarse veins its orthorhombic habit and marginal inversion to calcite may also aid rapid recognition (fig. 38.3A) (5).

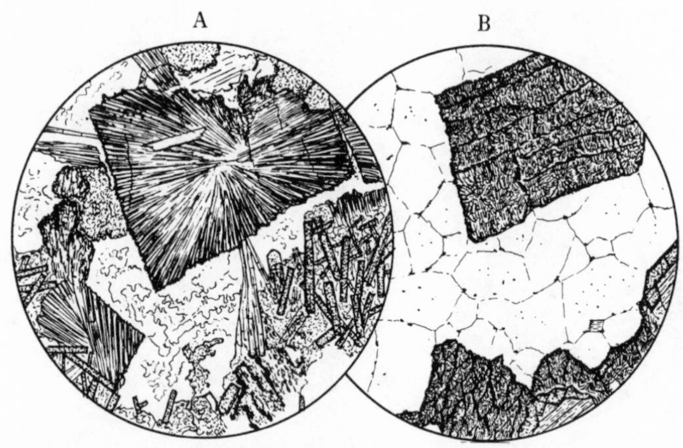

Fig. 38.2. Alpine blueschist facies metamorphism of acid igneous rocks; $\times$ 35.

A. Metadiorite, block in ophiolite breccia, Pic Marcel, Hautes Alpes. Euhedral plagioclase crystals metasomatically replaced by pale green pumpellyite and minor glaucophane. An original single crystal of quartz, strained by the volume changes in the rock, has begun to recover as a polycrystalline aggregate with highly convoluted boundaries. Lawsonite–chlorite aggregate to right.

B. Metagranite, Hercynian basement of Sezia root zone, Monte Mucrone near Biella. Essentially isochemical metamorphism, the euhedral plagioclase crystals being replaced by a fine-grained jadeite-quartz aggregate. Igneous quartz recrystallised to an annealed mosaic; biotite rimmed by garnet. K feldspar (not shown) is now a strained microcline.

Although we may regard glaucophane, jadeite, lawsonite and aragonite as characteristic of blueschists, the widely differing P/T fields of stability of these minerals makes difficult the definition of a useful blueschist facies and has led to boundary disputes in the petrological literature. Assemblages transitional between undoubted blueschist (glaucophane-lawsonite-aragonite) and undisputed greenschist (actinolite-epidote-calcite-albite-chlorite) predominate in some areas, as in the Sambagawa belt of Japan where glaucophane–albite–epidote–calcite assemblages, occasionally with an actinolitic amphibole, are common. Other transitional rocks lack a blue amphibole, containing instead the glaucophane equivalent albite + chlorite (cf. note 3a) coexisting with lawsonite. Such assemblages, developed in parts of the California coast ranges, in the South Island of New Zealand, and in Calabria, have been assigned to a *lawsonite–albite* facies (6).

Eclogites in association with blueschists pose problems of interpretation similar to those of amphibolite facies eclogites. Some may represent inclusions derived from greater depths and tectonically emplaced in their higher-level blueschist envelope. Many examples are however recorded in which cores of metamorphosed pillow lavas are composed of the eclogitic garnet–omphacite assemblage, the rims and matrix having developed the more hydrated assemblage glaucophane-chlorite-epidote-muscovite-carbonate-quartz (e.g. the Zermatt–Sans Fee ophilite zone of the Swiss Alps). The pseudomorphing of these obviously originally supracrustal pillow structures suggests such eclogites to represent recrystallisation under conditions of *P* and *T* identical with those of the rim blueschists – presumably however in deficiency of water (7).

Blueschist eclogites frequently show preferred orientation of an omphacite matrix, with scattered porphyroblasts of garnet; others may also show a pronounced compositional foliation, garnet and omphacite being segregated into distinct bands. The composition of blueschist eclogite appears to be more restricted than that of eclogite in the amphibolite environment, kyanite or hypersthene-bearing examples being seldom seen. As in the amphibolite case, however, primary hydrous minerals usually occur, here glaucophane, paragonite, or muscovite. The titaniferous mineral is rutile rather than the sphene characteristic of glaucophane schists (8).

Glaucophane, epidote, and chlorite may also occur as retrogressive products of omphacite and garnet. Commonly, however, alteration of both eclogite and of associated glaucophane schist is to coarsely-grained amphibolite or epidote–amphibolite assemblages. Thus glaucophane may be rimmed by hornblende in crystallographic continuity, or by hornblende–albite aggregates, the matrix consisting of tiny prisms of hornblende and epidote set in water-clear albite (fig. 38.1A). The change from blueschist to amphibolite facies conditions so recorded in particular rocks results either from re-establishment of normal thermal distribution (*T* rise) or from elevation to more normal levels (*P* decrease) following the initial recrystallisation at high *P* and low *T* (9).

Greywackes and pelites of the blueschist facies commonly develop in place of biotite the superficially similar layered mineral stilpnomelane (figs 38.1C, 38.3A) (10). In the higher-pressure echelons of the facies, lawsonite and jadeite are found in place of feldspars. Some of the best-studied greywacke assemblages are those of the Franciscan formation in the coast ranges of California, where transition from quartz-albite-lawsonite-stilpnomelane-muscovite-chlorite (with calcite) rocks may be traced through assemblages with albite-lawsonite-aragonite to the highest-pressure association jadeitic pyroxene-lawsonite-aragonite; thus in a field a 'jadeite isograd' may be mapped. In all of these stages

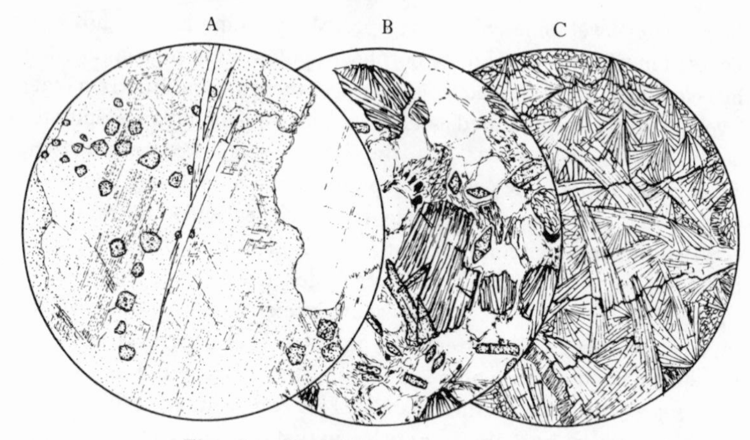

Fig. 38.3. Blueschist facies rocks; × 40.

A. Marble, Laytonville Quarry, Mendocino County, California. Large plates of aragonite show irregular lamellar twinning. The grain to the right of the picture with γ parallel to the analyser has inclusions and cleavages reflecting the orthorhombic symmetry. Spessartite garnet and stilpnomelane present.

B. Metagreywacke, Pacheco Pass, California. Jadeite pseudomorphs detrital plagioclase: prisms of lawsonite present.

C. Jadeite rock (jade), Tawmaw, Burma.

the minerals are fine-grained and preserve much of the original greywacke texture – notable are jadeite pseudomorphs after detrital plagioclase (fig. 38.3B).

Metacherts. The eugeosynclinal sedimentary association characteristic of regions of blueschist facies metamorphism often contains cherty layers rich in ferric iron and manganese. Quartz-rich rocks with hematite, magnetite, spessartite garnet, Na, Fe^{3+} amphiboles (riebeckite) and pyroxenes (aegyrine–acmite) result. The manganiferous cherts of modern sea-floors are reflected in the piemontite (Mn-epidote)–quartz rocks which occur interbedded with many glaucophanitic pillow-lavas, as in the Piedmont (Pennine) zone of the French Alps.

Metasomatism. While it is not yet proven that metasomatic effects are more common in the blueschist than in other facies, the products are certainly more obvious and spectacular. Such effects as the replacement of plagioclase by pumpellyite and glaucophane illustrated in fig. 38.2A may only involve diffusion over short distances; however, larger-scale migration of material is suggested by the abundant occurrence of monomineralic veins – especially of pumpellyite, aragonite, jadeite, and omphacite. Jadeite, often adopting a more fibrous habit than most pyroxenes, may occur in segregations as the matted, multicoloured aggregates of semi-precious jade (fig. 38.3A) (11). The occurrence of aegyrine in hematite–quartz rocks and of ferrostilpnomelane in magnetite–quartz rocks also seems to indicate mass transfer, in this case accession of alkalis: convincing evidence of sodium-metasomatism is provided by those originally soda-free cherty metasediments in which Na, Fe^{3+} pyroxene has grown at contacts between hematite and quartz.

References and notes

(1) These low gradients $\Delta T/\Delta P$ have been attributed to the depression, along subduction zones, of cold sediments at such a rate that heat flow is insufficient to maintain normal isothermal distribution. Ernst, W. G., 1973. Blueschist metamorphism and *P-T* regimes in active subduction zones. *Tectonophysics* **17**, 255–72.

(2) Compare the analogous substitutions for hornblende (p. 387) which however involve tetrahedral Al, viz. MgSi $\rightleftharpoons$ Al.Al and Mg, 2Si $\rightleftharpoons$ NaAl2Al.

(3) (a) $2Na_2(Mg, Fe)_3Al_2(Si_4O_{11})_2(OH)_2 + 2H_2O =$
glaucophane water
$4NaAlSi_3O_8 + (Fe, Mg)_6Si_4O_{10}(OH)_8$
albite antigorite

(b) $NaAlSi_3O_8 = NaAlSi_2O_6 + SiO_2$
jadeite albite quartz

(c) $(OH)_2CaAl_2Si_2O_7H_2O = CaAl_2Si_2O_8 + 2H_2O$
lawsonite anorthite water

(d) Pumpellyite: approx. $Ca_4Al_5(Mg, Fe)Si_6O_{21}(OH)_7$, but a wide range of solid-solution is found.

(4) Complex tectonics in regions of blueschist facies metamorphism make interpretation difficult. Some such transitions are tectonic contacts between glaucophane schists formed at different levels. Thus a section of the chaotic melange of the Franciscan formation, from the Pacific to the Great Valley of California, shows a regular increase in metamorphic grade from unmetamorphosed sediment to fine-grained glaucophane schist facies assemblages. Exotic blocks of coarse-grained blueschist facies rocks occur however at all levels in this prograde sequence. *See* Coleman, R. G. & Lee, D. E., 1963. Glaucophane-bearing metamorphic rock types of the Cazadero Area, California. *J. Petrol.* **4**, 260–301, also Turner, *Metamorphic petrology*, pp. 289–99.

(5) Kinetics of the aragonite–calcite inversion suggest that the preservation of aragonite indicates rapid rates of cooling and hence rapid elevation to surface conditions, see Brown, W. H., Fyfe, W. S. & Turner, F. J., 1962. Aragonite in California glaucophane schists. *J. Petrol.* **3**, 566–82.

(6) The Fe^{3+}-rich crossites will be stable at higher temperatures than glaucophane, so that sodic amphiboles may occur with albite and chlorite. Brown, E. H., 1974. Comparison of the mineralogy . . . of Blueschists from . . . Washington and Otago. *Bull. Geol. Soc. America* **85**, 333–44.

(7) Fry, N. & Fyfe, W. S., 1969. Eclogites and water pressure. *Contr. Mineral. Petrol.* **24**, 1–6.

(8) $CaTiSiO_5 + CO_2 = TiO_2 + CaCO_3 + SiO_2$
sphene rutile calcite quartz
The occurrence in eclogites of rutile, occasionally with carbonate and quartz, suggests the 'dry' character of such eclogites to result from an excess of CO_2 in the fluid phase.

(9) The example of fig. 38.1A may, however, represent polymetamorphism – the superimposition of a mid-Tertiary (Lepontine) amphibolite facies on the Cretaceous (Early Alpine) glaucophane schist. Frey, M. *et al.*, Alpine metamorphism of the Alps. *Schweiz. Min. und Pet. Mitt.* **54**, 247–63.

(10) Stilpnomelane is the only common silicate to show complete solid-solution between ferrous and ferric end members; most stilpnomelanes, however, probably crystallised as ferrostilpnomelane (approx. $K_1(Fe^{2+}, Mg, Al)_3\ Si_4O_{10}(OH)_4H_2O$) and were subsequently oxidised. Stilpnomelane + muscovite is the low-temperature, high-pressure equivalent of biotite + chlorite.

(11) Most commercial jades (including the classic ancient Chinese jades) are, however, composed of the actinolitic amphibole *nephrite* (not necessarily of blueschist facies origin) or of serpentine. The difficulty of propagating cracks in a fine felt of fibres gives toughness to aggregates of minerals not in themselves particularly hard or strong.

39
Very low-grade metamorphism

Although in some clay-rich sediments a continuous sequence of recrystallisation may be traced linking surface diagenetic recrystallisation with medium-grade metamorphism (p. 389), many rocks (especially igneous intrusives or extrusives) are coarse-grained or dry, and reaction rates are too low for palpable metamorphism to proceed until the relatively high temperatures of the greenschist or blueschist facies (200–300 °C) are achieved.

Volcanic rocks and greywackes containing highly reactive glass or clay-rich fractions may however show a transition from diagenetic mineral associations by the development of zeolites – notably analcite representing albite feldspar and heulandite representing anorthite. The classic region for such *zeolite facies* assemblages is the Taringatura district of Southland and contiguous areas of Otago in New Zealand. Here zeolite development and alteration can be directly related to stratigraphic depth in the Triassic sediments of the New Zealand geosyncline, hence the term *burial metamorphism* (1). Initially zeolite development is confined to micro-environments (glassy shards and to cementing along grain boundaries) the detrital feldspars remaining unaltered (fig. 39.1A). Increasing temperature, however, sees more pervasive alteration, with even the coarsest detrital feldspars producing zeolites. During this process the low-temperature layered minerals of the clay fraction – montmorillonites and mixed-layer associates such as celadonites – recrystallise to chlorite and muscovite or phengite, with new-formed quartz and sphene.

In the New Zealand type sequence two stages are recognised, heulandite-analcite and the higher-grade laumontite-albite. These are separated by two reactions occurring at virtually the same temperature: the replacement of the calcic zeolite heulandite by its higher temperature equivalent, laumontite, and (in the presence of quartz) the replacement of the sodic zeolite, analcite, by albite (fig. 39.1B) (2). The zeolite facies itself is terminated at the stage at which lime zeolites disappear by complex reactions with ferro-magnesian silicates to produce non-zeolitic assemblages in which prehnite and pumpellyite (3) coexist with albite, chlorite, quartz, and sphene. This *prehnite–pumpellyite* facies is widespread in vulcanogenic metasediments in a variety of geological situations, such as the Labrador trough of Canada (pre-Cambrian); portions of the Appalachian belt of Maine and in the Tasman geosyncline of eastern Australia (Palaeozoic); and in the Helvetic zone of the Alps. The rocks are universally fine grained and in hand specimen give little hint of metamorphism.

With rising temperature, prehnite and pumpellyite metagreywackes give way to the epidote- and actinolite-bearing assemblages of the greenschist facies – the New Zealand geosyncline again provides the premier example. Prehnite is consumed first, with the production of more pumpellyite and of actinolite; in favourable rock compositions stilpnomelane may also develop. At this stage, too, the rocks generally begin to develop textural indications of advancing metamorphism – penetrative fabrics and the progressive obliteration of original sedimentary features. Many examples of the prehnite–pumpellyite facies belong to distinctly high-pressure facies series, in which the subsequent

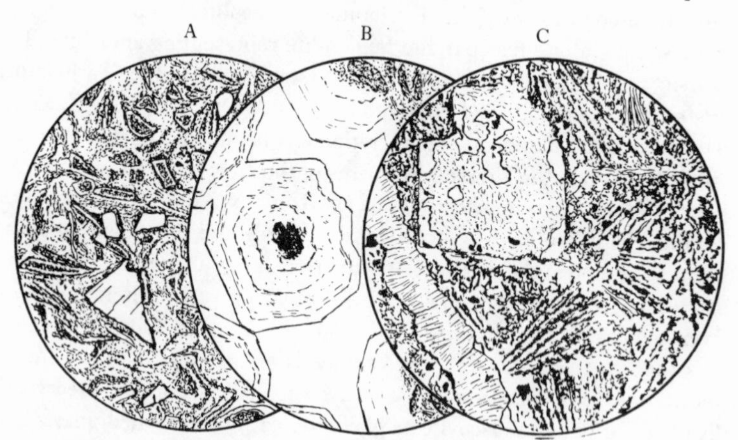

Fig. 39.1. Burial metamorphism; × 60.

A. Rhyodacite vitric tuff. Glass shards replaced by heulandite, the cores clouded with iron oxides. Fragments of andesine unaltered. Taringatura district, New Zealand.

B. Albite pseudomorphs after analcite, from vein in quartz-adularia-albite rock, Taringatura.

C. Basalt altered to chlorite and quartz. The variolitic texture of the original igneous rock is preserved and a phenocryst of feldspar retains some unaltered relicts. A monomineralic vein of chlorite traverses the lower left hand side. Carlsberg Ridge at $5\frac{1}{2}$° N, dredge haul.

greenschist facies itself grades into blueschist or lawsonite–albite facies assemblages. Archean examples, however, reflect higher geothermal gradients; thus in the Yilgarn Block of western Australia prehnite-pumpellyite facies rocks out-cropping over areas of some 1000 km^2 are separated by relatively narrow strips of greenschist facies culminating in amphibolite or even granulite facies rocks. Nickel deposits confined to the high grade areas represent original igneous sulphides metamorphically concentrated (5).

Iron-formations

Interesting products of very low-grade metamorphism of Archaean iron formations include sodic amphiboles, magnesio-arfvedsonite to riebeckite, possibly the products of reaction between NaCl and iron silicates or carbonates from the original sedimentary environment. Before evidence of health hazards caused prohibition of the industrial use of 'blue asbestos', fibrous crocidolite formed by the progressive filling of tension cracks in such rocks (fig. 39.2B) was of considerable economic importance.

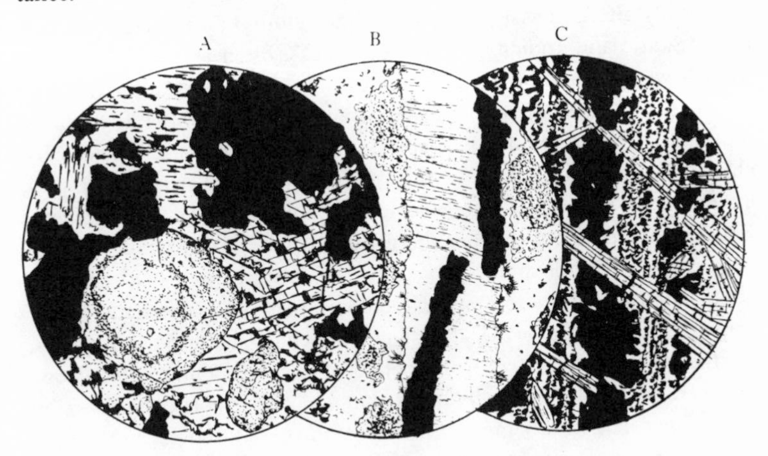

Fig. 39.2. Metamorphosed iron-formations; × 40.

A. Spessartite-rhodonite-braunite rock, Kodur Mine, Vizagapatam, Madras (Granulite facies).

B. Crocidolite (riebeckite) vein in siderite-quartz matrix, Hamersley Range, W. Australia. The growth of asbestiform amphibole normal to the edges of a tension crack opening parallel to the bedding is well attested by the disrupted layer of magnetite.

C. Magnetite-quartz banded iron formation, Hamersley Range, W. Australia. Note lack of disturbance of sedimentary lamination by riebeckite crystals growing across it.

Metasomatism

The composition of CaAlSi hydrates characteristic of the zeolite and prehnite–pumpellyite facies is chemically equivalent to that of mixtures of calcite, kaolinite, and water (6). A high ratio of $H_2O:CO_2$ in the fluid phase must therefore be inferred for the production of zeolite and prehnite assemblages; in some instances this may reflect the large quantities of water released during the devitrification of volcanic glass.

To be connected with this high flux of water inferred for the burial metamorphic environment must be the degree of metasomatism shown. Glass shards, which certainly were not originally of pure zeolite composition, may be completely replaced by heulandite (fig. 39.1A), and at higher grades we may infer extensive replacement of volcanic material by quartz and the highly disordered polymorph of potassium feldspar, adularia. This degree of metasomatism, and the unstressed 'burial' environment of many zeolite facies rocks, links them with the essentially greenschist (occasionally amphibolite) facies assemblages found in sub-oceanically altered basalts – including those found on the ridges of the mid-Atlantic and Indian oceans. Such rocks, often essentially spilitic, show a wide variety of alteration. A common feature is differentiation of a basalt into Na-rich and Ca-rich volumes: labradorite feldspar is replaced by albite, while associated pyroxene and glass is replaced by chlorite. The calcium released may migrate several metres before precipitation as the calcic phases epidote, prehnite, sphene, or calcite. Basalts dredged from the Carlsberg ridge show the alteration of some originally crystalline pillow lava cores to the greenschist assemblage albite–chlorite–actinolite–sphene (epidote) while the originally glassy rims have been converted almost completely to chlorite and quartz. Furthermore, much of the plagioclase of the crystalline cores shows replacement by chlorite or by quartz and chlorite, chlorite–quartz veins traversing the matrix being common (fig. 39.1C). Although the chemical changes in some instance were merely redistribution between adjacent rock units in the lava pile, rather constant chemistry of those 'fresh' basalts which retain their high-temperature mineralogy allows the inference that considerable quantities of at least calcium were lost from the metamorphosed basalts sampled (7). Since the samples studied were obtained by dredge-hauling, their provenance is by no means certain; however, laboratory studies on the stability fields of the minerals concerned suggest that the environment of such basalt alteration must have been some hundreds of metres below the rock–water interface, in regions of high (?1300 °C/km) geothermal gradient. Equilibration of basalts with convecting currents of sea water circulating within the sub-oceanic crust provides an agency of mass transfer (8). Such currents, perhaps also connected with the precipitation of the metals of many stratiform ore-bodies, may be compared with the modern hot brines of the rifting Red Sea and of the Salton Sea of California; the geologic setting is, of course, very different from the orogenic environment of most regionally metamorphosed rocks.

References and notes

(1) *Burial metamorphism*, large-scale recrystallisation of deeply buried rocks without perturbation of the geothermal gradient by igneous intrusions and without marked penetrative movements. Coombs, D. S., 1961. Some recent work on the lower grades of metamorphism. *Australian J. Sci.* **24**, 203–15.

(2) Coombs, 1961; Winkler, *Petrogenesis*, pp. 138–9.

$$\underset{\text{analcite}}{NaAlSi_2O_6 \cdot H_2O} + \underset{\text{quartz}}{SiO_2} = \underset{\text{albite}}{NaAlSi_3O_8} + H_2O$$

$$\underset{\text{heulandite}}{CaAl_2Si_7O_{18} \cdot 6H_2O} = \underset{\text{laumontite}}{CaAl_2Si_4O_{12} \cdot 4H_2O} + \underset{\text{quartz}}{3SiO_2} + 2H_2O$$

(3) (a) $$\underset{\text{laumontite}}{CaAl_2Si_4O_{12} \cdot 4H_2O} + \underset{\text{calcite}}{CaCO_3} = \underset{\text{prehnite}}{Ca_2Al_2Si_3O_{10}(OH)_2} + \underset{\text{quartz}}{SiO_2} + 3H_2O + CO_2$$

(b) $$\underset{\text{prehnite}}{8Ca_2Al_2Si_3O_{10}(OH)_2} + \underset{\text{chlorite}}{(Mg, Fe)_4Al_4Si_2O_{10}(OH)_8} + 2H_2O = \underset{\text{pumpellyite}}{4Ca_4(Mg, Fe)Al_5Si_6O_{21}(OH)_7} + \underset{\text{quartz}}{2SiO_2}$$

(4) Possible examples of the complex chemical reactions involved are given by Bishop, D. G., 1972. Progressive metamorphism from prehnite-pumpellyite to greenschist facies in the Dansey Pass area, Otago, New Zealand. *Bull. Geol. Soc. America* **83**, 3177–97.

(5) Binns, R. A. & Gunthorpe, R. J., 1976. Metamorphic patterns in Archean metamorphism of the Yilgarn Block *in* Windley, B., ed., *Early History of the Earth*. London: Wiley.

(6) $$\underset{\text{laumontite}}{2CaAl_2Si_4O_{12} \cdot 4H_2O} + 2CO_2 = \underset{\text{kaolinite}}{Al_4Si_4O_{10}(OH)_8} + \underset{\text{calcite}}{2CaCO_3} + \underset{\text{quartz}}{4SiO_2} + 4H_2O$$

In some regions (e.g. the Barrovian metamorphic zonal sequence of Scotland), the occurrence of non-zeolitic kaolinite–calcite assemblages within the P/T range of the zeolite facies has led to criticism of the status of this facies.
Coombs, *op. cit.*; Zen, E. An & Thompson, A. B., 1974. Low-grade regional metamorphism: Mineral equilibrium relations. *Ann. Review Earth and Planetary Sciences* **2**, 179–212.

(7) Cann, J. R., 1969. Spilites from the Carlsberg Ridge, Indian Ocean. *J. Petrol.* **10**, 1–19. Possibly under the totally unstressed conditions of burial metamorphism, nucleation rates are the critical rate-determining

factor, chlorite here being the readily nucleated mineral whose growth controls chemical migration. See also Miyashiro, *Metamorphism*, especially p. 419.

(8) Spooner, E. T. C. & Fyfe, W. S., 1973. Sub-sea-floor metamorphism, heat and mass transfer. *Contr. Mineral. Petrol.* **42**, 287–304; Miyashiro, *Metamorphism*, chapter 19.

INDEX

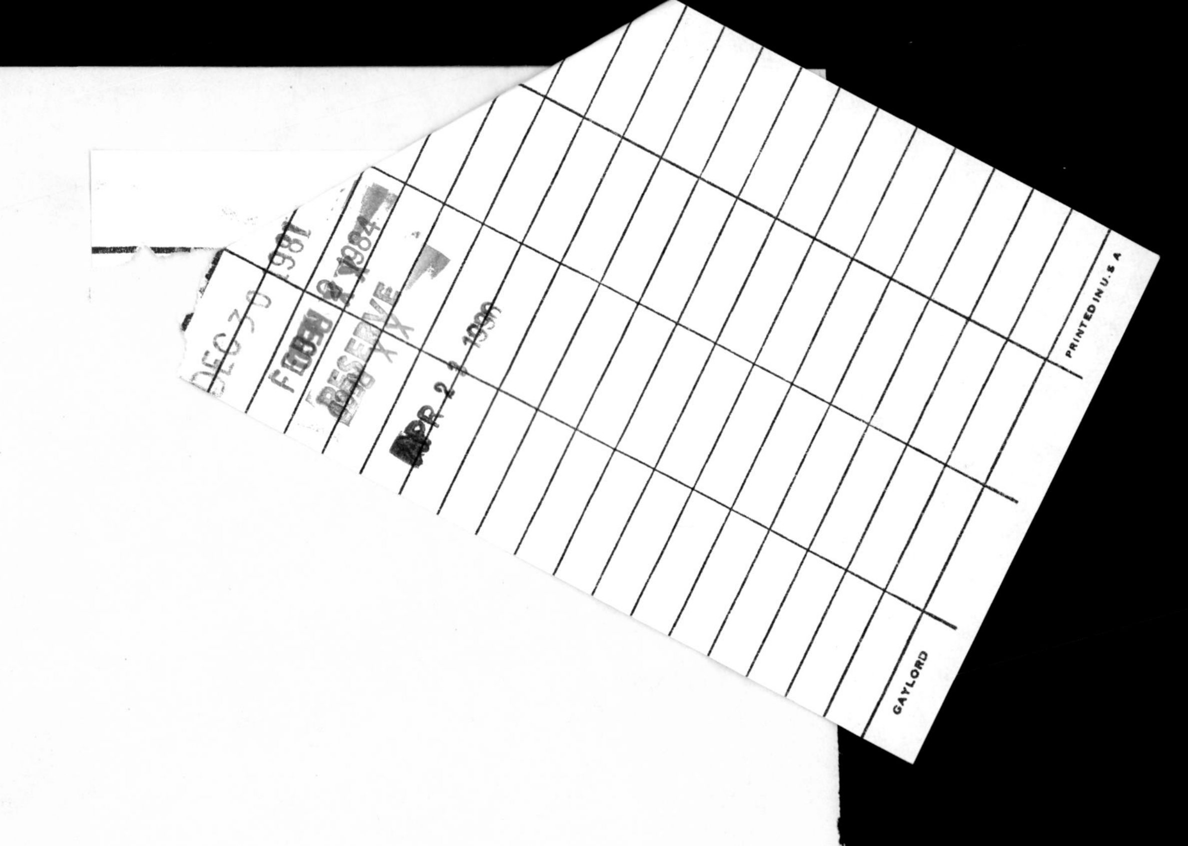